ADSL & DSL Technologies, Second Edition

WALTER **GORALSKI**

Osborne/**McGraw-Hill**

New York Chicago San Francisco
Lisbon London Madrid Mexico City Milan
New Delhi San Juan Seoul Singapore Sydney Toronto

Osborne/**McGraw-Hill**
2600 Tenth Street
Berkeley, California 94710
U.S.A.

To arrange bulk purchase discounts for sales promotions, premiums, or fund-raisers, please contact Osborne/**McGraw-Hill** at the above address. For information on translations or book distributors outside the U.S.A., please see the International Contact Information page immediately following the index of this book.

ADSL & DSL Technologies, Second Edition

1234567890 CUS CUS 01987654321

ISBN 0-07-213204-3

Publisher
  Brandon A. Nordin
Vice President & Associate Publisher
  Scott Rogers
Editorial Director
  Tracy Dunkelberger
Acquisitions Editor
  Steven Elliot
Project Editor
  LeeAnn Pickrell
Acquisitions Coordinator
  Alexander Corona
Technical Editor
  Gary Kessler

Copy Editor
  Darren Meiss
Proofreader
  Melissa Lynch
Indexer
  David Heiret
Computer Designers
  George Charbak, Elizabeth Jang,
  Melinda Moore Lytle
Illustrators
  Richard Coda, Michael Mueller,
  Lyssa Wald
Series Design
  Peter F. Hancik

This book was composed with Corel VENTURA™ Publisher.

This book is dedicated to all the firefighters, police, flight crews, innocent passengers, military personnel, and civilians who lost their lives as a result of the terrorist attacks on September 11, 2001. While they may be gone, we who remain will make sure they will never be forgotten. May they rest in peace.

ABOUT THE AUTHOR

Walter Goralski has more than 30 years of experience in the data communications field, including 14 years with AT&T. He is currently a Senior Course Developer with Juniper Networks and is the best-selling author of *SONET, Second Edition*, and several books on ADSL, VoIP, ATM, and frame relay. He has also written numerous articles on data communications and other related issues.

ABOUT THE REVIEWER

Gary Kessler is an Assistant Professor & Program Coordinator for the Computer Networking Program at Champlain College in Burlington, Vermont, as well as a Networking consultant, frequent contributor to *Network* magazine and, author of the bestseller *ISDN: Concepts, Facilities, and Services*.

CONTENTS

ACKNOWLEDGMENTS

With every book I write, there seems to be an ever-increasing list of people to thank for assistance and encouragement. Whether this is a good sign or a bad sign, I do not know. But this book would not exist without the following people guiding me through the whole process.

It was Dave Hill who first suggested a DSL course for Hill Associates, which later grew into this book. Nigel Cole of the ADSL Forum and Orckit first made me feel as if I knew enough about ADSL to fill a book and encouraged me to write it. His review of the manuscript was one of the most thorough and complete that I have ever endured in spite of time pressure, and the result was a much better book. Gary Kessler of Hill reviewed the first draft and made helpful comments, as always. Gary has been a good friend through the years and his help has been invaluable. Joe Charboneau and Mike Rude of ADC provided valuable information about HDSL and HDSL2. The ADSL Forum, in general, has been a valuable and unique source of information. Steve Elliot of Osborne/McGraw-Hill stepped outside his role as editor to gather valuable information in his travels.

I cannot thank my family enough either. Jodi gives me space and quiet support, Christopher has a first-born's pride in my accomplishments, Alexander makes sure I keep my writing totally in perspective, and Arianna often keeps me company while I write but never bugs me (see, I put it in the book).

PREFACE

...TO THE SECOND EDITION

When the first edition of the book was being written in late 1997 and early 1998, the importance of *Asymmetrical Digital Subscriber Lines* (ADSL) and related DSL technologies was by no means obvious. Many people I spoke to who were certainly as knowledgeable about the telecommunications industry as I was—or even more so—advised me to write about cable modems or wireless solutions or anything more promising than DSL. ADSL and other DSLs were, they assured me, at best stop-gap measures that were the last gasp of the traditional voice-oriented telephone company.

I must admit they had strong arguments. Cable modems were hot and running off with the broadband residential access market. Wireless was new and exciting. I could write about half a dozen stable cable modems, and only about a few ADSL modems, and most of these were mere prototypes pressed into service as shipping products only because the DSL industry had to do something, and fast, to counter the popularity of cable modems.

In the end, I allowed their arguments to influence my thoughts slightly. I hedged only a bit. I wrote on ADSL because, as I put it at the end of the original preface "Although many technologies can play a role in modernizing the voice network, ADSL is singled out in this text because ADSL has an edge over many others in terms of standardization, vendor activity, economics, and customer interest. The question is whether ADSL can maintain this edge. Other technologies will be mentioned in this regard, but ADSL is investigated in full."

When I read these words almost four years later as I sat down to prepare the second edition of this work, I realized that I was not as dumb as I thought. Most of the four pillars I chose to emphasize have held up pretty well. Standards for ADSL and its variants are in place all over the world; there are enough active and healthy vendors today to satisfy most; and customers are surely interested in high speeds on copper telephone lines with ADSL and its variations. Only when it comes to the economics has ADSL been vexed by troubling and complex issues (issues that are fully explored in the pages of this book).

One consequence of the complex economic factors surrounding ADSL in particular has been the inability for the broadband copper access industry to sustain anything like the truly competitive marketplace that many would prefer. For the time being, it seems that pricing has made it feasible for only an incumbent telephone service provider to offer ADSL and DSL with any hope of profit, even in urban areas. The several *competitive local exchange carriers* (CLECs) and specialized Internet access providers have retreated from the DSL marketplace, often leaving customers high and dry.

This second edition expands the coverage of other broadband access technologies as well as completely updating the state of DSL. Cable modems remain the most respectable competitor of DSL, and will be treated as such. But as the original preface concluded: "Whatever the outcome over the next 5 or 10 years, it certainly will be an interesting ride." It certainly has been just that in the past four years. And the broadband access ride has really only begun.

...TO THE FIRST EDITION

I wrote this book to do my part to promote change in the public telephone network. Although it usually comes with a price tag, change is good, and all networks must change if they are to grow and prosper. Networks that do not change will wither and struggle. It does not matter whether the network involved is the telephone network. All networks must evolve as the needs of the users change, as the technology on which the network is based changes, and as the entire economic, social, and political backdrop against which the network exists slowly transforms itself.

Consider the highway network of roads in the United States. From humble beginnings as simple "post roads" for mail delivery and stage coaches, the network adapted itself to automobiles by adding pavement and traffic signals. The system was transformed again when post–World War II prosperity encouraged automobile use for a wide variety of reasons, from commuting to the recreational Sunday drive in the country. The Interstate Highway System, with limited access and guard rails, better reflected this new environment.

It is not for nothing that telephone network terminology closely mirrors road system talk. Both systems have access links and bypasses. Both have tolls and interchanges. And more to the point, both have traffic and congestion. The talk about an Information Super-highway started out in discussions about an "interstate highway system for data." As will be shown, both telephone networks and highway networks share much more than terminology.

This book emphasizes one of the ways that may be used to transform the voice network in the United States (and for that matter, around the world) into a network better suited for the "automobiles" of the 1990s. If telephones are the voice network's stage coaches, modern personal computers attached to the Internet are its luxury cars. This being the case, perhaps local access lines using ADSL technology are better suited for users than the *plain old telephone service* (POTS) access lines of the last 100 years or more.

Although many technologies can play a role in modernizing the voice network, ADSL is singled out in this text because ADSL has an edge over many others in terms of standardization, vendor activity, economics, and customer interest. The question is whether ADSL can maintain this edge. Other technologies will be mentioned in this regard, but ADSL is investigated in full. Whatever the outcome over the next 5 or 10 years, it certainly will be an interesting ride.

INTRODUCTION

All successful technologies are successful because they solve problems. The problem could be as simple as getting around faster and more conveniently (airplanes and automobiles) or as complex as finding an alternative energy source by splitting an atom in a controlled fashion (nuclear power). This book is about ADSL technology, and the problem it attempts to solve is one that is becoming more critical as people change the ways that they use networks to communicate, work, and relax. The problem is easiest to explain through a few examples.

Pacific Bell (now part of SBC) is a major local telephone company servicing much of the West coast of the United States. Like most telephone companies, life had been good to Pacific Bell in the past. Revenues were plentiful, network usage predictable, and things pretty much revolved around the relatively routine tasks involved in supporting the typical three- to six-minute telephone call people made to order pizza, call in sick, or chat with friends. Over the past few years, however, especially since the explosion of the Internet and World Wide Web (or just "the Web" or even more simply "the web" today) in 1993–1994, life has gotten much more interesting for Pacific Bell and other local telephone companies. There was more intense local competition, increased interest in new services, and even heightened customer awareness of perceived shortcomings of the existing

Pacific Bell voice telephone network. But the biggest changes involved the interaction between the local telephone companies and the Internet. As it had done in many other fields, the Internet completely changed the rules of the voice networking game.

The Internet is not actually a network at all, but rather a worldwide collection of networks that are all interconnected, a "network of networks." This internetwork has been around for more than 30 years, but most of those years were spent as an obscure research and educational tool beyond the consciousness of the general public. This all changed in the mid-1990s with the appearance of the World Wide Web on the Internet. The Web is technically a subset of the Internet, which means that not everything on the Internet is part of the Web. The *Web* is collection of computers on the Internet (the Web site or Web server) with information available to almost anyone with Internet access, and Internet access did not have to include access to the Web. A special software package known as the *Web browser* or Web client was needed on the home PC (or any other computer system, for that matter) to enable users to get information from the Web sites on the Internet. The local telephone companies' lines were almost always used for Internet access, especially from home. This use of telephone lines changed the rules of the telephone company service game, although few noticed at the time.

But if anyone doubted that the local telephone company service game had changed for the good, these doubts were dispelled the afternoon and evening of January 6, 1997. To their credit, Pacific Bell had seen it coming, although there was little that technicians or anyone else could do about it. There was little that could be done because the problems that occurred on that date were the result of design decisions and engineering choices that were made many years before.

The root cause of the problem, oddly enough, was Christmas. For their holiday gifts—which included not only traditional Christian households, but also many non-Christians who exchange gifts at that time of the year—many people, especially elementary and high school students, found personal computers (PCs) under the tree. By 1996, many had discovered that college students were not the only ones who needed PCs to do assignments and research. The rise of Internet and World Wide Web popularity in 1993–1994 had completely changed education, for better or worse. It was no longer just a matter of typing a report with the word processor software on the PC. In more and more cases, the topic was explored, the material researched, and references checked and cross-checked all from the PC itself. In fact, in some cases, the assignment itself was distributed to the class with the help of the network. Attachment of the PC to the Internet and the Web made this all possible.

So during the Christmas season of 1996, many thousands of students nationwide and in California got brand-new computers that almost universally had built-in modem hardware and connection software. This was good news for PC vendors and companies such as Microsoft, which had bundled easy Internet/Web access into their products both in anticipation of and to encourage this network trend, but Pacific Bell officials were not as happy and were downright worried.

They fretted because the telephone network that Pacific Bell and all of the other 1,300 local telephone companies in the United States had built was not designed for the PC, nor the Internet, nor the Web. How could it be? The network that evolved to connect Alexander Graham Bell's 1876 telephone invention could not have possibly anticipated later inventions such as the PC, the modem, or the Internet. The network that had evolved was designed, engineered, and built for one basic purpose: two people talking to each other for relatively short periods of time over the telephone. This network is commonly called the *public switched telephone network* (PSTN), especially by those within the telecommunications field itself.

The PSTN is not the same as the Internet. True, the Internet essentially plays a role for PC connectivity similar to the role the PSTN plays in the voice world, but there are significant differences. For instance, as far as the PSTN is concerned, almost any device that generates the right electrical signals can use the PSTN to connect to a compatible remote device. The user devices attached to the PSTN do not have to be telephones; they might be fax machines or even PCs. Anything that makes noise can communicate over the PSTN. In fact, that is what a modem is for. A modem is needed on a PC to allow the PC to attach to some other remote device over the PSTN. The modem makes noise out of the digital bits that PCs and other computers understand. This noise is unlike human speech, but it is noise nonetheless. Anyone who has ever dialed a fax machine by mistake knows that such devices talk a very different language than people do.

There are more differences between the PSTN and the Internet—many more. The major components of the PSTN are called switches, and the major components of the Internet are called routers, just to cite one such difference. A more systematic examination of the PSTN architecture and Internet architecture is done in Chapter 2, as well as an attempt to sort out the differences between switches and routers. It should be pointed out, however, that the entire switch/router debate is an active one. Nevertheless, some more or less firm conclusions can be drawn about these differences.

So the PSTN can be used to connect a home or business PC to the Internet. All it takes is to plug the PC (by way of the modem) into a telephone line (known as an access line into the PSTN), dial up any one of the thousands of Internet Service Providers (ISPs) around the United States, and after a minimum of bookkeeping for payment options, you are on the Internet and cruising the Web in style. There are other ways to access the Internet, but you would be hard-pressed to find instances where the PSTN in one form or another is not used. This simple fact is both a blessing and a curse to the local telephone companies. It is nice to be so popular, but it is hard to do things in the same old way.

So the Pacific Bell people looked at the Christmas season of 1996 with some amount of fear. They were afraid of what would happen when all of those new PC users plugged in and logged on to the Internet and started surfing the Web. Because a large percentage of all the Web sites in the world were then in California, they had reason to be worried, but nothing much at all happened on Christmas Day.

It was not so much that the server Web sites were the problem. The Web sites almost always had dedicated bandwidth in the form of leased private lines connecting the servers to the Internet. The Internet backbone itself was similarly composed of dedicated bandwidth. The issue was that the scarcest resource between the user PC and the Web sites was the dial-up access lines or local loops used to connect these home users to the

Internet Service Providers. Because these dial-up links were designed primarily for intermittent voice telephone calls, long data calls made to the Internet could consume these scarce telephone network resources very quickly.

To understand the telephone company's dread, imagine that a new telephone was the prized gift of the holiday season. If everyone picked up their new phones at once and tried to make telephone calls, chaos would ensue because the PSTN is designed for a world of intermittent, independent telephone line use. There was no need to build a network where everyone used a telephone at the same time because, first of all, history had shown that people make calls more or less at random, and second, this was a basic assumption built into the very foundations of voice network engineering principles.

Apparently, there were enough other toys to play with Christmas Day to distract people, so the PSTN—the voice network—through which almost all of the analog modem calls traveled on the way to the Internet, survived. On Monday, January 6, 1997, the story was different. The students had returned to school and all seemed well, but around 3 P.M., the new PC users got home with their new homework assignments. A huge number of them turned on their computers, logged in to the Internet, and clicked their mouses, all at the same time.

Immediately, a PSTN *central office* (CO) where telephone lines come together to be switched in the East Bay area of San Francisco browned out. A brownout is not an absolute failure of the switch, which is basically nothing more than a big computer itself, but services are curtailed during a brownout, making it difficult to continue calls in progress and hard to establish new calls at all. Because both businesses and homes are serviced from the same central offices, both types of customers were affected. Throughout Contra Costa County, businesses lost their dial tones, and users heard nothing but silence when they lifted their telephones. Customers trying to dial a number heard fast busy signals (technically called *reorder*) that indicated too heavy a load on the PSTN trunking system, which links central office switches together. In frustration, many Pacific Bell customers had to use their cellular phones just to call up and report the service denials and outages.

The problems continued throughout the night as Pacific Bell became reacquainted with another networking truism: Congestion in any kind of network, once it occurs, is difficult to alleviate. It is much better to try to prevent congestion from occurring than it is to try to make it go away. Fortunately, things settled down, and the network more or less returned to normal. Until the next time.

And there will definitely be a next time. It has been estimated that some 10 percent of all central offices in California are chronically congested, and as California goes in many things, from fashion to networks, so soon will go the nation.

The Pacific Bell problem is by no means unique. All of the regional Bell operating companies (RBOCs: former pieces of the huge AT&T local network spun off in 1984 to go their own ways, but now oddly seemingly intent on somehow reassembling themselves in a fashion) have been dealing with problems like the one in California for a while as the Internet continues to grow in popularity, as the Web becomes a more essential part of people's lives, and as more people start or continue to work at home for a greater part of the work week. For example, a similar brownout (sometimes also called a brown down) occurred on the East Coast even before the Great California Brownout.

The winter of 1996 in the Northeast was harsh. Snowstorms were almost a weekly occurrence that January and February, and six inches or more of snow was not unusual for each storm. In the Washington DC area, a substantial blizzard made it nearly impossible for any government or other workers to venture in to the office. No problem: This is the 1990s. People who could not travel fired up the laptop and dialed in to the corporate network. Of course, many corporate networks are accessed today through the Internet as well, a practice sometimes called an *intranet*, although the term implies more than just this.

As one might expect, the same problems that occurred later in California appeared, this time in Bell Atlantic (now Verizon) territory. All over the service area affected by the storm, Bell Atlantic began to have overload and brownout problems with their switches. There were blocked 911 emergency calls, dial tone delays, and outright service denials. The effects were not as severe as in California, but the event sent a powerful message to network planners and designers.

Even if one knows nothing about networks, the Internet, or the PSTN in general, the problems posed by increased network usage—and changing network usage—are understandable. People are familiar with many systems that makes use of scarce resources. That is, the primary good that a system exists to deliver cannot be distributed to everyone at the same time. It makes no difference if the good is electricity, water, or phone calls. When workers in a downtown area hit the restaurants for lunch on a hot July day, the combination of air conditioning load and electricity in the kitchens dims lights all around town. During the Super Bowl halftime, water pressure drops so much that local fire departments hope that fires do not break out just then, because previously chair-bound TV spectators rise as one and make their way to the bathroom.

However, there is one major difference between water and electricity on one hand and the new PSTN loads on the other. The difference is timing. The power utility knows that it will be hot in July. The Super Bowl's low water pressure takes no firefighter by surprise. But network engineers never know exactly when the next mass login is going to happen, so planning for it in the short term becomes impossible. Problems have been triggered by events as diverse as the release of the Starr Report on President Clinton and the Victoria's Secret online fashion show.

This is not to say that planning ways to alleviate PSTN congestion cannot be done in the long term. After all, any system is designed for performance maximums, whether the key activity is air handling, flushing, or Internet access. Oddly, the PSTN was built for a number of maximum traffic days. Traditionally, heavy telephone traffic takes place on Mother's Day, the day after Thanksgiving, and New Year's Day. For the rest of the time, vast parts of the network more or less sit idle. The predictability of the traffic patterns is what has been shattered by the Internet and Web. The timing is off.

Another point that may strike one as odd is that in some senses it does not make much of a difference whether the increased traffic comes from talking. In California, everyone logged in to the Internet at the same time, but what if everyone had simply picked up the phone to call Grandma to thank her for the new PC (as if that would ever happen)? The effect would have been the same in some respects, but different in others.

The same sorts of service denials would take place. Someone once said that country star Garth Brooks has "crashed more central office switches than everything else put together." If tickets go on sale at 9 A.M., you can bet that a couple of thousand people will call at 9 A.M. Now the point has already been made that talk is just noise and so are the signals on the PSTN from devices such as fax machines and PCs, but the digital noise from PCs is very different from the noises made by people talking. In fact, these differences form one of the major points of this book.

It is somewhat reassuring that the examples in this section are several years old. As millions of people in the United States and around the world make the move to high-speed (broadband) Internet and Web access, the pressure on the PSTN has eased somewhat. However, because there are still many more people that do *not* have access to the Internet and Web (and yet very much want it and need it) than do, no one in the industry can afford to relax too much.

Because this fundamental point is so important to the book as a whole, some time should be spent exploring the differences between PC noise and people noise. Even those who have some familiarity with telecommunications may benefit from this section because this very familiarity sometimes leads to a kind of blurring of important distinctions.

ANALOG AND DIGITAL

There is no more profound or basic a distinction to be made in telecommunications today than the distinction between analog and digital. When two humans talk, their voices are *analog* signals carrying *analog* information content. When a user sitting at a PC employs a modem to link a PC to an Internet Service Provider (ISP), the link uses *analog* signals to carry *digital* information content. Consider that there are two aspects to these conversations: the information content and the signal (a less ambiguous term is *line code*) itself. The information content is what is being carried, and the signal, or line code, is how it is being carried. Either or both may be analog or digital. In other words, the information content being carried across a network may be analog or digital, and the underlying signal into or on the network itself may be analog or digital as well.

Before considering information content and signaling in detail, some words about the differences between analog and digital are in order. Even within the telecommunications industry among those who should know better, there is much debate about the correct use of the terms *analog* and *digital*. For instance, sending 0s and 1s over a modem attached to the PSTN is "analog transmission," but sending exactly the same string of 0s and 1s over a satellite channel with a different type of modem (but a modem nonetheless) is considered a form of "digital TV" (to be generous, the term "digital TV" in this context is just shorthand for "digitized TV").

To develop consistent digital and analog terminology for the rest of this book, consider the information content itself. Never mind how the information gets across the network, just look at the information being sent and received. This information may itself be of an analog or digital form.

If it is analog, the information may take any value between an allowed minimum and maximum. There is no forbidden value—any value between the maximum and minimum is just as good as another. If the value of the information were plotted against time, the resulting graph would form a single continuous line with no jumps or discontinuities. Examples of analog information content include things such as temperature, air pressure, weight, and almost all common physical measurements. The air temperature cannot go from 50 degrees to 70 degrees without passing through all of the values in between. The same is true when the weight of a person increases from 190 to 200 pounds, although it may seem that way, especially around the holidays. The point is that the human voice is an analog quantity; human speech can be described by a number of variables such as *amplitude* (signal strength) and *frequency*, all of which are analog variables that continuously vary over time between a minimum and maximum value.

The upper portion of Figure I-1 shows how the amplitude of an analog variable might be plotted against time, giving a smooth curve.

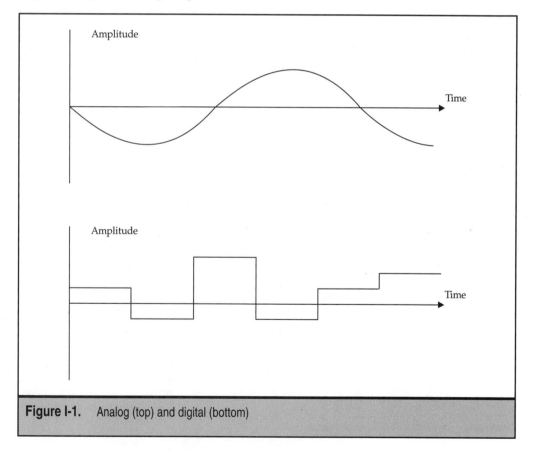

Figure I-1. Analog (top) and digital (bottom)

Of course, the information content does not have to be analog; it might also be digital. A variable is considered digital if it can take on only one of a limited number of discrete, or separate, values. There may be the same concept of minimum and maximum, but these limits are just the boundaries of the allowed values. Everything else outside this set of values is strictly forbidden or represents some type of error condition. The simplest example of digital information is the set of digits themselves. That is, the integers 1, 2, 3, and so on represent the only values that an integer can take. Binary digits, or *bits*, can take on only one of two values, 0 or 1, which is what makes them binary in the first place. A graph of digital information would show many discontinuities in value, just as one counts from "1" to "2" without worrying about values in between, such as 1.5. Modern computers are the prime example of devices that generate and consume digital information content.

The lower portion of Figure I-1 shows how the amplitude of a digital variable might be plotted against time, giving a disjointed curve. In fact, the only reason that sharp vertical lines are included between the levels is for the human eye to better interpret what is going on. Technically, there is no "jump" between one value and another. There is a time when the value is X, and a time when the value is Y, with no values between X and Y at all. As it turns out, this is hard to do in the real world because most electrical phenomena—and electricity is most often used to represent information traveling across a network—cannot be made to instantly jump from one value to another. Note also that digital values do not necessarily have to be binary. That is, there can be more than just two values for a digital signal. The figure shows at least four values.

This point about the electrical representation of information content on a network brings up the next aspect of the analog/digital world of telecommunications. It is clear that the information to be sent and received across the network may be analog (voice, for example) or digital (a PC file, for example) in nature. To convey that information, some method of signaling must be used on the network itself. This use of the term "signaling" should not be confused with other meanings of the term in telecommunications. For instance, signaling is said to be used on the PSTN to establish, maintain, and release circuits. For this reason, many in the industry prefer the term *line code* instead of *signaling* for the way that information is represented (coded) on the telecommunications medium (line) to avoid any chance of confusion. But this text uses the term *signaling* for this information representation process because this is the term much more commonly used. This use of the term *signaling* refers to the way that information content is transferred and nothing else.

The same logic that was applied to information content to define analog and digital can now be used to define how that information is signaled across the network. If the actual signal generated by a sender or transmitter and received by a receiver can take on any value at all within a defined range, the signal is analog. Note that the signal is analog regardless of whether the information sent is analog or digital. Sending analog information across an analog network becomes an almost trivial process, especially where analog voice is concerned. All that needs to be done is to convert the pressure waves of ordinary voice into waves of electricity using a simple transducer. The process is reversed at the other end of the network, naturally.

The underlying signaling network may also be digital in nature. In this case, as before, either analog or digital information content may be carried across the network. The digital information to digital signal process is relatively simple, but the conversion process from analog information content to digital signaling is not. Recall that an analog variable can take on any value within a range, but digital variables can only take on a limited number of values. How can one be converted into the other? In other words, if analog information can take on any value between 1 and 2, including 1.5, how is this to be represented as a digital signal which allows only the values 1 and 2? Obviously, there can be no simple one-for-one mapping of analog values to digital signals.

This dilemma has an answer, but not one that leaves everyone feeling satisfied. The representation of analog values by digital signals is known as *quantization* (sometimes the term *quantizing* is used). It turns out that there is always some *quantization noise* that creeps in during the analog-to-digital conversion process. The term *noise* is just used here to indicate the inaccuracy of mapping analog into digital that interferes with the desired result. *Noise* is just a general term often used to indicate the absence of needed information, or the presence of unwanted information on a telecommunications link. In the simple example I presented, the analog 1.5 value must either be represented by a 1 or 2. Perhaps an engineer could make it a rule to round off all values equal to or greater than 1.5 to 2, and all other values would round to 1. Even this simple example shows how quantization noise works. If 1.5 rounds to 2, then so will 1.6, 1.8, and even 1.555438. Now, there was obviously some meaningful difference between an analog 1.6 and 1.8, or there would be no point in generating the values in the first place, but after each has been changed to a digital 2, a 2 they must both remain. In other words, the receiver, which might want to convert the received digital signal back to analog information content, would have no way to distinguish between a 2 that started out as a 1.6 and a 2 that started out as a 1.8. Noise has been introduced into the analog-to-digital conversion process.

This quantization noise may be greater or lesser, but it never goes away. A network could try to convert analog information into 10 digital values instead of only 2, so 1.6 analog is 1.6 digital (digital values do not have to be whole numbers, just discrete numbers). Now the problem shifts to values like 1.555438, which must become either 1.5 or 1.6. The noise is ten times less, but it is still there. The question is how much time and effort should be expended to address the quantization noise issue. Quantization noise is always an issue when converting analog to digital, especially when analog voice is converted to digital signals for transfer over portions of the PSTN. Analog voice, in fact, usually uses 255 quantization levels for its digital representation.

Two other points about analog and digital should be made. First, when speaking of analog, the term *bandwidth* is used to indicate the frequency range that the analog signal operates in, so it is common to say that "analog television signals occupy a 6 MHz bandwidth." The MHz just means millions of cycles per second, where a *cycle* is the total range of analog values that the signal can take on. Here the term *bandwidth* means the same thing as frequency range, and this usage is common in the analog world.

But what about analog information content and analog signaling? Don't each of these have a characteristic bandwidth? Yes. If the bandwidth of the analog information (perhaps 6 MHz) is the same as or less than the bandwidth of the analog signal (perhaps 7 MHz), there is no problem. But what if a 7 MHz analog information stream is trying to make its way onto a 6 MHz analog signaling network? Something has to give. And give it does. In this case, it is said that the information content is *passband limited* to the 6 MHz signaling limit. The term *passband* can be taken to mean the amount of analog bandwidth that an analog signal will pass through the network. In this example, the 7 MHz of information can use the 6 MHz signaling channel by "chopping off" the frequency range between 6 and 7 MHz, chopping off the frequency range between 0 and 1 MHz, or any place in-between, as long the result is a 6 MHz bandwidth.

In the PSTN, analog local loops will passband limit analog signals in many places to the frequency range between 300 and 3,300 Hz. That is, nothing below 300 Hz or above 3,300 Hz will arrive at the other end of the wire. Outside of the United States, this upper limit is frequently 3,400 Hz, but the point is the same. The choice of the 300 to 3,300/3,400 Hz passband is no accident. Some 80 percent of the power of the human voice is within these limits, so modest expansions of the passband has only marginal effects on perceived voice quality. The passband limit is enforced by passband filters in electronics gear. This analog bandwidth limiting is an important point and plays a large part in ADSL discussions to come.

But when applied to digital, a second meaning for the term *bandwidth* is common. It is commonly said that "this Internet access line has 64 Kbps bandwidth." The Kbps means thousands of bits per second, where a *bit* is a binary digit and must be either a 0 or a 1.

As might be expected, there is a characteristic bandwidth associated with digital signals as well, but it is measured in bits per second rather than cycles per second. What if the bandwidth of the digital information content coming out of the back of a PC's serial port is 128 Kbps and the bandwidth of the telecommunications line is only 64 Kbps? Should 64 Kbps just be "passband filtered" out? Well, maybe not. If the 64 kbits just discarded are part of the file, the missing bits will cause a error condition at the receiving end and will probably result in the need to retransmit the whole file across the network. Why not just delay the extra bits in a special place called a *buffer*? In this context, a buffer is just a special area in memory used for communication purposes.

Why would a buffer help when PC communications are concerned? Only because of one of the major differences between using the PSTN for human-voice conversations and using the PSTN for computer communications. When people talk on the phone, a constant stream of information usually flows both ways. Even silence, which occurs when one party is listening or the other pauses to collect their thoughts, is represented by the full 3 KHz bandwidth (300 to 3,300 Hz) on the network. In other words, the bandwidth assigned to a voice conversation cannot be easily taken away and given to another conversation. In fact, this process defines what is called a circuit. A *circuit* can be defined in this context as bandwidth continuously dedicated to a conversation for the duration of the conversation. The duration of the conversation in turn defines the *holding time* of the call, as was already alluded to earlier.

But data, as defined by computer-to-computer file transfers, Internet Web browsing, and the like, is different. There is no longer a constant stream of information from source to destination. Instead, information is organized into units called *packets*. Packets have a minimum and maximum size (which is usually quite small) and must conform to a standard structure and set of rules called a protocol. On the Internet, packets conform to the *Transmission Control Protocol/Internet Protocol* (TCP/IP) structure, now officially known as the *Internet protocol suite*.

Data is different because most data applications are bursty in nature, which means that when packets are sent from a source to a destination, many packet are sometimes generated in a second, but sometimes no packets at all are generated in a given second. For example, when an Internet user downloads a Web page, many packets are generated to carry text and graphics across the network to the destination PC. However, as the user mulls over the new information now presented on the PC monitor screen, no packets are sent back and forth at all. Ironically, the PC is typically linked to the Internet by way of an ordinary telephone circuit, which dedicates all of the bandwidth to the circuit all of the time. The bandwidth is still there, but it is not being used by the reflective PC user. This whole issue of "packets on circuits" (which is really shorthand for "bursty data packets carried on circuit-switched networks designed for voice conversations") is explored more fully in Chapter 4.

So it is possible, and prudent, to buffer bursts of packets above the bandwidth limit of the communications link until there is a lull in the packet stream from the information source. The buffer can then be emptied in a more leisurely fashion without loss of packets and the resulting errors from the lost information in the packets at the receiver. Of course, it may be desirable to increase the bandwidth on the communications link if the buffer itself is constantly full or is in danger of overflowing and losing packets.

Note the tradeoff implicit in buffering versus bandwidth. More bandwidth may cost more, but it may be the only way to efficiently handle packets in the buffers, which is important because packets sitting in buffers are not going anywhere at all. Buffer time adds to the end-to-end delay that packets must endure as they make their way across the network. In this case, more bandwidth will cut down on the packet delay, make the transfer of Web information occur more rapidly, and potentially cut the time needed to access the information needed at a given Web site. This interplay between bandwidth and delay is an important one and is explored more fully later in this Introduction.

The discussion up to now has focused on analog/analog and digital/digital situations. The other possibilities should be discussed a little as well.

Because there are two alternate forms of signaling and two alternate forms of information content, it is easy to construct a matrix to show how these two aspects of communication interact with each other. Please note that whether the information content is analog or digital, or whether the signal is analog or digital, communications can still take place as long as both ends of the conversation employ the same techniques. This general rule has some exceptions, but these complications are beyond the scope of the present discussion.

When interfacing a specific form of information content to a specific form of signaling, a specialized interface device is usually necessary. If no special interface device is needed, one could say that the network is specifically engineered for a particular type of signal and information content. That is, the reason no special device is needed is that the network fully expects to perform with this type of information and signal.

So the analog/digital matrix would show the name of the interface device in the intersections of the matrix. If none were necessary, it would be because the network was engineered for that particular type of information content and signal. Obviously, there would only be one instance of this "default" network operation. A network cannot be engineered for two different things at the same time. Such an analog/digital matrix is shown in Table I-1.

The table shows the names of the interface devices needed for all combinations of information content and signaling, whether analog or digital. Looking at the left column, it is immediately noticeable that no special interface device is needed to send analog information content across a network using analog signaling. Actually, this is not strictly true, as was mentioned previously. A transducer is normally required to convert analog sound waves to analog electrical waves, but this conversion is so accepted and transparent that it is seldom mentioned in any special way. When applied to the PSTN, the equivalent statement would be "no special interface device is needed to hook up a telephone to the analog local loop linking the premises to the telephone switching office." This is also just another way of saying that the PSTN is "engineered for voice." Analog voice information content is the "default" mode of operation for analog local loop signaling in the PSTN.

However, there may also be digital forms of information that need to be sent on the PSTN, such as would be needed to link a home PC user to an ISP over the analog local loop, as pointed out previously. In this case, the information content is digital (the PC's 0s and 1s), but the local loop remains analog. The special interface device needed in this case is the *modem*, which stands for *modulator/demodulator*, with *modulation* being a technical term for this operation. The coupled terms mean that modems must be able to both modulate digital content to analog signals and also demodulate analog signals to digital content back again. Modems may be external boxes separate from the PC or built-in on a board which is inside the PC chassis.

Continuing the matrix on the right side, a device known as a *codec* is needed to send analog information content over a network employing digital signaling. The term *codec* comes from coder/decoder, and *coding* is the technical term for expressing otherwise analog information as a stream of bits. A codec device is needed when analog voice conversations are coded for transmission on a digital link using digital signaling. An example is

		Signaling	
		Analog	*Digital*
Information content	*Analog*	NA	Codec
	Digital	Modem	DSU/CSU

Table I-1.   Interface Devices for Analog and Digital

when analog voice from an ordinary telephone is to be sent through a corporation's or carrier's digital network.

The last device is listed as a DSU/CSU, which stands for *Data Service Unit/Channel Service Unit*, but its purpose is quite simple. Note that no special interface device was needed to go from analog information content to analog signaling in the PSTN network example using a telephone. At first glance, it would seem that the same should be true if the information content is digital and the signaling is digital, but this is not the case. The network can only be engineered for one type of signaling, and when talking about residential access, this is analog signaling. But as it turns out, there are several kinds of digital signaling methods that can be used on the local access line (local loop). Digital information content (for example, a PC file) can also be represented in many ways, which means that a special interface device must be used to convert between a specific form of digital information content and a specific form of digital signaling and back again. Although in a very real sense, the DSU/CSU performs the same simple type of conversion for digital/digital links that the transducer performs for analog/analog links, the DSU/CSU plays an important role in this book and so is listed in the table whereas the transducer is not. There are many forms of DSU/CSU devices, as many as there are digital content forms and digital signaling formats. This book introduces some of them and places them in the context of DSLs.

BROADBAND

Just as no book about ADSL can be understood without a clear definition of analog and digital, so no book about ADSL should start without a definition and discussion of *broadband*. The telecommunications field is sometimes criticized—and rightfully so—for being a field inundated by professional jargon. To be taken seriously, certain terms must be used over and over and acronyms tossed about with reckless abandon. Unfortunately, the key terms in the telecommunications field are used over and over again with slight or radical changes in meaning, depending on context, often to the dismay of those seeking to understand just what all the excitement in telecommunications is about. A simple example already given was the use of the term *signaling* for both call control and line coding. The term *asynchronous* is one of those overused and underdefined terms, and the term *broadband* is another.

A definition of broadband is needed because ADSL is sometimes referred to as a method of providing *broadband access* to residential users. The *access* is just the way a person at home would link up to the PSTN and is usually through an analog local loop consisting of a pair of copper wires. This is not the only form of PSTN access, of course, but ADSL primary addresses this form of access line.

In this book, the term *broadband* refers to a telecommunication link that has a *latency* less than that of a 2 Mbps (2,000 Kbps) telecommunications link used for digital voice communications. The term *latency* is usually defined as a delay in the network, but there is more to latency than that. A few words have been said about the interplay between bandwidth and delay. Now is the time to explore this relationship more completely, with an eye toward defining latency and applying the definition to broadband access networks.

Both delay and bandwidth play a crucial role in networks. Today's networks must support not only traditional voice and data applications, but many applications that fall into the category of "interactive multimedia," so not only the network delay (for interactivity) but also the network bandwidth (for huge video and audio files) must be adequate for these services. Some applications on a network are inherently *delay bound*, such as voice. Assigning more bandwidth to voice does not make it better or more efficient. All that matters is adequate bandwidth and delay. People will not talk if the delay between speaking and listening is 5 seconds, or starts at 1/2 second and rises to 3 seconds.

But other applications on a network are *bandwidth bound*, such as most data applications. Assigning more bandwidth to these applications makes them run better and more efficiently. Assigning more bandwidth to a PC file transfer makes a big difference. What matters here is adequate delay and bandwidth. A file transfer does not care whether the beginning packets of the file transfer make their way across in 1/2 second and then rises to 3 seconds for the packets at the end of the file. Even a delay of 5 seconds might be okay, but sending a huge Web page over a circuit with very small bandwidth can cause delays of many minutes and will definitely make users angry.

Delays vary according to the number of bits sitting in a buffer somewhere, the current congestion load on major network components (switches or routers), or both. Most people have an almost intuitive grasp of the effects. A telecommunications network is mathematically equivalent to a queuing system—so are banks. It makes sense that it takes longer to cash a check either when many people are in the bank, when each teller is taking longer than usual because each person has multiple transactions, or both. A banking line is just a buffer, and a teller with a lot of transactions to sort out is congested. The delays in both situations are highly variable. A person may get in and out of the bank in 5 minutes in the morning, whereas someone else may languish for 30 minutes in the afternoon. Likewise, a packet may arrive 20 milliseconds after it is sent (a millisecond is one-thousandth of a second) whereas the next packet makes its way in a leisurely 40 milliseconds. On the Internet, delay variations on the order of hundreds of milliseconds are not uncommon. On the PSTN, the delays are much lower and are generally stable.

Interestingly, the *International Telecommunication Union* (ITU), which is in charge of all international telecommunications standards, defines a term it calls *transfer time* for networks. Transfer time has a little different meaning than what one might expect because the ITU frequently specifies network services in terms of latency, instead of a combination of bandwidth and delay, as one might expect. Yet this latency specification is totally adequate. Because most people use the terms *latency* and *delay* interchangeably, something else must be going on here. And there is.

Figure I-2 shows how network delay and bandwidth can be calculated for a data application on a network. Information transfer takes place as a series of transmission frames (which contain the packets), which are some number of bits long (X bits in the figure). Because this is the data world, packets and frames always contain bits. The transmission path is pictured as just a passive "bit pipe" that in no way alters, stores, or converts the

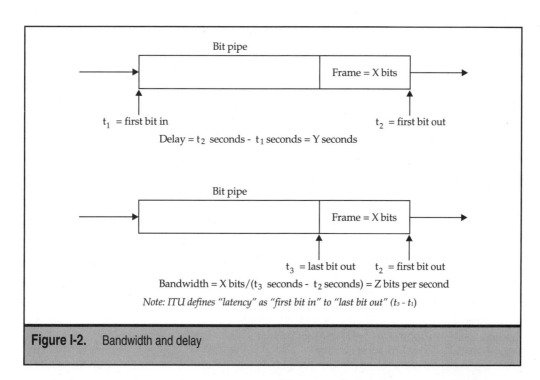

Figure I-2.   Bandwidth and delay

bits. Now, by ITU definition, the time elapsed from the moment the first bit of a given frame enters the network until the moment the first bit leaves the network is the delay. Note that as a frame makes it way through the network, the delay may vary, have a maximum and minimum, an average value, a standard deviation, and so on. The delay could be measured end to end from sender to receiver, between two particular points on the network, or a number of different points along the way.

In contrast, bandwidth is defined in terms of bits. Note that because this concerns data applications, the definition uses the digital version of the bandwidth term given earlier and not the analog version of frequency range. This bandwidth is in bits per second, so digital bandwidth is defined as the number of bits in the frame divided by the time elapsed from the moment the first bit of a frame leaves the network until the moment the last bit leaves the network. Actually, this is only one possible measurement point. Note that as frames make their was from an access link to other parts of the network, and so forth, the bandwidth the frame has to work with may vary considerably.

The whole point is that the ITU defines "latency" of "transfer time" as the interval from first bit in to last bit out. As should be obvious from the figure, this concept neatly combines the effects of delay and bandwidth on frame transfer. In fact, it allows networks with lower delays to compensate somewhat for restricted bandwidths, and it makes no difference at all whether the bits of digital information content are sent with analog signaling (modem) or digital signaling (DSU/CSU). The definitions of bandwidth and delay are still valid for all forms digital information content.

Keep in mind that this example ignores the effects of errors or overhead on frame transfer time. Typically, if a frame and the packet it contains is received in error, it must be retransmitted over the network, which adds to the overall latency of the file transfer, of course. Also, some of the bits in a frame or packet are not used for information content, but for the internal use of the network itself. These are overhead bits. Factoring in the effects of errors and overhead would give the overall *throughput* of the network. The lower the error and overhead penalties, the closer the throughput is to the transfer time.

Again, it should be noted that no distinction is usually made between delay and latency. The somewhat special definitions used here are peculiar to international standards and used primarily for illustration. Many working telecommunications experts might consider this whole discussion overly "picky," but the idea here is to introduce the concepts precisely before working with them.

With firm definitions of network delay, bandwidth, and latency in hand, it is now possible to take a second look at the term *broadband*. Broadband was defined previously as "a telecommunication link that has a latency less than that of a 2 Mbps telecommunications link used for digital voice communications." It should be clear that this latency applies to both the bandwidth and delay of a 2 Mbps link. The PSTN itself, designed and optimized for voice, has minimal delays to work with in the first place, so broadband access usually refers to a bandwidth greater than 2 Mbps, purely and simply.

The current interest in broadband access lines is due to the increased use of the Internet from residential users. If someone dials in to the Internet over a modem attached to an analog local loop, chances are they are getting about 33.6 Kbps of bandwidth or slightly more with some types of modems. Obviously, this is a long way from the 2,000-Kbps minimum of a broadband access line. ADSL is a way to make a broadband access line out of a local loop without any major expense in re-engineering the wire in the loop or replacing it. In fact, ADSL and related techniques go far above and beyond this 2 Mbps minimum. ADSL itself can operate at 8 Mbps, and some related technologies can push bits at a rate up to 50 Mbps through the access line.

This definition of broadband is still completely valid. But oddly enough, the success of DSL and other technologies such as cable modems has made it necessary to redefine the term broadband when applied to high-speed residential access lines. This new, working definition of "broadband" was needed because of the throughput speeds of what are called "broadband" services but do not meet the strict definition of "broadband" given here.

In other words, a service provider cannot offer broadband services unless the service actually meets the definition of broadband. This only makes sense. So a customer orders up a "broadband" Internet access service from a telephone company or cable TV operator that offers cable modems, and the customer finds that the link runs to their home not at 2 Mbps or more, but a more pedestrian 640 Kbps or less. This might be very fast compared to dial-up modems, but it's just not broadband, And in many places there are strict rules about businesses accurately describing what they are selling. So service providers needed a definition of broadband that more accurately reflected the speeds that their customers were actually getting.

Some service providers just called it "high-speed Internet access" and let it go at that. But "high-speed" is really not defined at all, and so meaningless. After all, 33.6 Kbps is "high-speed" compared to a 9600 bps modem.

So many "high-speed" services now characterized as "broadband" use a newer form of definition of just what constitutes a broadband service. The working definition is "any service with enough bandwidth and low enough delays to support streaming (sustained) digitized (or true digital) video services in one direction." Advances in video systems currently put this speed at about 400 Kbps and above. So most "high-speed" services that struggle with the older "more than 2 Mbps" easily pass the "streaming video" test.

More details on the differences between analog and digital information, circuits and packets, phone calls and Web browsing, and the PSTN and the Internet are discussed in full in the first few chapters of this book. Understanding and appreciating these differences are the keys to understanding and appreciating what ADSL does, and what it is for.

This introduction has introduced and defined terms that are used constantly throughout the remainder of this book. This might be a good place to summarize the major points examined in this section of the book. Briefly, the PSTN is designed, engineered, optimized, maintained, and operated primarily for analog voice conversations. Even though the PSTN is primarily digital today, the engineering is geared toward short time duration voice calls from human to human. But more of the information content, voice and otherwise, is being transported in the form of digital bits. Perhaps the time has come to make some accommodation to this longer lasting digital information, especially with regard to the limited bandwidth (passband) local loops, and the congested switching and trunking networks on the PSTN that these loops provide access to. This accommodation by using ADSL is the basic theme of this book.

CHAPTER 1

Welcome to the Information Superhighway

For a while in the early 1990s, it was fashionable to use the term *Information Superhighway* to describe a kind of ultimate network for residential and business use. This network would enable people at their desks at work or in their living rooms to have access to all of the information they needed to accomplish business tasks, to do their homework, or even to plan a family vacation. Press conferences were held, federal programs were announced, and many companies and vendors made big plans, but today if anyone mentions the idea of an Information Superhighway, it is usually in the context of the "good old days."

Ironically, it now appears that the Information Superhighway actually exists, at least in a rough, preliminary form. This would be the Internet, of course. From relatively humble beginnings as a research and education network confined to government agencies and major universities, the Internet has exploded in the mid-1990s to become an almost indispensable part of ordinary people's lives.

The desire that people have to be able to interact with network content, whether manifested as talking to a TV show or mumbling to e-mail, has never been so strongly demonstrated as with the Internet and World Wide Web. The Web is a gold mine of multimedia, and applications such as "interactive video chat rooms" form a kind of rudimentary video conferencing package that adds an element of interactivity to the mix.

The lure of interactive multimedia and the Internet is amply illustrated in the phenomena of "work at home"—most people are familiar with the idea of working at home by remotely accessing the Internet or intranet at the office.

The concept of "home at work" reverses the trend: Take the whole family to the office. Why? Faster access for the Web, of course! The Web is a multimedia paradise, and many family educational and recreational needs revolve around the Web. The children can do their homework, the latest movies can be previewed and reviewed, Dad can work on his golf game with an instructional video, and Mom can plan the family vacation (and see what the view from hotel room 422 at the resort is), all on the Web.

So in spite of the targeting of many broadband services for residential users, many of the users of video services are right in their cubicles at work. According to a Jupiter Media Matrix study publicized in 2001, 24 million people in the U.S. have multimegabit connections to the Internet at work, but only 8.6 million enjoy broadband (as defined informally as "fast enough for streaming video" in the Introduction) connections at home. People seem to check out the movie previews at work and see the show at night. Although most businesses seemed resigned to workers using the Internet for more than work-related activities during business hours, some have attempted to block all but internal videos to the cubicle.

And it is not just ordinary residential users that need adequate Web access bandwidth for multimedia and busy Web pages. The speedy lines used by large businesses are often too expensive to cost justify in small offices with just a handful of employees. This whole contingent of *small office/home office* (SOHO) workers and their families is one of the key market segments that ADSL is aimed at.

The Web has invaded almost every aspect of modern life today. Once people realized what the Web could do, they quickly came up with new and innovative applications to take advantage of the new medium. Table 1-1 shows some of the activities the Web is used for today.

Activity	Advantage	Comment
School homework	No "book is out" problem at library	Web assignments routine at all educational levels
Shop at home	Online catalogs always up-to-date and largely typo- and error-free	Retailers have no worries about shopping season weather
Banking	No lines at the bank	Large percentage of transactions will be done on Web soon
Bookstores	Author information, reviews available	Largest volume on Web
Buy a car	Easy to get best prices around	Second largest dollar volume
Product information	Cuts down on printing/ distribution costs	Potential customers can access only the information they need
News	Almost instantly updated	No wait for print journalism
Marketing	New products introduced immediately	Less need for costly ad campaigns
Research	No travel to special centers	Much historical materials, scientific, and classical literature online
Software distribution	No expensive mailings needed	Convenient and simple

Table 1-1.   Ten Things to Do on the Web

This list is neither exhaustive nor limited in any other way. For instance, adult enter-tainment and related euphemisms are a big money maker on the Internet and Web. How-ever, because many observers try to downplay the role of this segment, the significance of this market is merely noted here. All of the activities listed are current and firmly en-trenched. Newer activities that have been proposed for the Web, such as television pro-gramming distribution and voice, have not been included in spite of enormous interest, but only due to their current sporadic use.

This table has been reproduced unchanged from the first edition of the book. What was valid then is still valid now, but with a few interesting notes. The "adult entertainment industry" has emerged from the backroom just on the strength of economics to take its place as one of the driving forces behind Internet evolution. Many of the most exciting developments on the Web, from improved streaming video to instant credit card validation for authorized access had their origins in the adult sector. And what works for the adult entertainment industry can be applied to the e-commerce and research sectors of the Internet as well. After some vestigial reluctance, many equipment vendors and high-speed link service providers candidly admit that some of their biggest customers are in the adult sector, and money is never an issue. Sex sells, and always seems to sell very well. None of this should be interpreted as an endorsement of this aspect of the Internet and Web, which has posed many problems for society as a whole, from a resurgence of child pornography, which had all but disappeared by the 1980s, to "cyber-sexual harassment." This is just an acknowledgement of the reality of the human condition.

Other notes are of interest as well. The first edition of this book badly missed the sudden popularity of online music services such as the widely hailed and at the same time widely attacked Napster. And in spite of the widely reported issues about copyright and distribution that such services have raised, a report by Jupiter Media Matrix in May of 2001 showed more than a 33 percent rise in the number of music media players in use between January 2000 and January 2001, from 31.3 to 41.7 million users.

The same timeframe showed a sharp rise in the use of tax-related Web sites in the U.S. Not only does it seem to be easier to gather tax information on the Internet (no surprise there, as a visit to the post office or IRS office will show in the spring), but sound reasons and incentives are in place to encourage the filing of both state and federal tax forms online over the Internet in a totally electronic fashion, right down to the taxpayer's electronic signature.

So this table could easily be extended to include "adult entertainment," music, and taxes as more things to do on the Internet.

It should also be noted, but hardly needs to be proven, that home Internet use continues to grow. Nielsen//NetRatings, for example, has documented a 40 percent increase in home Internet use in Philadelphia between March 2000 and March 2001, from 40.4 percent to 56.5 percent of the population. Denver was up 34 percent, Boston 32 percent , and Atlanta also up 32 percent. Other U.S. cities show less dramatic rises, but substantial nonetheless. And the lowest increases in the survey—such as 13 percent in San Francisco and 16 percent in Washington, D.C. during the same time period—came in cities with considerable Internet usage to begin with. And for at least the last six months of this period, the entire telecommunications industry in the U.S. seemed to be in some kind of doldrums as Internet-related stocks fell and layoffs were rampant.

Nielsen//NetRatings also tracked Internet user growth by occupation in the same March 2000 to March 2001 time frame. The results are of direct interest here, and the figures published by Nielsen//NetRatings Web as reported in several sources (this version comes from *Internet Week*, 5/7/01, p. 72) in May 2001 are shown in Table 1-2.

Results are ranked by percentage growth rather than absolute numbers, but several trends are clear. Growth on the part of the more traditional Internet user groups such as students and technical professionals has been overshadowed, in some cases by a wide

Occupation	March 2000 Users	March 2001 Users	Percent Growth
Factory workers/laborers	6.2M	9.5M	52%
Homemakers	1.6M	2.4M	49%
Service workers	2.1M	2.9M	37%
Salespeople	4.2M	5.6M	34%
Clerical/administrative	4.2M	5.6M	34%
Craftspeople	3.2M	4.1M	29%
Educators	3.8M	4.8M	28%
Retirees	6.6M	8.5M	28%
Military	1.3M	1.7M	28%
Self-employed	7.4M	9.2M	24%
Professionals	14.9M	18.5M	23%
Full-time students	1.8M	2.2M	23%
Technical workers	7.2M	8.8M	22%
Executives/managers	11.9M	14.4M	21%

Note: Numbers are for heads of U.S. households. Source: Nielsen//NetRatings.

Table 1-2. Occupations of Internet Users

margin, by the growth of the new breed of Internet user such as ordinary workers, salespeople, and homemakers. Obviously, the Internet is now for everyone.

Without adequate bandwidth, all of the multimedia information available on the Internet and Web comes with a steep price—usually, a required high degree of patience. Some files available on the Internet are up to 45 megabytes (for example, a virtual reality tour of Notre Dame, complete with hunchback). It is worth remembering that computer people like files to appear small, and so measure them in bytes (8 bits), whereas network people like networks to appear fast, and so measure bandwidth in bits ($\frac{1}{8}$ of a byte).

So using the office Internet connection rather than a dial-up link at home is one way to get the bandwidth people need to access the Internet and Web. After all, it takes a lot of bits to preview a major motion picture, access a multimedia encyclopedia, and so on. Books are the number one dollar item sold over the Internet, but why stop there? Why not put the whole book on the Web and let you read it there (for a small fee)? Text is the smallest and most compact form of information, so why not just download the whole book to your home PC? Copyright issues aside, distributing even something as small and simple as book text over the Web creates problems just in terms of bandwidth.

Consider sending a compressed, 300-page book containing a modest 20 figures (sort of like this one). How long would it take to send this book from one site on the Web to another at various speeds? The results are shown in Table 1-3. (The table ignores the slowing effects of line overhead and errors, but the comparison is what is important.) But perhaps people are not yet ready for electronic books. Widely publicized efforts to promote electronic books involving both popular authors and specialized viewers and Web sites have floundered. A book, it seems, is not totally defined just by the information it contains. Then again, modern books have been around for 500 years more than the Internet.

Because the table introduces some of the characteristic speeds at which many technologies operate, including but not limited to ADSL, this is a good time to look at these technologies in general. Several of them are examined in more detail later on in this book.

The table begins with a modem that operates at 2400 bps. That's right—this modem could download a whopping 300 bytes, usually just text characters, in a single second! In the 1970s, this was considered a high-speed modem. Most modems at the time ran at only 110 or 300 bps. This was state of the art at the time, and fortunately not many computers that existed in those days could handle information arriving or leaving much faster than that.

Today, modems operate at 33.6 Kbps and above, potentially all the way up to 56 Kbps. However, finding any modem that actually operates at 56 Kbps is difficult, and speeds above 33.6 Kbps are only in one direction. Moreover, many access lines still might operate with digital content well below even the more modest 33.6 Kbps, which remains the highest purely bidirectional analog speed. So 33.6 Kbps is used as a sort of "average." When the Web first entered the public consciousness in mid-1994, the most common modem speed was 9600 bps or 9.6 Kbps. This was not bad, but downloading the Japanese national anthem (for instance) could take about 10 minutes or so. Of course, the song was a lot shorter than 10 minutes to listen to, so no streaming real-time operation was possible.

Speed	Time	Amount
2400 bps (old modem)	1.5 hours	0.0625 pages/second
33.6 Kbps (new modem)	6 minutes	0.875 pages/second
64 Kpbs (ISDN)	3.5 minutes	1.67 pages/second
1.5 Mbps (T1, HDSL2)	7.7 seconds	0.13 books/second
6 Mbps (ADSL)	1.925 seconds	0.52 books/second
45 Mbps (T3, OC-1)	0.25 seconds	4.0 books/second
155 Mbps (B-ISDN)	0.07 seconds	14.3 books/second

Table 1-3.   Downloading a Book from the Web

In the recent past, a lot of talk has centered around delivering higher speeds to residential users and small office users by using an *Integrated Services Digital Network* (ISDN). A full description of ISDN would fill a book by itself, and some aspects of ISDN are investigated later. For now, it is enough to point out that much of what ADSL is intended for, ISDN was intended for first—that is, a kind of early "Information Superhighway" for access to all kinds of digital information content. Unfortunately, ISDN proved very expensive to implement using 1980s hardware and software and was priced way over the heads of most users. And ISDN still used the same existing and congested switches of the public switched telephone network (PSTN—this is the normal acronym used to describe the global voice network used by all telephones today). Oddly, the Web seems to incorporate many of the best features of ISDN plans. That is, ISDN services included plans for text messages (e-mail), information services (news), and even video (just as the streaming video sites do today on the Internet), features that are all available on the Internet and Web today. In any case, ISDN operates at a basic speed of 64 Kbps, although higher speeds are routinely used by combining two (and sometimes more) 64 Kbps channels to make them act as one higher-speed channel.

Next in the table is something called T1 and HDSL2. A T1 is a form of digital, leased private line, which means that a company (or even a residential user with deep pockets) can lease a point-to-point circuit at a flat monthly rate from a telephone company (others can and do offer these services today as well) and use it to their heart's content without concern for additional usage costs. This approach differs from the approach of the previous speeds listed in that the others so far have been "switched" network services; that is, a modem or ISDN user can connect to many sites over the access line, just not more than one at a time (the Internet is still just one place to these services). Although a T1 can run at a full 1.5 Mbps for a single source and destination, the line is designed primarily be used in the form of 24 channels (each an individual circuit) running at 64 Kbps each. Of course, all must start and end at the same physical locations, because no switching is provided—just a point-to-point link. Alternative ways of providing T1 service, such as wireless radio or fiber optic cable, have been around for some time. But the "new" way of delivering, transporting, and providing T1 is known as HDSL2, which is related to ADSL, but not closely. It should also be noted that the ISDN *Primary Rate Interface* (PRI) runs at 1.5 Mbps or 2.0 Mbps, but no separate entry in the table is provided for this line speed because it is so very close to the T1 speed.

ADSL itself is the next speed listed and is the subject of this book. The speed listed is 6 Mbps, but ADSL runs at a whole range of speeds, and admittedly, this listed speed is toward the high end. Much will be said about ADSL in a whole series of later chapters.

Following ADSL, the speed jumps up through the T3 and OC-1. Although it is possible to obtain speeds between a T1 and T3, these intermediate speeds are not common and are of no concern here. A T3 is a faster T1 and runs at 45 Mbps. However, like T1, a T3 is most often broken down into 64 Kbps channels, 672 of them to be exact. A company with a T3 running at 64 Kbps may only be able to give any user 64 Kbps at any one time, which can cause a problem when speedy Web access is the goal. An *Optical Carrier level 1* (OC-1) is basically a T3 running on a fiber optic cable, although a full description of OC-1 would go far beyond this simple relationship. The point is that both operate at about 45 Mbps.

The last entry runs at 155 Mbps, which is the speed for a newer and better version of ISDN known as *Broadband ISDN* (B-ISDN). One of the shortcomings of ISDN was that by the time ISDN was ready to roll out in large markets, the needs of users and capabilities of PCs had gone far beyond a mere 64 Kbps. In 1988, the ITU essentially added a lot of video services to ISDN and called it B-ISDN. Popular technologies such as *Asynchronous Transfer Mode* (ATM) and *Synchronous Optical Network/Synchronous Digital Hierarchy* (SONET/SDH) were part of this 1988 effort, as well. Unfortunately, B-ISDN still suffered from many of the other shortcomings of ISDN in terms of expense and awkwardness, but many of the intended B-ISDN features have found their way onto the Web.

It seems obvious from the table that a network operating at about 1.5 Mbps or above to everyone's house would be just fine for interactive multimedia. As mentioned, ADSL can operate above even this speed. ADSL would enable people to stay home on the weekend and leave the company parking lot empty for a change.

Notice something else odd about the preceding table. At least two of the speeds, ISDN and B-ISDN, were introduced not just as a speedy network technology—although that was a key feature—but they also included a whole suite of services that users at home or at work could access. Yet the point has been made that much of what ISDN and B-ISDN are supposed to do is now available on the Internet and Web. Both ISDN and B-ISDN were the products of the ITU and the most respected networking companies in the U.S. and around the world. The Internet started as a research and education network funded by the U.S. military.

What happened? Why are people around the world clamoring for faster Internet access instead of services based on ISDN and B-ISDN? The Internet has been around in one form or another for almost 30 years now. What happened in the mid-1990s to catapult the Internet into mainstream consciousness?

The answer to all of these questions is the World Wide Web. In the middle of 1993, a new type of Internet application appeared on the Internet scene. A year later, this new Web concept was undoubtedly the most popular aspect of the Internet. Soon many people's only contact with the Internet was through the Web. Without unraveling the whole history of the Internet, a few words about the early Internet are necessary to explain why the Web caused such a revolution when it did.

Probably the best way to understand the rapid growth of the Internet and Web and its transformation from just a giant, specialized computer network to a network everyone needs and uses is to trace the development of the Web browser and the resulting impact on the Internet itself.

THE INTERNET AND THE WEB

The Internet officially began operations in 1969. While some young people living in the present gathered at Woodstock, other young people living in the future were sweating and straining to get four sites connected by the end of the year. Of course it was not called the Internet then, but rather ARPANET, named for the U.S. military Advanced Research Project Agency, which funded it. Throughout the 1970s, the "Internet" continued to grow and expand as it added computers from many universities and research centers.

Strangely enough, although this giant network was a decent way for computers to communicate, no tools were specifically designed to navigate it. It was a fairly difficult environment for the users to use their various text-driven interfaces to communicate and exchange information on the network. But as awkward as it was, by the early 1980s, the ARPANET had grown so big that to better oversee its development, it was split into two parts. One formed links between the military networks (MILNET), and the other linked the academic networks. By the mid-1980s the academic part was overseen by the National Science Foundation and was called the NFSNET.

Today someone looking at the Internet of the late 1980s would hardly recognize it. Even with this continued growth and impressive progress, the greatest use of the Internet was for e-mail, but as resources and data availability expanded as each university was added to the NFSNET, the potential for wider use grew every day. The Internet remained an untapped gold mine unless a user was familiar with their own site's system, knew specifically where to go, how to get there, and what to do once there. The problems of the early Internet can be referred to as the "messy room" syndrome. Because many of the Internet users at that time were college students, this was certainly an appropriate image. There may be valuable resources buried somewhere under that pile of clothes, or then again, there might not be. Where to start looking for anything was more a matter of luck than skill, and finding one sock did not guarantee that the other one would turn up.

This is not to say that the Internet in the 1980s shares nothing in common with the Internet or Web today. Quite the contrary. All networks have an enormous inertia that must be overcome before change can take place. The more radical the proposed change, the greater the inertia, so only small changes are possible in short periods of time. As it turned out, the rise of the Web was based on a relatively small change, but one that had enormous implications and applications. Some of the milestones in the history of the Internet are shown in Figure 1-1.

Then, as now, the Internet was based on the TCP/IP protocol. Then, as now, people could dial in for access to the Internet. Then, as now, the Internet was based on what is known as a client-server architecture. Because this is such an important feature of the Internet and Web, more should be said about the overall concept of a client-server architecture.

According to the client-server concept, all applications running on a computer are either client or server versions of the application. A single computer can run both client and server versions at the same time, just as a PC can run a spreadsheet and word processing application at the same time, but it is more common to use a single machine to run all client versions of an application or all server versions of an application. Not surprisingly, a PC running only client applications is called a client, and a PC running only server applications is called a server. The Internet exists basically to link clients to the servers they need to access.

All interactions between computers on the Internet follow this basic client-server model. Consider a fairly common Internet activity—transferring a copy of a file from a remote computer to a home PC. In this case, the home PC is the client and the remote computer is the server. Both computers must be running file transfer software that is compatible with the TCP/IP protocol and with each other. The easiest way to ensure this

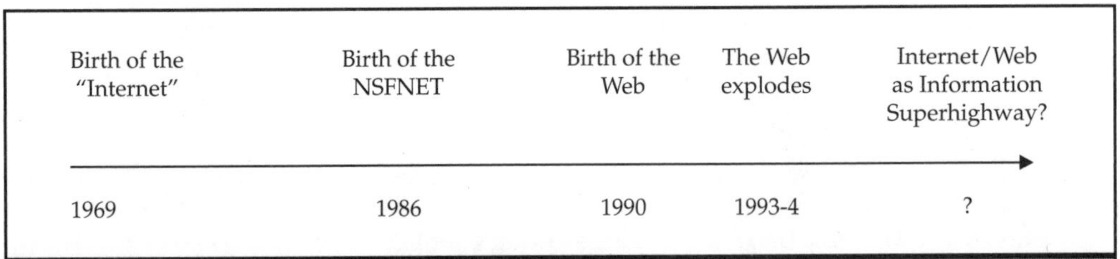

Birth of the "Internet"	Birth of the NSFNET	Birth of the Web	The Web explodes	Internet/Web as Information Superhighway?
1969	1986	1990	1993-4	?

Figure 1-1.   A (very) brief history of the Internet and Web

application-level compatibility is to base both client and server portions on the Internet's *File Transfer Protocol* (FTP) standard. There is an FTP client software package and an FTP server software package that embody these standards. The home PC runs the client FTP process, and the remote server runs the server FTP process. The home PC user accesses the remote FTP server over the Internet and transfers the file (actually a copy of the file; the original remains where it is). This client-server system is easy to implement and quite efficient, as long as the server and the server FTP application are up and running whenever a home PC user accesses it. A lot of effort is needed to keep heavily accessed servers up and running 24 hours a day, 7 days a week.

This very client-server model goes a long way toward explaining why the Internet was as organized as a messy room. The fact that no single company or entity ever controlled the Internet certainly helped as well: Getting one child to clean their room is easier than trying to organize three children and get them to coordinate the task. On the Internet, there eventually turned out to be thousands and then millions of children.

But with regard to client-server, suppose a user needs to send e-mail to a colleague, fetch a file from a remote server, and log in to a remote computer to run some custom application written to run under TCP/IP. Obviously, three different client applications would be needed—these are usually mail, FTP, and telnet, respectively—to enable the user to accomplish these tasks. In the old days, around 1984, each of these client applications would have to be run one at a time. In 1984, this was just as well. Few desktop computers or workstations could run more than one application at a time because the familiar Macintosh and Windows multitasking operating systems and computer systems powerful enough to run them remained in the future.

So the PC user ran the e-mail client to talk to the e-mail server, then ran the FTP client to talk to the FTP server, and finally ran the telnet client to talk to the telnet server. The whole process was awkward and slow, and by the time people got to the telnet phase, they forgot what they were looking for in the first place. Even worse, each of the client processes was command driven, meaning that a user had to type in a terse string of letters that had to be looked up or memorized before the server could perform the necessary task. For example, the FTP command **mget *.txt** would fetch all of the text files present in the current directory on the remote server, but **mget** was useless as a command in other applications.

Two ways were tried on the Internet to work around the twin problems of separate clients and text commands. Sometimes these approaches were known as *hierarchical* and *hypertext*. Their differences are considerable, but the important point is that the hierarchical approach was tried by an application called Gopher, and the hypertext approach was tried by an application called a Web browser. Both appeared around the same time, and the Web proposal actually pre-dated the Gopher idea. After some initial success in the early 1990s, the Gopher concept of hierarchical client and server organization pretty much disappeared, while the Web concept of hypertext clients and servers flourished. Cynics are quick to point out that the change in 1993 occurred at about the same time that the inventors of Gopher began to charge for Gopher server software while the Web server software remained free, but there was more to it than that.

The Internet Gopher was developed by the University of Minnesota and was released on the Internet in late 1991. Named for the school mascot and state nickname, as well as for every low-level worker who at one time had to "go fer" coffee, Gopher was designed to allow servers to be set up as information providers. Gopher servers could offer files with readable lists (instead of terse file names) which were organized hierarchically. Gopher menus contained locations and resources from across the Internet that enabled users to "go fer" it simply by pointing and clicking a PC mouse instead of typing a command.

By 1992 and early 1993, *graphical* (only a mouse is needed to coordinate a pointer on the screen) Gopher clients were available for both the Macintosh and Windows platforms as well UNIX. Because of its ease of use and installation, and Gopher server availability across multiple platforms, such as UNIX computers and Macintosh and Windows PCs, Gopher quickly caught on and became a popular tool for navigating the Internet. The Gopher was an important early step from an Internet "for some people" to an Internet "for all of the people." Today, Gopher is mainly of historical interest to all but the most dedicated Gopher site and user. Today belongs to the Web.

Early mammals lurked in the shadows of the hulking dinosaurs until changing environmental conditions killed off the dinosaurs and led to the age of mammals. In the shadow of Gopher's popularity was a piece of the Internet using a different method of data organization. In opposition to the hierarchical approach of Gopher, this method was designed to use hypertext for navigation and linking of data. The actual term "hypertext" was invented in 1981 by a computer scientist named Ted Nelson in his book called *Literary Machines*. Hypertext is based upon the idea of following a nonlinear path through data and documents through a series of linked "nodes" of information.

The concept of hypertext traces back to 1945 when its methods were proposed by Vannevar Bush, who wanted to design a computer information database he called a *memex* (fortunately, the term never caught on). A user of a memex computer would be able to read data on a subject and from within that document be able to link to related information throughout the database. In essence, the user would follow a trail of links throughout the database that was not limited to a specific topic but allowed for connections across boundaries of classification. Just as any good scientist knows, for example, that although biology and chemistry are separate fields of study, they are in reality so interrelated that to truly understand one, a scientist must have an understanding of the other (biochemistry). Bush therefore surmised that knowledge, in any field, needed to be explored, accessed, and found by users taking different pathways in their pursuit of information. This concept became the basis for the Web.

In fairness to the Gopher approach to the two systems of Internet organization, it is important to realize that neither method is right nor wrong in any real sense. Hierarchical is based primarily on organization and classification: a place for everything and everything in its place. To counter the messy room syndrome, hierarchical cleans up the room and puts everything in a drawer. If someone needs socks, they're in one drawer, pants in another, shirts in the closet, and so on. It is easy to get dressed if the person knows where to find socks, pants, shirts, and so on to complete the outfit.

On the other hand, hypertext views the *relationship* of data as the guiding factor because this method accepts that information needs the ability to cross boundaries and classifications to be useful. This method leaves the messy room messy, but allows that finding a pair of socks will likely lead to finding a pair of jeans, which leads in turn to a pile of shirts (ones that actually go with the jeans), and even though the person did not need anything else, hypertext can lead to some ties, which make the outfit perfect. Of course both hierarchical and hypertext's strengths are the weaknesses of their counterparts. In a hierarchical system a user needs to climb up and down the hierarchical tree before different but related subjects can be found.

There is one other aspect of hypertext that makes the Web a frustrating experience in some ways. Following links can be a confusing thing. Finding socks leads a person to pants and then to shirts, but in with the shirts is the disco shirt last worn for a Seventies retro party, which leads to a collection of disco records from someone's misguided youth, which leads to the entertainment cabinet, where also happens to be located the manual for programming the VCR. What does the VCR have to do with getting dressed? Maybe nothing, maybe something. But perhaps that is the important point that has lead to the success of hypertext as the navigation method of choice for many Internet users.

Through hypertext a user has the flexibility to organize and link information any way they see fit (even hierarchical!) and also link information together in a way so the user can go from one concept to another without limitations. Life does not always fit into nice categories, nor can human beings be expected to think in a completely linear fashion.

Hypertext in a very real sense is the information organization idea that just would not go away, even while experts both praised and condemned its methods and implementation. Although many associate hypertext with the Internet, it would eventually have its first productive application outside the Internet. In 1987, Apple Computers released a software package called HyperCard that came bundled free with the Macintosh and its operating system. It not only allowed the linking of documents, it also enabled the user to create links to sounds and images within those documents. Although it was limited to restricting its links to the same system, its popularity would go a long way in setting the stage for users to adopt its methodology on the Internet. In a very real sense, the Web is just the old Mac HyperCard system with links allowed to HyperCard applications on other computers.

The first efforts to bring the concept of hypertext to the Internet began in the late 1980s when Tim Berners-Lee was employed as a software engineer at CERN, the European Particle

Physics Institute, in Geneva, Switzerland. The problem facing Berners-Lee was the familiar "messy room" that the Internet had become. The high-energy physics community was spread throughout various universities and industries in Europe. Information and results of various experiments and the research data it produced was spread over many computers and LANs. It was believed that there was great potential to accelerate research if a way could be found to correlate and share all this information electronically.

The problem of separate clients was discussed previously. Users might have to send e-mail to a colleague, then remotely log onto their computers to examine text, and finally transfer a file with the graphs and images, recording the results of the experiment they needed to look at. Moreover, they were forced to do each step from a different client program with different commands and usually one at a time. Berners-Lee was seeking a kind of "universal client" for the Internet that used hypertext to link files, e-mail, and graphics all in one.

By March, 1989, Berners-Lee proposed a way of linking text documents with other documents using "networked hypertext." Working with colleague Robert Cailliau, they eventually produced a design document and published it in November, 1990, as a proposal for developing a system based on their idea of networked hypertext. "Hypertext is a way to link and access information of various kinds as a web of nodes in which the user can browse at will," the document stated. "Potentially, hypertext provides a single, user-interface to many large classes of stored information, such as reports, notes, databases, computer documents, and online systems help."

The term "World Wide Web" was introduced in this document, although without the capitals. It proved to be a name that would stick, although often just shortened to "the Web" today. After passing a few documents between computers, they found that the documents needed a standard format that could be interpreted by each computer and still convey the information contained within. With this in mind, they developed a language called the *Hypertext Markup Language* (HTML) modeled after, and basically becoming a subset of, a powerful, but much more complex formatting language called *Standard Generalized Markup Language* (SGML). Today HTML and SGML have both been adapted into a powerful tool called the extended markup language (XML).

The resulting Web software would be a mixture of Web *servers* (now just Web *site*) and Web *browsers* (the "universal client" terminology has become the Web browser). The browser is software that runs on the user's client computer and "talks" to the software running on the Web server to request certain files. These files could be any type representing the various resources a user needs to access. If written in HTML, the files can contain information the browser can display in addition to the names of files of related resources and their locations (the hypertext links).

By May, 1991, the World Wide Web was being used successfully at CERN. The first browser developed was a "line-mode" browser that was really just an advanced version of a telnet session, as shown in Figure 1-2. The term *line-mode* refers to the line-by-line output of this first browser. Berners-Lee then officially presented the World Wide Web to the world in December, 1991, at the Hypertext '91 conference in San Antonio, Texas, where it caused somewhat of a sensation among the crowd, which included some influential Internet people. In January, 1992, the line-mode browser was made available to anyone with an Internet connection.

```
Welcome to the World-Wide Web
THE WORLD-WIDE WEB

For more information, select by number:
A list of available W3 client programs [1]
Everything about the W3 project [2]
Places to start exploring [3]
The first International WWW Conference [4]

This telnet service is provided by the WWW team
at CERN [5]
1-5 Up, Quit, or Help
```

Figure 1-2.   The original Berners-Lee Web browser

In the true spirit of Internet cooperation, the ideas of the World Wide Web spread and were developed through Internet discussion groups on many newsgroups and separate conferences. In July, 1992, the University of California at Berkeley became the birthplace of the first modern-looking browser. This browser was developed by a post-graduate associate named Pei Wei. The browser, which he named Viola, was available on UNIX systems using X-Windows, which is roughly the UNIX equivalent of Microsoft Windows. It was the first browser to introduce distinctive browser features such as distinguishing hypertext links by colors and underscoring. Viola's other key feature was the capability to allow simple mouse pointing-and-clicking to activate the hypertext links. Unfortunately, Viola remained only an interesting experiment because Pei Wei had no interest in developing Viola further, not even as a commercial product.

So by early 1993, the World Wide Web had a small but solid foundation of interest throughout the Internet community, but Gopher remained king. Gopher was generating most of the Internet community interest. Gopher, with its hierarchical and menu approach, had more Internet traffic, more servers, more interest, but fewer features than the Web. During discussions within the Web community in order to try to figure out how to compete with the popular Gopher, Berners-Lee and his group of developers at CERN detailed the major limitations of the early Web.

The two major points that came out of these discussions were based on two fundamentals of the Web itself: the Web server and the Web browser. In the case of the Web server, the installation and maintenance of the server software was quite difficult even for experienced system administrators. As for the browsers, many users who accessed the Internet were going through a terminal, such as the popular VT100, attached to their company's or university's mainframe computer, and not through a PC.

The VT100 was and is a "line-mode" display device used with many DEC mini-computers. Not an intelligent device like a PC and thus unable to run its own programs, the VT100 family of devices consisted of a keyboard and a simple green-on-black monitor. The monitor was incapable of displaying anything except text, one line of 80 characters at a time. The screen was usually 25 lines high. If a line scrolled off the top of the VT100, it was lost, because the VT100 had no memory to store information from the central computer. Those who remember the DOS operating system will recognize that a DOS monitor behaved exactly like a VT100 terminal.

DEC later extended the product line to include the VT102, the VT-200, and so on. All differed in support for color and number of rows and columns on the screen, but all remained *dumb terminal* devices. Today, many PCs still run software that makes the intelligent and powerful PC behave like a VT100 dumb terminal. These terminal emulation packages are popular methods for accessing network devices such as routers and switches for configuration purposes, and sometimes still used to access the Internet, especially in odd situations where Web support is not critical.

The problem with browsers and terminals was that no reliable text-mode VT100-based browser existed that could deal with the many limitations to the full screen features that a VT100 client offered. More importantly, for a PC market that was already growing in leaps and bounds, a simple, full-screen, color, point-and-click browser was needed to compete with the already successful Gopher clients.

The first problem with regard to improving the Web server software was immediately addressed by Berners-Lee and his group of developers at CERN, as well as others around the Web, and resulted in much improvement the next year, 1994. The second problem, improving the browsers, was once again addressed in true Internet fashion. A VT100 text-mode browser called Lynx (for following hypertext "links") was modified from an existing client software project. The Lynx parent software had been developed a year before, in 1992, by Lou Montulli for a campus information system for the University of Kansas. Lynx would soon develop into the standard for text-mode Web client software for those without true graphical capabilities. In fact, because users concerned with the slowness of loading complex Web graphics prefer a method that offers only the text on a Web page, the simple text-mode browser refuses to die. Today, the text-mode browser has given birth to other forms such as the *Wireless Application Protocol* (WAP), designed to bring Web text to the world of cellular telephones.

THE WEB EXPLODES

Everyone who saw Pei Wei's Viola Web browser was instantly dissatisfied with text-based browsers like Lynx. The reaction was similar to those who saw early PCs instantly recalculate spreadsheet rows and columns: "I want that!" Again, Pei Wei had no interest in developing Viola and was not interested in product development.

But others were. They quickly realized that as PC power grew, graphical Web browsers would not need the powerful UNIX operating system and X-Windows to perform adequately. Perhaps even Macintoshes and Windows-based PCs would do just as well—if not now, then quite soon. Of course, they were right.

Events were unfolding that would not only meet the requirement of providing a graphical Web browser, but would also transform the Web from a little-known corner of the Internet into being the dominant presence on the Internet and a significant presence in ordinary people's lives. This browser would also be the major catalyst resulting in truly opening the Internet and *cyberspace* to the general public.

In February, 1993, a new Web browser called Mosaic was announced for release by the National Center for Supercomputing Applications (NCSA), part of the University of Illinois at Urbana-Champaign. The NCSA is subsidized by the U.S. government to develop tools and software that provide the means for researchers to be able to work collectively and share data and resources. Berners-Lee's project at CERN and the resulting development of the World Wide Web utilizing HTML on the Internet had generated interest in the researchers at NCSA as a promising medium for the interchange of information on a national, if not global, scale.

Marc Andreessen, a graduate student at the University of Illinois at Urbana-Champaign, with the help of Eric Bina, a programmer at the NCSA, developed the new Web browser. It was originally designed to run in a UNIX environment (with X-Windows, naturally). Because of its popularity and despite being limited to a UNIX server, the MS Windows and Macintosh versions of Mosaic browser were released in the fall of 1993.

Although the Web community had specified that a *graphical user interface* (GUI) browser was important to the further development of the World Wide Web, the initial success of Mosaic was based primarily on the fact that it was easy to obtain, install, and use. Of course, this would not make any true difference if a software package was terrible, easy to use or not. Many powerful and productive software packages have failed because no one wanted to expend the effort in obtaining, learning, or just plain using them. Not only was Mosaic easy, it was flexible. It was designed as more than just an HTML browser and could handle links to other Internet services such as FTP and Gopher.

More than that, Mosaic could "learn." If Mosaic met a file it did not know how to handle, the browser allowed the user, within Mosaic itself, to set a link to an application (referred to as a Helper Application or plug-in) that could handle the file at once and then in the future. Mosaic did not stop there. Marc Andreessen wanted to expand the graphical browser and HTML beyond just text and hypertext. Mosaic was designed with the capability to display in-line images and pictures within the same document as the hypertext. Furthermore, images were taken another developmental step to exist as image maps. Image maps allow for different sections of the picture to operate just like a hypertext link and access another source. A user can click anywhere on the image and the browser interprets its corresponding link. This extensible feature of Web browsers has come in handy over and over again, as new forms of Web animation such as Flash, new forms of network applications based on languages such as Java, and new forms of audio and video encoding have made their way onto the Internet.

With all its new features and flexibility, Mosaic was not just an innovative Internet application, it was the tool, the "universal client," that opened the portal to the Internet for the regular person. Once there, many realized that HTML was not a terrible, scientific, complicated computer language, but rather a relatively simple format that even those who had previously only used a word processor could master in a short time. Today, many word processors allow users to build pages within them and will generate the HTML code so that the users never even have to learn any of the HTML code. The users just keep using the word processor as they have been all along. With its initial success, Mosaic would be the first of many Internet browsers to come.

Marc Andreessen left the NCSA to co-found a company called Mosaic Communications, which eventually became Netscape Communications Corporation with the release of their Netscape browser, a supercharged redesign of Mosaic. Andreessen jokingly called Netscape a "combination of Godzilla and Mosaic," and the merged code-name of "Mozilla" has crept into the Netscape world in a variety of forms.

Netscape's battles with Microsoft for the world browser market in the mid- to late 1990s have been well-documented and need not be repeated here. Microsoft even held off the release of its much anticipated Windows 95 operating system software so that it could incorporate its Internet Explorer Web browser into the package. To keep ahead of the pack, Netscape continued to offer extensions to the HTML standards to enhance the interaction and function of its Web browser. The goal is to continue to offer a more creative, simpler, interactive medium through the browser.

This brief history of how the Web invaded the consciousness of the world could be extended to include the development of Java, cookies, and other concepts that have encouraged the Web to be used for e-commerce and business, as well as simple Web surfing for pleasure. But the road to Netscape and Internet Explorer was the path that needed to be blazed. In some sense, the addition of these new Web features has just been the icing on the cake.

THE RISE OF THE INTERNET SERVICE PROVIDERS

Mention has already been made of the watershed years between 1984 and 1986. In that time, the National Science Foundation found itself heir to much of the Internet infrastructure as the U.S. military sought to sever its ties to what came to be officially called the Internet. The NSF now ran the very backbone of the Internet, now called the NSFNET, which linked various regional networks together into one unified whole. There were a lot of reasons for the breakup of the military and science relationship that the Internet was founded on, including international criticism and the reluctance of international standards committees to consider protocols and applications developed for a "military network" as seriously as other proposals. So for better or worse, the federal government began to funnel funds to the NSF instead of the military to operate the equipment that comprised the Internet and pay the people who kept the network up and running.

Another watershed year took place in the 1990s. In 1994, about the same time that the Web was making a big splash on the Internet, the federal support dollars for the NSFNET were starting to be phased out as planned. In 1993, the NSF declared the experimental high-speed NSFNET network begun in 1984 a success, but not the type of research network needed for the future. So the NSF invested in a newer study of high-speed computing and networking and announced that "the Internet" would have to eventually fund itself commercially. NSF continued funding to certain portions of the Internet, cutting the funds 20 percent per year from 1994, until 1998.

The idea was that the federal government should provide taxpayer dollars to encourage new technologies that might improve the life of citizens (this is outlined in the Constitution), and, in any case, the Internet did not set a precedent in this regard. Samuel F.B. Morse, inventor of the telegraph, was an art professor, not a professional inventor (although he did explore more than the telegraph in this regard). Only with the help of federal money was the first telegraph line strung between Washington, D.C. and Baltimore, Maryland. Early use of this line was free, but by the time the money ran out, it was obvious that people had embraced the telegraph and money from users would support the whole industry and assure profits for all involved.

The same was true of the Internet. By the 1990s, it was obvious that the Internet formed an important part of university and college life. Withdrawing the subsidy from the NSFNET just meant that funds for continued operations had to be gathered from the users of the network, so the Internet was restructured into a small number of *Network Access Points* (NAPs) run for the most part by major telephone companies. Any part of the Internet could reach any other part of the Internet through these NAPs if the separate parts did not have direct connections to each other.

No one need think of the NSF as abandoning the Internet. When the NSF built the NSFNET, they were looking to run a network to support high-speed computing and high-speed data communications (such as they were in those days). Their intent in 1994 was the same. The result was a new backbone network and the NAP structure to make sure both older research and newer commercial backbones could communicate. In both cases, the NSF goal was always higher speeds for computing and communications.

Of course, any regional network providing Internet service (which boiled down to access to servers on the Internet) had to pay for the lines running to the NAPs and other charges as well. How did they get the money needed to pay for NAP interconnections? They simply reached out beyond the university and college community and went public. Now calling themselves *Internet Service Providers* (ISPs), the smaller, regional portions of the Internet began advertising and signing up as many people as they could for Internet access. Some enterprises were so successful that they grew into national ISPs very quickly, while others either accepted their smaller, regional role or at least were content with it.

It helped that there was already a precedent for such online services. In the 1980s, the availability of PCs and the appearance of these PCs in people's homes, as well as on their desktops at work, led to the appearance of various specialized networking companies.

The appearance of the IBM PC in 1982 caused a revolution not only in business computer use, but also in home computer use. For the first time, ordinary people at home had access to the enormous computing power previously available only to a select few with home computer terminals. And those few users had no real computing power in their homes. The computing power was still in the office mainframe or minicomputer; the terminal provided access only to the office computer over a telephone line. A person with a home PC could, however, run applications and programs formerly restricted to a corporate environment. In fact, when it came to word processing, spreadsheets, and even simple graphics, the humble PC often outperformed many a mainframe and minicomputer, due to advances in technology embodied in the new PCs.

The appearance of the PC had an interesting side effect as well. Everyone could have their own personal computer, but the effect was to isolate people from one another and limit their access to hardware and software to what was installed and functioning on their local PC itself. Someone may have written a stirring piece of prose, but it could only be read by those who had direct access to the PC it was composed on. Of course, PCs had floppy drives (and most had only floppy drives back then) that someone could use to copy the writing to and distribute, if one knew who had the writing and how to get it. This "sneaker net" process was not different from Medieval scribes copying manuscripts by hand and distributing them to select libraries and repositories one at a time. What was needed was the PC equivalent of the printing press.

By the mid-1980s, the PC community had their printing press. It appeared in the form of a variety of online services, some large, some small. The smaller ones consisted of a single PC running special "server" software similar in nature to Internet server software, but not based on the same protocols in most cases. A few telephone lines connected to modems allowed others to simply dial in with their terminal emulation software packages to become "clients." This simple but powerful arrangement launched a cottage industry of *bulletin board systems* (BBSs) that lasted well into the 1990s and that still linger in some portions of the networking world. These small electronic bulletin boards usually addressed the needs of a small, contained group (a telecommunications workers union local, for example) and offered meager features, such as e-mail, among users, as well as a common point to exchange files and programs. Many were staffed on a voluntary basis, and there were no charges to users beyond telephone calling time charges, which were usually minimal, except for users outside the local calling area.

The larger systems addressed the needs of the public at large for such services. Generally, the services were much the same, with the addition of some special *user groups* for those with shared interests, such as military history or house plants. These larger online services benefited from special rules that the federal government laid down in the U.S. in 1984, specifically to encourage the growth of such services. The larger services had many points of presence (POPs) throughout the U.S. so that local users had to constantly dial long-distance numbers and pay high rates and the consequent high bills for access to these systems. Membership was typically based on a flat monthly rate and usage charges, although some offered totally flat rate services, which users embraced wholeheartedly. Naturally, they concentrated on major metropolitan areas, but many were in almost every state in one place or another.

By the time the Web exploded onto the world in 1993 and 1994, the world of the large online service providers was filled by three companies: CompuServe (CSi), America Online (AOL), and Prodigy. This list arranges them in order of their first public service offerings.

CompuServe began in 1982, virtually at the beginning of the PC industry itself. They quickly gained a reputation as the service of choice for PC enthusiasts. That is, the more a user cared about things such as operating system optimization, writing their own utility programs, or adding their own hard drive to their system, the more the user liked CompuServe. Oddly, for all its reputed sophistication, CompuServe was one of the last to realize that the Internet and Web were important to its users and lagged behind in offering easy and integrated Internet and Web access to its members.

America Online began offering services to the public at large in 1985. Their reputation was that their members did not need a great deal of PC sophistication or have to be a computer science wizard to access their various chat rooms and sample their offerings. AOL's aggressive marketing techniques turned some people off, but certainly raised the consciousness of average PC users who at least tried the online service to see whether they would like it. Many did. AOL was among the first to see the Internet and Web not only as an adjunct service to their bulletin board-based system, but potentially as a replacement service.

Prodigy began as a joint effort between IBM and Sears in 1984 to encourage PC users to communicate online. Public service was launched in the U.S. in 1988, and Prodigy seemed determined to "out-AOL AOL" in terms of user friendliness and services, such as home shopping and electronic newspapers. However, Prodigy alienated some with insistent advertising from sponsors (virtually absent on other service providers' systems) and suffered from some early negative publicity. While Prodigy saw that the long-term future of networking for ordinary people would revolve around advertising, sales, and marketing, CompuServe and AOL people at the time tended to treat these key aspects of the Internet today with disdain.

By the early 1990s, PC users in the U.S. were comfortable with the idea that an online service could put them in touch with other users with similar interests and offer them access to information that would otherwise be difficult to obtain. When the ISPs were faced with a loss of government funds, many of them saw a solution in the simple repositioning. Rather than provide Internet access to colleges and universities, why not offer Internet access to anyone willing to pay the price? This involved installing TCP/IP client software on the user's PC, but software installation by this time was not the headache it once had been.

In the mid-1990s, three separate threads came together to encourage the expansion of Internet and Web access in the U.S. The first was the idea that anyone with a PC could benefit from networking with other PC users through an online service. The second thread was that the ISPs were faced with having to make their operations financially self-supporting. The last thread was the Web itself, with its colorful graphics and multimedia in place of the stodgy commands of the recent past. The Internet had the Web. ISPs had Internet access. The end.

Of course, the online service providers, especially the "Big Three" of CSi, AOL, and Prodigy, saw the writing on the wall. When the Web broke onto the world, Prodigy became the first of the Big Three to offer a Web browser as an integral part of their client software package in January of 1995. AOL quickly followed and did Prodigy one better. In 1995, AOL went out and bought the assets of a company that had begun operations in 1990 as Advanced Network & Services (ANS). ANS was the principal operations arm of the NSFNET Internet backbone and was backed by a consortium composed of IBM, MCI, and a few other organizations. Naturally, with the new Internet emphasis being to make money rather than spend government grants, this ANS purchase made sense to both AOL and ANS. ANS got money and AOL got Web access for their users. By the time CompuServe realized that their members might want access to the Internet and the Web, many had already decided not to wait and signed up with AOL or another ISP (CompuServe members generally shunned what they perceived as the blatant and crass commercialism of Prodigy).

Today, there are still thousands of mostly small ISPs around the world. Only about 20 or 30 have an extensive national backbone of their own (the uncertainty is due to the lack of a standard definition of "national," "backbone," and "own" when it comes to ISPs). The others basically link to other ISPs, who might link to other ISPs, and eventually everything is linked together at the major NAPs. So the Internet is not one network, but more like thousands of interconnected ISP networks, each with a large or small number of Web servers (Web sites) maintained by their individual members (customers) or members' organizations.

WOE TO THE ISPS

This discussion of the Web, online service providers, and ISPs might seem to have little to do with ADSL, but it actually has everything to do with ADSL. The Web explosion has led to dissatisfaction with the speed and efficiency at which the Internet operates. To see why, consider the structure of the Internet at the beginning of the Third Millennium, shown in Figure 1-3.

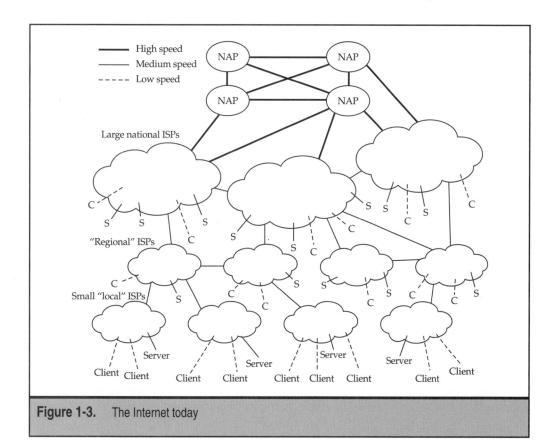

Figure 1-3.   The Internet today

At the bottom of the figure, millions of PCs, minicomputers, and mainframes act as either clients, servers, or both on the Internet. Although all attached computers conform to this client-server architecture, many of them are strictly Web clients (browsers) or Web servers (Web sites) as the Web continues to take over more of the form and function of the Internet at large. Only at this bottom level are the terms *client* and *server* spelled out. At the other levels, members of each ISP's network are only represented by C and S. The figure ignores important details such as LANs and routers for the sake of simplicity, but it is important to realize that many of the clients and servers are on LANs, and that routes are the network switching points of the Internet. The figure is a little misleading because the number of clients actually exceeds the number of servers many times over, but this is not apparent in the figure.

Moving up one level, the figure shows the thousands of ISPs that have emerged in the 1990s, especially since the Web explosion of 1993–1994. Usually, the link from the client user to the ISP is by way of a simple modem-attached, dial-up telephone line. In contrast, the link from a server to the ISP would most likely be a leased, private line, but there are important exceptions to this simplistic view. Note that there may also be a variety of Web servers within the ISP's own "cloud" network. For instance, the Web server on which an ISP's members may create and maintain their own Web pages would be located here.

Moving up again, the smaller ISPs link into the large backbone of the national ISPs. Some may link in directly, whereas other are forced for technical or financial reasons to link in "daisy chain" fashion to other ISPs, which link to other ISPs, and so on until an ISP with direct access to a NAP is reached. Note that direct links between ISPs, especially those with older Internet roots, are possible and sometimes common. In fact, the NAPs were once so congested that most major ISPs prefer to link to each other directly today, a practice called "peering" because it can bypass the need to use the NAP hierarchy to deliver traffic.

The NAPs themselves are fully *mesh connected* (all link directly to all other NAPs). Figure 1-3 shows the structure of the U.S. portion of the Internet. However, a large percentage of all inter-European traffic passes through the U.S. NAPs. Most other countries obtain Internet connectivity by linking to a NAP in the U.S. Some large ISPs even link routinely to more than one NAP for redundancy purposes. The same is true of individual ISPs, except for the truly small ones, which rarely link to more than one ISP, usually for cost reasons.

Also, speeds vary greatly in different parts of the Internet. For the most part, although there are exceptions, client access was by way of low-speed dial-up telephone lines, typically at 33.6 to 56 Kbps. Servers were connected by medium-speed private, leased lines, typically in the range of 64 Kbps to 1.5 Mbps. The high-speed backbone links between national ISPs ran at higher speeds still, sometimes up to 45 Mbps. On a few, and between the NAPs themselves, speeds of 155 Mbps (known as OC-3c), 622 Mbps (OC-12c), 2.4 Gbps (OC-48c), and now even 10 Gbps (OC-192c) are not unheard of. High speeds are needed both to minimize large Web site page transfer latency times and to the fact that the backbone(s) concentrate and aggregate traffic from millions of clients and servers onto one network.

With the information presented in the figure, coupled with the information presented at the beginning of this chapter, the problem facing the "Information Superhighway" concept can be easily understood, which is the problem that initial implementations of

ADSL have addressed. The Internet exists to provide connectivity between clients and servers. Increasingly, this translates to home PCs running Web browsers accessing remote Web sites.

But the speeds that home client PCs are dealing with are the simple dial-up modem speeds that are limited by the passband frequency ranges that these telephone circuits through the PSTN allow. Moreover, the sending and receiving of bursty packet application traffic needlessly ties up bandwidth on these circuits. The end result is that the World Wide Web is sometimes called the World Wide Wait, and the ISPs are under a barrage of criticism for not doing more to alleviate the situation.

Although residential access to the Internet is the most important aspect of the concern about limited bandwidth on the access line, the Internet today is used for much more than consumer activities. The use of the Internet has expanded in the past few years yet again, mostly through a collection of specialized browser applications and Web sites. The Internet today is also used by businesses to a far greater extent than this chapter—with its emphasis on residential use—would seem to convey. However, in many cases DSL can be an important technology not only for Web site access, but also for the following Internet-related activities:

▼   **Intranets**   An intranet is typically formed by an organization that wishes to cut down on—or cannot afford to pay for—the many leased private lines needed to link all of their business sites together. The Internet, with its global connectivity, provides a nice alternative to the mesh of private lines required. All any site needs is a single link to an ISP, although two links can provide better reliability. Intranets can be especially effective for companies that need to go global—and many do just to stay competitive today. A private line from the U.S. to Europe can be prohibitively expensive, but a simple dial-up or short access line to an ISP at both ends can use the Internet for the desired connectivity.

■   **Extranets**   An extranet can be viewed as a logical extension of an intranet. Intranets are used to link sites belonging to a single organization over the Internet. An extranet extends these links to other organizations that are best characterized as *trading partners.* So car makers can link to tire suppliers, medical centers can link to insurance carriers, and so on. Both entities must agree to this type of network, of course, because the Internet provides no security in and of itself. Extranets work best when the ISP can also provide a type of closed user group or some other form of security to prevent unauthorized access, such as by forming a virtual private network (VPN).

■   **VPNs**   A *VPN* is an intranet or extranet with security added by the companies involved, the ISPs, or both. The added security can be provided through a number of methods, from simple virtual circuits over the Internet to very elaborate schemes of encryption and user validation. The intent is to provide as much security as could reasonably be expected from a network of leased private lines, but over the Internet. Hacking and unauthorized access are the main targets of the VPN, but VPNs can also guarantee performance levels, often with the inclusion of a managed IP service.

■ **Managed IP** Many ISPs, especially the large regional and national ISPs, offer what are called *managed IP* services. The service manages not the customer's network, but the resources available to the customer on the ISPs access and backbone network. For example, an ISP might offer Gold, Silver, and Bronze levels of service to their VPN or general users at various prices. Each service level would have a guarantee of bandwidth, delay, availability, and so on attached to it. The ISP is responsible for managing their IP resources so that each customer gets the service level they are paying for and making sure that the customer is satisfied. Generally, for this reason there are no guarantees *between* ISPs, only within a single ISP's network. There are many ways to add these service guarantees to the global public Internet, and no way is expected to prevail in all cases. But some way of managing IP resources is critical as more and more voice and video is making its way onto the Internet.

▲ **Voice and video** It might seem odd that voice is often packetized as well as digitized so that the voice can travel over the Internet. Circuits designed to carry voice are now used to carry not only data packets, but voice packets as well. Why not just use the voice circuits for voice? Well, for one reason, packet networks can be used for voice and data at the same time. Also, local links to ISPs are less expensive than making long distance calls all over the world—and the link is needed for Internet access anyway. Why not use it for packetized voice also? The idea of voice over DSL (VoDSL) is explored in Chapter 11. The video referred to is usually corporate videoconferencing, but the Internet can also be used to distribute video entertainment. The video aspect issues are almost the same as the voice issues, except that the bandwidth need for video is much greater.

Although this book emphasizes the use of DSL for home and small office access to the Internet, it is worth remembering that DSL can also be a vital part of an intranet, extranet, VPN, managed IP, or voice-and-video-over-DSL network as well. Whenever clients and servers need to be connected, DSL can play a role.

Note that Web servers send much more information to clients than clients send to servers. This is not true of many Internet applications, such as FTP or telnet, where client-server interactions can be fairly symmetrical when it comes to traffic. In many cases, Web browser (client) output consists of simple mouse clicks and brief text strings, such as names and addresses or credit card numbers. Web server output, sent in response to client requests, may be thousands or even millions of bytes (a byte is 8 bits). Yet most of the technologies used for dial-up and leased lines offer the exact same speeds in both directions. What is needed is a cost-effective, broadband solution for residential and small office access to the Internet, and this is what ADSL is for.

In fact, ADSL solves a number of tricky problems plaguing not only the ISPs, but the local telephone companies themselves. Chapter 2 takes a closer look at the structure of the global PSTN, which is the root cause of the problem.

CHAPTER 2

The Public Switched Telephone Network

A ny discussion of ADSL technology, especially considering how it is currently used in the United States, must include details not only about how the Internet and Web are structured and function, but also how the Public Switched Telephone Network (PSTN) in the United States is structured and functions. Only then can the major design decisions and implementations of ADSL be understood. Some of the details about the structure and functioning of the Internet and Web were explored in Chapter 1. This chapter is intended to supply the same degree of information about the PSTN itself. Although the discussion emphasizes the PSTN in the United States, there are numerous similarities with other telephone networks around the world, perhaps not in detail, but certainly in general characteristics.

It has often been observed that "network people" who work in the telecommunications industry tend to fall into two main categories: those trained in and experienced with "data networks," and those trained in and experienced with "voice networks." The structures of many organizations who must deploy and employ both types of networks reflect this distinction. This dichotomy arose at the very start of computer networking as the very different requirements between voice and data networking became obvious. However, some of the latest advances in voice technology make voice, as well as audio and video, look pretty much like any other stream of data bits in networks. The enormous implications of this simple statement will become clear later in this book.

ADSL is concerned with both traditional voice and newer data applications that essentially share a link to and from a residence, home office, or small business office. Much of the task of explaining the role and functioning of ADSL in this environment consists of explaining the PSTN to the "Internet people" and explaining the Internet to "voice people" familiar and comfortable with the PSTN. The previous chapter discussed the Internet and Web; this chapter looks at the PSTN.

A NETWORK IS A NETWORK IS A...

The whole task of investigating networks is made much easier by the simple fact that all networks look pretty much the same. That is, all networks share certain structural and architectural characteristics that make them appear quite similar to each other. This is not to say that there are no significant differences in their function and operation because there are, but all networks share characteristics that make them networks in the first place.

The discussion of the Internet introduced the familiar network "cloud." No details were offered as to the functioning of network components inside the cloud, which is why it appears as an amorphous cloud in the first place. The idea behind the cloud is that users need not concern themselves with the inner workings of the network. All users need to worry about is whether they can reach other users through the network (or internetwork of linked clouds). The rest of the task is handled by the appropriate hardware and software.

The good news is that all networks are basically simple in overall structure. They have to be or they would not work at all. After all, computers can do networking, and anyone who has had a computer foul up their bills knows how intelligent computers are in and of themselves. The bad news is that each network type has its own jargon and acronyms for network components that do much the same thing.

One word of caution: The following discussion emphasizes the similarities of many types of networks. However, this does not mean that there are not significant differences between various network types. For instance, the PSTN employs *circuit switching* and the Internet employs *packet switching*. These differences will be dealt with elsewhere.

The PSTN is a network just as much as the Internet is a network, if not more so, given the PSTN's priority in age. The PSTN is almost 100 years older than the Internet, and some aspects of the PSTN reflect the older technology that was state-of-the-art in its day, but is now considered aged or aging. These network discussions need not be limited to the PSTN and the Internet—the arguments extend to all networks in general. To see why, consider the network shown in Figure 2-1. This time, some of the details within the cloud are presented.

Figure 2-1 shows that within the cloud are devices known as *network nodes.* Outside the cloud are other devices usually called *user devices.* This is how one knows whether they belong outside the cloud or not: network nodes go inside, user devices go outside. Usually, but not always, a user device links to one and only one network node. Network nodes, on the other hand, may have multiple links to other network nodes, but again not universally.

User devices link to network nodes by a link known as a *user-network interface* (UNI). There is nothing magical about the properties of the UNI link. It typically can run at a variety of speeds; is supported on a number of different media—from coaxial cable to fiber; and runs adequately up to some standardized or designed distance. Network nodes link to each other with what is sometimes called a *network node interface* (NNI), although *interswitch interface* (ISI) or *internode interface* (INI) are much more accurate terms. No magic here either. These links also may vary by speed, media, and distance.

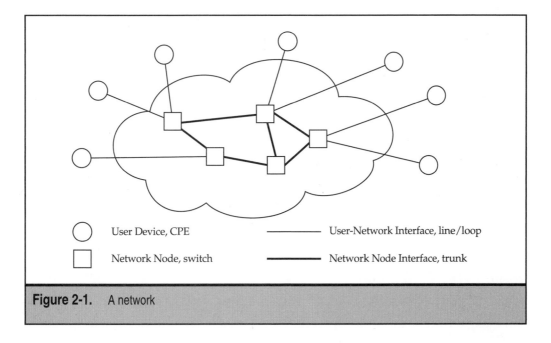

| ⭕ | User Device, CPE | ⎯⎯⎯⎯ | User-Network Interface, line/loop |
| ⬜ | Network Node, switch | ▬▬▬▬ | Network Node Interface, trunk |

Figure 2-1.   A network

That is pretty much all there is to networking, but one exception must be added to this general network diagram. Older *local area networks* (LANs) did not conform to this generic *wide area network* (WAN) structure. Since the early 1990s, most newer LANs do indeed look like this, although the entire cloud might encompass only a single building or even a single floor.

All users can be reached through the network. Different networks play around with different hardware and software used by users, network nodes, UNIs, and network node interfaces, but most of the differences would be lost on those not intimately familiar with the various protocols. Remember that user devices do not send to network nodes directly, but rather to other users, and some of the network nodes do not link directly to users, but only to other network nodes. It is common to call network nodes with user interfaces *edge* devices and network nodes with links only to other network nodes' *backbone* devices, but this is only by convention. The key is that these internal network nodes exist only to link network nodes to each other, not directly to users.

The network nodes, whatever they may be, typically accept traffic through an *input port*, decide where it goes by some rule or set of rules, and put the traffic out through an *output port*. The easiest way to determine where the traffic goes is to look up the destination in a table maintained and updated in the network node itself. Many variations of this theme exist, but this discussion is concerned with basic operations.

To this point in the discussion, general data network terminology has been presented. Both UNIs and network node interfaces are usually leased telephone lines, but there are variations. Figure 2-1 also shows the terms that would be applied if the network figure were of the PSTN specifically. From the PSTN perspective, the user devices would be telephones, fax machines, or computers with modems, which all fall into the category of *Customer Premises Equipment* (CPE). In the United States, the form that the CPE must take is completely up to the customer. Any approved device by any manufacturer may be used, as long as it conforms to some basic electrical guidelines. The CPE in the United States is owned and operated by the user and is beyond the direct control of the network service provider. In other countries, the CPE can be provided and owned by the service provider under a strict set of regulations. The process of giving users more control over the CPE is part of the general movement toward deregulation. This process does not concern ADSL directly, at least not yet, in many parts of the world outside of the United States.

The network node from the PSTN perspective would now be a voice *switch*. A switch is just a type of network node that functions in a particular way in terms of how traffic makes its way from an input port to an output port, the way the table that tells the traffic appearing on an input port which output port to use is maintained, and so on. At least that is how most network people see it. A continuing controversy lingers between what is meant by the term *switch*, especially between the voice and data networking communities. The controversy extends to the companion term *router*, which is another type of network node found in the Internet. Fortunately, the router-switch controversy is not a concern in a book about ADSL. Suffice it to say that a router is a type of network node found on the Internet that performs networking tasks differently than a switch (such as packet switching in routers and circuit switching in voice switches), which is the basic network node type of the PSTN.

Instead of the edge and backbone structure of data network nodes, the PSTN uses terms such as "local exchange," "toll exchange," or "long distance" to distinguish voice switches having links to users or not. All users link to a local exchange, usually called a *central office* (CO) in the United States, and often a *local exchange* (LE) elsewhere. The local exchanges link to *toll offices,* or *tandems,* or *long distance switches,* with the actual terms varying depending on the detailed structure of that portion of the PSTN.

Moving on to the interfaces, the UNI is now the *access line* or *local loop.* Use of these terms varies, and some prefer one over the other in all cases. Some reserve the term *line* for digitized loops where analog voice is represented as a stream of bits, and others reserve the term *loop* for purely analog user interfaces. This book uses the term *analog local loop* to mean a PSTN user interface carrying only analog signals and analog information. This book uses the term *digital subscriber line* (DSL, as in ADSL) to indicate that some accommodation or optimization (in terms of handling digital information) has been performed to make the loop more efficient for high-speed digital information content. Note that this usage does not distinguish loops or lines based on analog or digital *signaling* (information encoding) at all. Either type of signaling may still be used. In this usage, *loops* carry low-speed digital and analog information content (modem traffic and voice), whereas *lines* carry high-speed digital information content and may also carry analog content (as ADSL does). This definition and usage may offend purists and engineers, but its usefulness will become apparent later on.

In any case, loop or line, the user interface is not normally a leased, point-to-point, private line. Rather, it is a *switched,* dial-up connection that is capable of reaching out and touching almost every other telephone in the world by dialing a simple telephone number. One such destination may be the local Internet Service Provider (ISP) if the loop or line is connected to a PC with a modem or other specialized network interface device. Naturally, the ISP needs to be connected to a PSTN local exchange for this to happen at all, which is how PC users use the PSTN to access the Internet.

In the PSTN, the network node interface has become the "trunk." In the PSTN, there is little to no physical difference between loops, lines, and trunks. The difference is in how their physical facilities are used. Lines and loops are used to connect users to the network. *Trunks* connect network nodes (voice switches) to each other, but suppose a link runs between the voice switch on the PSTN to the ISP router connected in turn to the Internet? Is this a line or trunk? To the PSTN, the ISP router is CPE as much as a telephone is, so the access line directly to a router is still a line or loop to the PSTN. But the router is certainly a network node to the ISP, and links between one network node (the PSTN switch) and another network node (the ISP router) are by definition trunks to the ISP. Fortunately, telephone companies and ISPs alike have started calling such PSTN-ISP links "trunks." Technically, the PSTN should consider these links to ISP routers to be lines or loops to CPE devices.

Trunks are typically high-speed links for one main reason. Trunks must aggregate a lot of traffic from thousands of users and ship it efficiently around the network to other network nodes. The same is true in general for network node interfaces for the same reason.

Before closing this section, it might be a good idea to summarize the differences in network terminology, not only between the PSTN and the Internet, but among many new technologies in general. These differences are shown in Table 2-1.

Network	Network Node Called a	User Device Usually a	User-Network Interface Is a	Network Node Interface Is a
PSTN	Local exchange switch	Telephone	Local loop	Trunk
Internet	Router	PC client or server	Dial-up modem or leased line	Leased line
X.25 packet switching	Packet switching	Computer	X.25 interface	(undefined)
Frame relay	Frame relay switch	Router, FRAD	UNI	(undefined)
ATM	ATM switch	Router, PC	UNI	NNI
LANs since ca. 1990	Hub	Router, PC	Horizontal run	Riser, backbone

Table 2-1.   Differences in Network Technology

A few words are needed about networks other than the PSTN and Internet. X.25, frame relay, and ATM all fall into the category of packet-switching data networks, just like the Internet. Their network nodes are switches, not routers; oddly enough, however, the router can be a user device on frame relay or ATM networks. It should be noted that X.25 networks can use a special X.75 interface as a network node interface, but this is not mandated or universal. Frame relay also defines a specialized device for the network known as a *Frame Relay Access Device* (FRAD), although it is not really a user device in some senses at all. Note that of the three, only ATM defines a network node interface. (Oddly, frame relay does define an NNI acronym, but as a *network to network interface*, which handles the interface from one frame relay network to another.)

The brief tour of the PSTN, X.25, ATM, and frame relay cannot begin to do justice to the real differences and similarities among the technologies. For instance, ATM and frame relay switches have much more in common with routers than is outlined here. Without going into a long debate on just what is a switch and just what is a router, this brief introduction will do for the purposes of this chapter. The emphasis here is on use of terminology, not functional details.

Before 1990 or so, LANs looked very different from WANs. Pre-1990 LANs were mostly shared-media, distributed networks, and some still are. Today, most LANs conform to the network node model by way of the *hub*. There are no firm equivalents for lines and trunks, however, and the "horizontal" and "riser" terminology in the table is used for convenience only.

The discussion has so far been a little abstract with regard to the PSTN. The next section explores the general structure of the PSTN in more concrete terms. It puts the PSTN in evolutionary perspective so that decisions made many years ago do not appear to be haphazard; it addresses the needs of the PSTN in the best way possible, given the technology available at the time.

THE PSTN: THE FIRST NETWORK FOR THE PEOPLE

The telephone (and PSTN behind it) has become such a part of life that like other inventions, such as televisions and airplanes, it is hard to imagine life without it. Yet such a world existed before 1876. Of course, people still had access to a network that they used to exchange information in a highly cost effective manner. It was called the postal system.

Back then, people wrote letters to communicate. Lots of letters. During the week, mail was delivered in the morning and evening (many newspapers included this "evening mail" concept in their titles because they were distributed this way), and there was another delivery on Saturday (some people still fondly remember the "Saturday Evening Post" magazine). People needed to communicate the same things as today, only instead of reaching for the telephone, they reached for the pen.

There was a global telecommunications network, too—the telegraph, invented and perfected by Samuel F. B. Morse in 1838. By 1876, the telegraph was in every major city and was routinely used by news agencies, government departments, and ordinary people to exchange messages deemed more urgent than those entrusted to the mail. If someone's relative was suddenly taken ill, this news could be distributed to the rest of the family by going to the nearest telegraph office, usually at the local train station (telegraph lines closely followed railroad rights-of-way from the very beginning). For a few cents a word, the telegraph operator would click the message in Morse code to another operator (humans were the "network nodes" in many cases in the telegraph network, although automated message-relaying equipment did exist in major metropolitan areas). The message was relayed to the telegraph office closest to the destination and printed out. From there, a messenger, typically on bicycle, would rush to the home of the recipient, who didn't have to wait for the mail. Letters took a day or so. Telegrams took hours.

Many of the techniques taken for granted on telecommunications networks today were first pioneered on the telegraph system. These included compression ("hw r u?"), early data terminals (which printed Morse coded messages on paper tape), and even an early form of telemarketing (Sears began this in 1886 when a railroad conductor named Richard Sears quickly disposed of a watch shipment by advertising to other railroad people over the telegraph).

The drawbacks of the telegraph system were threefold, although the benefits of the system made these drawbacks relatively minor in most cases. First, the system was closed to end users themselves. No one had a telegraph wire running directly to their home and tapped out Morse code directly. Second, although the telegraph system could distribute messages much faster than the mail, there was still a need to send really urgent messages much faster, within minutes. Finally, a telegraph wire could only handle one message at a time. There was no way to stop a message in the middle and replace it with a more urgent one. Urgent messages got in line—the telegraph form of a "buffer"—behind routine ones.

This last liability was the most damaging. In metropolitan hubs, the telegraph office backed up with messages on busy days and during busy hours of any day. Putting in more telegraph lines was an expensive solution because these lines would sit idle most of the time. Clearly, whoever could invent a way to "multiplex" several telegraph messages over a single line stood to make a lot of money.

Alexander Graham Bell was one of these dreamers who wanted some extra cash. His intended bride was quite above him in Boston society, and he thought that if he struck it rich with a multiplexed telegraph technique, he could marry. What happened instead is that Bell accidentally invented a way to expand the useful bandwidth of ordinary telegraph wire enough to actually carry an adequate amount of the voice spectrum (range of frequencies) to produce intelligible speech some distance away. Actually, it was more of a way to convert the analog voice into electrical signals (and back again) suitable for the telegraph line that Bell discovered. Only Bell's work with the deaf enabled him to realize what it was that he had found. Bell first called it the "harmonic telegraph." Fortunately, Bell eventually called it the *telephone.*

Bell used his research into the third telegraph problem (lack of multiplexing) to solve the second telegraph problem (delay). Voice communication was almost instantaneous. There was no delay for coding, relaying, or bicycling after the intended party was reached. It came as a surprise to Bell that the telephone solved the telegraph's first problem (no telegraph to the home), but this last development needs some explanation because it has implications for ADSL.

Bell saw the telephone primarily as a tool for business and similar uses. For example, factories could call suppliers instantly, and a train crash resulted in doctors being called to the scene faster than ever before. The telephone was embraced for these purposes.

Bell never could understand why ordinary people would want a telephone at home. To the day he died, Bell refused to keep a telephone in his office because he found the ringing quite distracting—people wrote letters. What Bell missed was that people can pick up inflections of voice when listening to speech and that the act of talking to someone, no matter how far apart they were physically, made the interaction that much more intimate. What Bell discovered is that there are social consequences to technology.

The early telephone "systems" were elaborate intercoms. They were point-to-point affairs where every telephone needed a direct wire to every other reachable telephone. This system worked well enough for the factory-supplier or train-crash-doctor applications of the telephone, but it would hardly do if everyone wanted a telephone at home to call whomever they pleased whenever they pleased.

To better accommodate the growing residential population using the phone, the local exchange using a switchboard was invented. An operator sat before a huge array of lights. When a person wanted to call someone in the community, they took the telephone off the hook and turned a crank, which generated electricity on the wire and lit a light above their circuit position on the switchboard. The operator plugged a cord into the position and asked who the person wanted to speak to. If that line had no other cord plugged into it (a "busy" line), then the connection was made by hand and the operator "cut through" the call and no longer listened in. A second light above the position showed when the conversation was over and the cords could be unplugged from both ends.

Early switchboards had no numbers, but numbers were added because new operators could not become familiar with everyone's location on the switchboard fast enough to be trained quickly. Of course, users could not be expected to remember everyone's number, so a directory of subscribers and their numbers was published and distributed.

The first such local exchange, or central office, was established in Hartford, Connecticut, in 1878. The United States already had 1,000 telephones. With the exchange "switch," anyone in the local area needed only one telephone line to reach anyone else. The advantage of the central office is shown in Figure 2-2. Without the local exchange, 6 telephones would need 15 lines to interconnect them. The formula giving the number of point-to-point links for any N number of telephones linked in this fashion is $N(N-1)/2$, so $6(5)/2 = 15$. Once the central office was in place, the number was only N lines for N telephones. This method came just in time. By 1880, there were 50,000 telephones in the United States.

Initially, male college students were hired as operators, but this did not work out. The young men had a "relaxed" attitude about working hours, required spittoons at all work locations, and sometimes used harsh language with customers. When replaced by young women from good homes, all three problems went away. Mark Twain was one of the first customers and so also one of the first people to have a telephone in his home. Supposedly, Twain was one of the first people to engage in another telephone tradition: the complaint about service quality.

It is certainly understandable that anyone in those days would complain about the quality of the voice delivered through the telephone. It was pretty horrible. The standard greeting was a shouted "Ahoy!" that competed with the crosstalk of other conversations and the clicks and squeaks of adjacent telegraph signals. Users claimed that they could hear the rumblings of thunderstorms thirty miles away.

The main problem was with the local access line or local loop itself. Initially, these were single strands of iron wire, uninsulated and exposed to the elements. These loops were based on the same technology and physical plant as the telegraph system, and early telephone systems just rented spare telegraph lines from the telegraph companies. Unfortunately, iron rusted and single strands of wire acted as long antennas, picking up signals from near and far. Due to the nature of telegraph signals—simple "digital" electrical pulses representing dots or dashes—neither of the problems with rusty wire or antennas

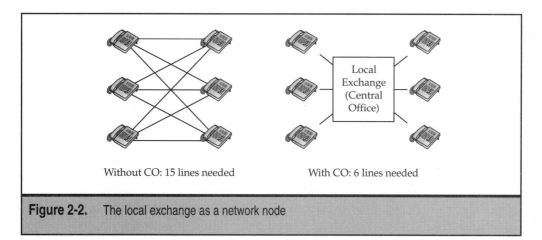

Without CO: 15 lines needed   With CO: 6 lines needed

Figure 2-2.   The local exchange as a network node

affected Morse code to any great degree. In fact, it was the other way around. The pulsing "digital" Morse code signals interfered with the analog voice signals on nearby wires when they were run together on utility poles or in trenches. Incidentally, it will be shown that this same effect has implications for ADSL implementations.

In any case, some basic changes improved voice service dramatically. In 1881, single-strand telephone wires were running the 45 miles between Providence, Rhode Island, and Boston, Massachusetts. These long wires were constant sources of complaint, until one day, a technician named John Carty accidentally hooked up two wires at the same time. Suddenly the voice quality improved dramatically. Carty had invented the two-wire pair for voice *circuits*. The two wires formed a closed circuit on which the electricity could flow. Another breakthrough came in 1884 when a new method of "hard drawing" copper wire made copper affordable for telephone wiring. Everyone knew that copper had better electrical properties than iron, but up until 1884, copper had been too expensive. With copper, longer runs for telephone lines were possible, all the way to New York from Boston (292 miles), for instance.

Crosstalk from adjacent telegraph signals and even conversations in cable bundles was still a problem, however. Some early efforts were made to *shield* the wires from these effects or keep the wires widely separated by a foot or more, but both methods were expensive. When the Brooklyn Bridge opened in New York in 1883, all of the telephone wires between Brooklyn and Manhattan had to be jammed into one small conduit pipe. The crosstalk was unbearable. However, even crosstalk became manageable after pairs of copper wire were twisted together, usually three or four times per foot. The new *un-shielded twisted-pair copper wire* (UTP) made cables consisting of many wire pairs possible and lowered network expansion costs. Even when jammed together inside these cable bundles, the analog signals would not interfere and cause crosstalk between each other. The cables offered more environmental protection as well.

By 1915, it was possible to place a telephone call from New York to San Francisco. The call was switched by hand from switchboard to switchboard and it took 23 minutes to complete the process. The cost was $20.70. As expensive as this sounds even today, it was much more so back then when a complete lunch cost about a nickel.

This section has positioned the local exchange or central office as the network node of the PSTN; however, each line connected to the central office has so far ended up at a telephone. Yet mention was just made about New York to San Francisco conversation being hand-switched from switchboard to switchboard across the country. How could this happen if the local exchange is only connected to local telephones? The answer to this problem is to invent *long distance*.

The trouble with the local exchange is that they connected only local users through the switchboard. The term *local* turned out to be a relative one, and in the densely popu-lated Northeast, most users were only a few miles from the central office. In the Midwest and Far West, things were different. It could be many miles just into town, where the local exchange was usually located. Providing service was expensive and voice quality was relatively poor. For this reason, the early *Bell System* generally shunned rural areas, and it was in these areas that *independent telephone companies* first sprang up around the turn of the century. As the Bell network grew, it became common to refer to local telephone companies as *Bell Operating Companies* (BOCs). The BOCs together formed the Bell System, although this was not an official designation in any way.

So just how did a local user in one community call another hundreds of miles away? By interconnecting the local exchanges themselves. Calls within the *local calling area* could be switched by any operator. Calls outside this area had to rely on the services of the *long distance operator*. Not all operators had access to the links leading to other central offices; they did not have long distance switchboards. That was fine because few calls actually went outside the local area. But whenever someone did want to make a long distance call, the call was made to the long distance operator. The links that connected to the long distance switchboard did not actually terminate at customer telephones. In this case, there was another long distance switchboard at the other end. A call was switched from "board to board" until the local exchange servicing the called party was reached—only then did the telephone ring at the other end.

As was mentioned previously, the process was long and tedious. In most cases, the long distance operator obtained the city and local number information from the caller, and then said, "We will call you back when the call is completed." Sometimes, the call back indicated that the intended party was on the telephone themselves and could not be reached.

Of course, every local exchange in the country did not have a direct link to every other local exchange in the country. This was as impractical with central offices as with individual telephones, and the local telephone companies faced different challenges when installing and maintaining telephone lines that were not only a few miles long, but tens of miles long. And that was all these links were: regular unshielded twisted-pair copper wire, the same as that used for local loops. Special amplifiers were invented for long links to periodically boost the power of the voice so that there was something to listen to by the time the signal reached the recipient.

This long distance interconnection problem had a solution as well. The early Bell System, the largest provider of local exchange service in the country in the 1890s, founded The Long Distance Company to link local exchanges together all over the United States. While displaying a singular lack of imagination, the name was nothing else if not descriptive. For all intents and purposes, The Long Distance Company was the property of the Bell System. That is, The Long Distance Company existed to link Bell System local exchanges together and saw no reason to link any of the newly formed independents to their system. This practice generated a lot of controversy because the new and independent telephone companies could not readily afford to copy the nationwide coverage of The Long Distance Company. There were under-funded attempts made periodically, but The Long Distance Company remained the only game in town. Ultimately, The Long Distance Company did begin to connect non-Bell local exchanges, but usually only in areas where Bell had no interest in expanding their own local services.

When the long distance network was added to the network of local exchanges, it gave the telephone system a distinctive, two-tiered structure, as shown in Figure 2-3. Note that the local exchanges could be Bell-owned telephone companies or independents, which

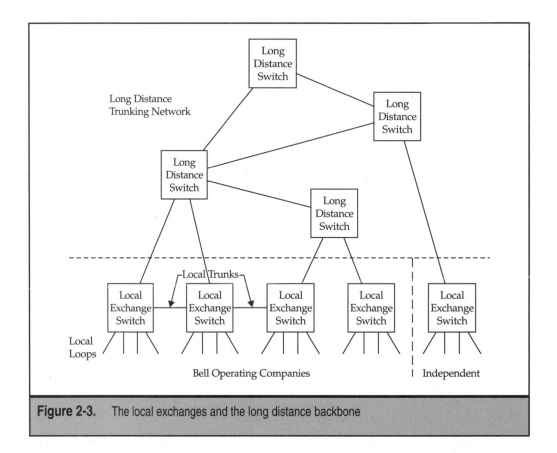

Figure 2-3. The local exchanges and the long distance backbone

allowed nearly all of the telephones in the United States to be linked, as long as a caller knew the city and local number they were calling. Keep in mind that in these early days, all of the telephone switching was still done by hand, board-to-board, and there was nothing special about the trunks used to connect any of these switches, local or not. Back then, the trunk designation referred to usage, not special construction or engineering. Some long trunks needed amplifiers, but so did some spectacularly long local loops. In the rural West, these local loops sometimes used barbed wire fences to carry analog telephone signals.

Figure 2-3 also shows that "local trunks" could be used to connect neighboring local exchanges. This made sense only because there were many more telephone calls made within a community than across the country.

A special problem soon arose on these trunks. Sometimes, an operator on one end would attempt to "seize" an idle voice circuit for use while an operator on the other end was also trying to do the same thing. There were other differences as well, but the whole point is that different "signaling" (call control) methods were needed on trunks as opposed to local loops. In the tradition of telecommunications terms, this use of the term "signaling" is distinct from the same term when applied to the form that information takes when traveling on a telephone line (analog signaling versus digital signaling). In this case, the term applies to a form of control information and supervisory processes that prevent the special trunk problems from occurring. In this context, signaling means *call control procedures,* which govern the setting up, maintenance, and taking down of voice conversations on circuits. Saying that signaling on a trunk is different than signaling on a local loop is the same as saying that call control procedures on a trunk are different than call control procedures on a local loop.

The technology used in this basic local-long distance telephone network evolved rapidly, especially in the switching and signaling (both kinds) environment. A detailed discussion of the technology is the major topic of the next chapter. This chapter will continue to deal mainly with the social and political developments that make the PSTN what it is today. As concerned as ADSL is with technology, these social and political factors cannot be overlooked in any discussion of ADSL.

Long distance calling greatly enhanced the value of the telephone. Instead of waiting days for a letter to arrive, people could call each other and speak immediately. In spite of this value, telephone service in many cases was perceived to be expensive and was not available in isolated areas. Even as late as 1941, only 40 percent of Americans had telephones, but those who did generated an enormous amount of revenue for the telephone companies.

This had been true practically from the beginning, especially with regard to long distance. The Bell System basically tried to keep local service costs low to encourage the popularity of the telephone, while at the same time pricing long distance comparatively high. The result was that by the late 1890s, the parent company of the local Bell operating companies was struggling, and only the subsidiary known as The Long Distance Company was on solid financial footing. The solution was to have the subsidiary buy the parent, so the Bell System entered the 1900s as *American Telephone and Telegraph* (AT&T), with The Long Distance Company becoming a business unit called AT&T Long Lines. Some 20 or so BOCs handled local services, organized mostly along state lines (New York Telephone, Illinois Bell, and so on).

THE END OF THE BELL SYSTEM

The Bell System endured with a form of state and federal government supervision known as *regulation* from 1913 until 1984. Each state regulated telephone service quality and rate structure for calls that were initiated and terminated within the boundaries of their individual states. For interstate telephone calls—calls going between two different states—regulation was handled by the federal government. Before 1934, this was done by the Interstate Commerce Commission, but after the Telecommunications Act of 1934 was passed, control and regulation passed into the hands of the *Federal Communication Commission* (FCC).

In 1984, as a result of a decades-long battle between the FCC and AT&T, and with the new competitive long distance companies such as MCI joining in, a federal judge and the United States Department of Justice split the Bell System up into effectively AT&T Long Lines and seven newly organized *Regional Bell Operating Companies* (RBOCs). The local independents more or less continued as they were, but there were major changes in how the RBOCs and independents handled long distance calls.

The RBOCs could still carry local calls end-to-end on their own facilities. For all other "long distance" calls, the RBOCs had to "hand off" the call to a long distance carrier, which could not be an RBOC. Furthermore, the RBOCs and independents had to let their subscribers not only use AT&T for long distance service, but any of the competitive long distance carriers such as Sprint and MCI. In fact, any long distance carrier that was approved by the FCC could offer long distance services in any local service area if they had a switching office close enough.

The problem was that there was no firm definition of what a "local call" was or even what "close enough" was, so the court and Department of Justice provided one. The entire United States was divided into about 240 areas with about the same number of calls within each. Calls inside these areas, known as *Local Access and Transport Areas* (LATAs), could be carried on facilities wholly owned by the RBOCs. All calls that crossed a LATA boundary had to be handed off to long distance companies, which were now called the *Interexchange Carriers* (IXCs, or sometimes IECs). The local companies, RBOCs and independents alike, were collectively called the *Local Exchange Carriers* (LECs). This whole structure neatly corresponded to the low-tier, local-long distance structure already in place.

In order to carry long distance traffic from an LEC, the IXC had to maintain a switching office within the LATA. This switching office was called the IXC *Point of Presence* (POP). The POPs formed the interface between the LECs at each end of the long distance call and the IXC switching and trunking network in between. For the most part, LATAs were contained within a single state, but there were exceptions. Any subscriber served by a LEC had to be able to route calls through the IXC of their choice, as long as the IXC maintained a POP within the originating LATA through a rule called *equal access*. If the chosen IXC did not have a POP in the destination LATA, the IXC could decline to carry the call (rarely) or hand off the call to another IXC with a POP in the destination LATA. Naturally, the second IXC charged the first for this privilege. It soon became apparent that there were just too many LATAs anyway, and as late as 1993, only AT&T had a POP in every LATA in the United States. But the system was in place, and cynics noted that the LATA structure closely mirrored AT&T Long Lines switching office distribution. With the breakup of the Bell System in 1984, it became common to speak of the entire system of telephones and switches in the United States as the PSTN.

The efficiency of the PSTN depended in large part on connecting calls as quickly as possible. There were two reasons for this concern. First, if calls could be put through faster, more calls could be handled per hour or day, customers would be more satisfied, and potentially more revenues could be made if the service were *metered* by number of calls and connection time. Second, the state regulators and the FCC, charged with approving telephone service rate increases, were themselves concerned with the quality of service the telephone companies provided to their customers.

Naturally, the fewer switching offices and switching steps needed from source to destination, the faster the calls could go through. Also, it could hardly be expected that each local exchange could maintain a full set of trunks to each and every other local exchange office in a given area. Of course, this would just reproduce the point-to-point telephone loop problem discussed earlier on a grand scale, at the switch level. And just as the point-to-point telephone problem was solved with a switch, so too was the trunk-to-trunk switching problem solved with a switch.

In many cases the practice developed to run trunks not directly to other local exchanges (although this practice also continued based on calling patterns), but to a more centrally located local exchange. Often, this local exchange received a second switch, but one that switched only from trunk-to-trunk and not from loop-to-loop or loop-to-trunk. These trunk-switching offices were called *tandems,* and the practice of switching trunks without any loops was said to take place at a *toll office.* Keep in mind that these definitions were never spelled out in any great detail. This was just the way telephone people talked.

Usually, a call routed through a toll office was a "toll call." A *toll call* is exactly analogous to a toll road, which is simply a road that one must pay a fee to drive on, above and beyond the road-use taxes assessed against drivers. In the same fashion, a toll call is just a telephone call, but it costs more to make it, above and beyond whatever the subscriber pays for local service. The amount of the toll depended on the distance and duration of the call. Remember that these calls were distinct from long distance calls, which crossed a LATA boundary. A toll call stayed in the same LATA, but just cost more (there were a few odd LATA arrangements, but these need not be of concern in this general discussion).

Also, the tandem/toll office arrangement offered a convenient way for IXCs to attach POPs to the LECs' networks. Instead of running trunks from a POP to each and every local exchange, an IXC could just link to the area's tandem or toll office. Because the tandem or toll office existed to tie all of the local loops in the area together, this guaranteed that all subscribers would be able to make *inter-LATA* calls through that IXC's POP, at least on the originating end. From the IXC perspective, this preferred point of trunk connectivity was called the *serving wire center* because the POP was served from this switching office. Again, this term was used from the IXC perspective. To the LECs, a wire center was just a big cabling rack (which they called a *distribution frame* where trunks and loops connected to the switching office) in the local exchange. In other words, a wire center is nothing special to the LEC, but is quite important to the IXC. Many IXCs maintain trunks to several wire centers in a LATA, all in the name of efficiency.

This is a good place to introduce the concept of access charges. The LECs had a monopoly on local service, but inter-LATA calls had to be handled from POP to POP by the competitive IXCs. These inter-LATA calls were often listed on the bill sent to customers by the LEC, which collected the entire amount on the bill. (Today the trend is for IXCs to send separate long distance bills to their customers, but the idea of access charges still applies no matter who sends the bill.) The IXCs paid for this billing service, which avoided the need for the IXCs to bill the customers directly. But how did the IXCs recover their rightful portion of the LEC bill? And how did the LECs charge the IXCs for the use of their facilities from POP to customer? The answer was through a system of *access charges,* which are still an item of discussion with regard to Internet access and the ISPs.

A local call is carried over the facilities owned and operated by the local carrier for just this purpose. A local call may be a toll call, but that just means it costs more. Local calls, toll or not, are precisely defined: both endpoints must be within the same LATA. Whenever a LATA boundary was crossed, this was a "long distance" call, even if it was only a few miles from end-to-end across a state line (very few LATAs crossed state boundaries). These were handled from IXC POP to IXC POP on interexchange carrier facilities.

The whole point about access charges was that all long distance calls included two local calls as well. There was a local call on each end, and an IXC call in the middle. The local calls went to the POP on the originator side, and from the POP to the destination in the other LATA. But if the IXC in the middle gained all the revenue from the long distance call (even though the call was billed by the originating LEC in some cases), how could the LECs at each end be compensated for the use of their lines, trunks, and switches? After all, these were in use for the duration of the call and could not be used for local calls or to make money for the LEC in any way.

The answer was to install a system of access charges in 1984 with the breakup of the Bell System. *Access charges* are paid by the IXC in the middle to the LECs at each end to compensate them for the loss of the use of these local facilities during the duration of the call. The IXCs pay the LECs for the use of their facilities to complete local calls. Now, the trunk to an IXC POP might be carrying 24 voice calls. That is not subject to access charges. The LEC voice switch ports that link to the trunk are, however. So the trunk is still a leased line, but each call that is carried to and from the POP consumes LEC resources and is subject to a per-call access charge system.

All entities that can cross LATA boundaries for services must pay access charges, except for one major category service provider in most cases: the ISPs. Data services were explicitly exempted from paying access charges according to the rules put in place in 1984. This was done to make sure that the data services segment, just emerging in 1984, would not be crushed by costs that they could not afford. This also reinforced the basic split between *basic transport* services provided by the protected LECs and *enhanced services* provided by the ISPs. This principle meant that the monopolistic LECs could only transport bits through their network, not store, process, or convert them in any fashion. Changes to bits were enhanced services and were reserved for competitive entities like the ISPs.

Recent regulatory rulings have made it clear that there is no reason why ISPs should not pay access charges (Internet traffic does cross LATA and state boundaries, no one denies that), it is just a matter of what policy best serves the public interest. For now, the reactions of regulators at the federal and state level to proposed ISP access charges have ranged from what is called *forebearance* ("please don't charge the ISPs for now") to a more hands-off approach ("LECs and ISPs must work this out on their own") to a more proactive regulatory stance ("pay up" or "don't bother").

If anything seems clear, it is that sooner or later the ISPs will have to pay some form of what most of the ISPs see as an "Internet tax" to the LECs in the form of access charges. The only question remains how much.

THE ARCHITECTURE OF THE PSTN

Today, the PSTN in the United States has a structure similar to the one shown in Figure 2-4. The local exchanges and toll offices inside the LATA make up the first tier of the PSTN, the LEC portion. Since the Telecommunications Act of 1996, service providers may be any entity approved or "certified" by the individual states to become a LEC. Newer companies are *Competitive LECs* (CLECs) and the former service provider in a given area

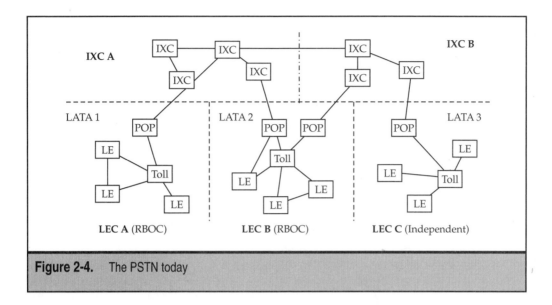

Figure 2-4.   The PSTN today

becomes the *Incumbent LEC* (ILEC). Terms such as *Other LEC* (OLEC) are sometimes used as well. ILECs, CLECs, OLECs, or some other exotic alphabetic combinations may still be RBOCs, independents, ISPs, or even power companies in various parts of the United States. There are some 1,300 LECs operating in the United States today, but many of these LECs are quite small and have only a few thousand subscribers in an isolated area.

The second tier of the PSTN is comprised of the IXC's networks. The IXC POP in the LATA could handle long distance calls for all subscribers in the LATA. The IXC had to have its own backbone network of switches and links, as well. The acknowledged leaders in this arena are AT&T, MCI, and Sprint. Sprint remained an oddity for a while because Sprint was also a LEC in some parts of the country, a rare mix of local and long distance services. Some 700 IXCs are still in operation in the United States today, but most of them have POPs in only a handful of LATAs. Many still handle calls to almost anywhere that originate from people within a LATA where the particular POP appears, but frequently only by handing the call off to another IXC.

A few points about Figure 2-4 should be emphasized. All lines shown on the figure are trunks, not local loops, because they connect switches rather than user devices to switches. Although shown as a single line, each trunk may carry more than one voice conversation or *channel*. In fact, many can carry hundreds or even thousands of simultaneous voice conversations. Also, IXC B, because it does not have a POP in the leftmost LATA, cannot carry long distance calls from anyone in that LATA. Of course, the same is true for IXC A in the rightmost LATA. In the center LATA, customers will be able to choose either IXC A or IXC B for long distance calls, either under equal access arrangements or *presubscription*, which automatically sends all calls outside the LATA to a particular IXC. Recently, the practice of deceptive IXC presubscription switching—known as *slamming*—has been universally condemned by regulators and all public-spirited IXCs. As a minor point, note that a POP need not be linked to a toll office switch. The actual trunking of the POP depends on calling traffic patterns, expense, and other factors.

At the risk of causing confusion, it should be noted that LEC B and LEC C, for instance, could both exist within the same LATA, depending on state ruling and certification. In this case, one would be the ILEC and the other the CLEC (or OLEC), and both would compete for the same customer pool within the LATA for local service. Finally, there is no significance at all to the number and location of LEs, POPs, and so on, nor the links between them. The figure is for illustrative purposes only.

The PSTN, at both the LEC level and the IXC level, is what is known as a *circuit-switched* or *circuit-switching* network. Much more will be said about circuit-switching in the next chapter in order to compare and contrast this practice with packet-switching, but any discussion of the current PSTN architecture would be incomplete without introducing one of its most distinctive features.

All trunks and local loops are divided into *voice circuits* or *voice channels*, both usually abbreviated VC. Technically, a voice channel in each direction is needed to carry one voice circuit, but this distinction is rarely made consistently. Sometimes, the voice channel outbound and inbound shared the same frequency range; this was usually employed on the local loop and was called *full-duplex* operation in the United States. At other times, the voice channel outbound and inbound used different frequency ranges. This method was usually used on the trunks and was called *half-duplex* operation. Because many conversations consisted of local loops on each end and trunks in the middle, a means of converting from full-duplex local loops to half-duplex trunks was needed at each end of the circuit. This interface arrangement is called a *hybrid* and introduces some nagging problems with echoes on the voice circuits. Echo on voice circuits is covered more fully in subsequent chapters.

A voice circuit is just a two-way connection between two compatible end-user devices (telephones, fax machines, or modems), so that communications can take place over the PSTN. The PSTN switches, just like any other network node, exist to map the incoming voice channels to the outgoing voice channels for the *holding time,* or duration, of the call. That is, the voice switches on the PSTN switch circuits on behalf of customers, who then interact as long as they care to.

While a voice circuit is in use, the network resources, in terms of bandwidth and switching capacity, cannot be assigned by the network to anyone else. Circuit-switching is sometimes called an "all the bandwidth, all the time" approach. The network resources tied up on a voice circuit in a circuit-switching network cannot be used for other purposes. For example, even if one of the parties of a telephone call put the other on hold for three hours, and there were no voices going back and forth at all, the bandwidth that the call represented could not be used by the network for any other purpose (such as carrying Web pages).

This approach made sense in the early telephone system. People called, and if they were not talking, they were listening. When they finished talking, they hung up. The length of their interaction defined the holding time of the call. It did not even make sense to attempt to develop technology to try to use the "quiet time" while people were listening on one half of the voice circuit for other uses in those days. The periods of silence were usually quite short, and no one really had any idea how to attempt such an undertaking anyway, so circuits were assigned a voice channel in both directions, and that was that.

This concept of voice circuit holding time brings up an interesting point of paying for telephone service, both local and long distance—how to charge a telephone call to the customer's bill.

The early telephone system experimented with a variety of rate structures. By 1923, some 206 different rates for local telephone service were in place in the Bell System alone. Some locations paid three times what others paid. The actual rates were based on a combination of the community's overall wealth, historical reasons, and simple greed. Obviously, the longer the distance involved in a voice call, the more trunks and switches were tied up and dedicated to the customer for the holding time of the call, so it seemed only reasonable that the long distance calls should cost more than local calls. And naturally, the longer the call, the higher the amount charged.

But what about local service? After all, the local loop had to be there whether it was in use or not. The same applied to the switch. The same is not true of trunks, which are not dedicated to any particular customer when not in use and can carry anyone's voice circuits. In this way of thinking, after some initial installation cost for local loops and the local switch, the telephone company basically had fixed monthly costs for maintenance, salaries, and so forth, so perhaps the customers should have a flat rate, or fixed monthly cost, for local service. With flat rate service, the amount of a monthly telephone bill was fixed, regardless of the number of telephone calls or their duration.

The alternative to flat rate service was *metered,* or usage-based, service. A monthly telephone bill using metered service varied based on the number of calls and their duration. More technical aspects of flat rate versus metered service are detailed in the next chapter. It need only be pointed out now that customers overwhelmingly preferred flat rate service to metered service whenever such a choice was offered, and with good reason.

Early telephone service was very expensive and pretty much limited to the businesses and wealthy families in a given location. As late as the start of the 1940s, only 40 percent of Americans had telephones in their homes. As the telephone companies tried to expand services and encourage everyone to have a telephone, metered service was a problem for families of limited means or on tight budgets. This is hard to explain simply with telephone service, so perhaps an analogy would help.

Suppose the monthly payment for the family automobile was not fixed, but varied based on how many miles it was driven in the previous month. Some months, when the car was driven particularly long distances, the monthly payment might be quite high, forcing families with tight finances to scramble around to scrape up the cash needed to keep the car. The question "How much should be set aside for next month's car payment?" has no easy answer and could be far off the mark. The result is added stress and concern. People end up thinking long and hard about buying a car versus using fixed rate public transportation.

Now suppose that the automobile company says, "We know that monthly auto use varies widely, but on average, each car we sell is driven about 1,000 miles a month, so we will just average this all together and charge everyone with a car a fixed monthly payment."

Now, some pay-by-the-mile car owners will end up paying more per month, and some will pay less, depending on where the fixed-rate average is set. But if the process is handled correctly, there should be no impact on the auto company's total revenues from this new "flat rate" monthly car payment system.

With flat rate payments, more families with restricted finances could figure out if they could afford a car. Payments could be budgeted consistently. The car company potentially benefits by having to perform fewer repossessions, handling fewer customer complaints over billing accuracy, having a stable and reliable revenue stream, and not having to decide whether any mercy should be granted those who cannot pay up in a particular month. There can be technical benefits as well, such as the elimination of complex accounting systems to track usage, but these are explored later.

Basically, the same arguments extend to flat rate telephone service. A lot of early telephone service was flat rate simply because no one had any idea how to invent technology to track calls by distance and time. As late as the 1960s, it was still common to charge for a toll call by filling out a toll ticket, which was just a card that an operator filled out based on connection time and distance to the destination. These toll tickets were collected daily and used to compile the subscriber's bill (which was typed by a clerk using a typewriter).

It might be argued that by the 1960s, American families had become wealthy enough that the monthly telephone bill was a bargain and not a burden, whether based on flat rates or not. This may have been true for domestic calls, but international calls were still a problem. In the late 1960s, thousands of young people in the United States were serving in the Armed Forces around the world, especially in Southeast Asia. When someone was wounded or—even worse—killed, it was not unusual for the monthly bill for calls to hit $700 or more. With many families at the time earning $10,000 per year or less, this was a real concern. In some tragic instances, the bill arrived on the day of the funeral. In a few cases where the family was close or well known to a telephone employee, the toll tickets would mysteriously become lost.

Flat rate local service avoided many of these social issues with respect to local calls. Once invented by the telephone companies and embraced by their customers, flat rates were mandated by many state regulators. Flat rate local service remains entrenched, in spite of perfectly valid telephone company studies that show that metered services would save a lot of people at least some money, and a few people a great deal of money (some others, of course, would pay more—much more). But in this case, the financial reasons proved to be secondary to the social reasons behind flat rate local service. There are even some who have seen Marxist overtones in the state's regulation and mandating of flat rate local service. There is no need to do this, and even Social Security was accused of being a Marxist invention at one time. There is only one fact that matters: People prefer flat rate services of all types, whether for bus fares or telephone calls.

THE PIECES OF THE PSTN

The PSTN is a fascinating place to explore. All types of loops, trunks, switches, and equipment have been incorporated into the PSTN since 1876. Before looking at some of the more technical issues that provide a basis for ADSL in the next chapter, this might be a good place to summarize the key elements of the PSTN. The emphasis is not historical, but more contemporary.

The entire PSTN was put in place to handle basic telephone-to-telephone voice communication. The telephone just provides an inexpensive way to convert mechanical, acoustic, analog waves into electricity and back again. The electrical signal modulates the pressure waves of voice and sends them through the PSTN via copper wires. The electrical signals need to be periodically strengthened or amplified to carry them through thousands of miles.

All of the basic components of the PSTN are shown in Figure 2-5.

In its simplest form, the telephone *handset* contains a small microphone powered by a small charge of electricity sent through the local loop itself. For those with at least a small knowledge of basic electricity, you may know that the handset microphone is filled with carbon granules that vary their electrical resistance under the acoustic pressure of the voice. The acoustic wave compresses and releases the granules, allowing the electrical circuit to mimic the voice waves. The electricity on the local loop in turn obeys basic electrical laws regarding resistance, voltage, and current.

The electrical signal can be sent over the PSTN to another telephone handset, where the signal is sent to the speaker. This tiny speaker relies on the varying current coming in to alter the strength of an electromagnet. The magnetic variations vibrate a thin metal disk. These vibrations model the acoustic wave generated by the person talking at the other end of the voice circuit and enter the receiver's ear where it is (presumably) understood. So the handset has the transducer needed for acoustic-electrical conversion. The trick of the telephone is that although the acoustic wave is in its electrical form, it can be transmitted over great distances.

The telephone handset is connected to the local exchange or central office by a local loop. The loop is usually dedicated to one subscriber, but there were exceptions known as *party lines* (the term reflects multiparty use, although people could have a lot of fun with party lines). Technically, the local loop is only one form of access line from the subscriber to the PSTN. In any case, whenever a customer wants to make a call, the local loop is

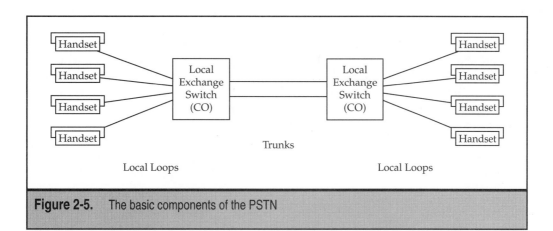

Figure 2-5.   The basic components of the PSTN

available unless another person at an extension (or on a party line) is using the loop. In this instance, the customer will overhear the current conversation.

In its simplest form, the local loop consists of two copper wires twisted around each other. Twisting was done to provide voice quality at a reasonable minimum and to reduce cost. With two wires on the local loop, the signals propagate in both directions. This is okay for voice conversations because usually only one person talks at a time. This two-way, but alternating, conversation is called half-duplex. However, the local loop can handle two-way simultaneous voice (try it), which is full-duplex. However, there are problems when attempting to do this with digital content, such as the output from PCs and modems. Special techniques are needed to make full-duplex transmission over two-wire local loops possible for digital information content.

At the other end of the local loop is the local exchange or central office switch. This is the PSTN network node that sets up, maintains, and terminates the temporary connections, known as calls, between any two end devices. The local exchange must terminate and switch local loops, trunks, and service circuits (for call control signaling and the like) to complete the task. In fact, a lot of effort is spent in providing *call progress* information to the user, which includes providing dial tone, accumulating the dialing digits, playing recorded announcements, generating ring back so the user hears the remote telephone ringing, and so forth.

The switch could be as elementary as a cord switchboard, as complicated as a solid-state computerized device running computer programs, or one of several varieties of *stepper, relay,* or *crossbar* switch in between. The switch connects loops and trunks in any combination needed to complete the call. Naturally, all of the loops, trunks, and service circuits must have ports on the switch itself. Some overall control is needed, which might be the human operator, but today is more likely to be a computer program running in the switch itself.

The switches are themselves connected by interoffice trunks. These can be thousands of miles long and are shared sequentially by users. Note that the user has no control at all over the trunks assigned to handle their call. Due mainly to their length and number, it was necessary and economically reasonable to provide much better voice quality on trunks than loops, so trunks use four wires (two twisted pairs) that provide a separate transmission path for each transmission direction. The main benefit is the ability to put amplifiers on the trunk that are able to operate in only one direction instead of both.

Note that in the simple network shown in Figure 2-5, there are more loops than trunks. In the PSTN, there are far fewer trunks than loops. This was possible and became an abiding principle of telephone network design because not everyone should be on the telephone at the same time. As was already mentioned, if it seems like everyone wants to use their telephone at the same time, chaos results on the trunks and within the switches. The time has come to take a closer look at this issue.

CHAPTER 3

Loops and Trunks

T he previous chapter introduced the structure of the PSTN from a historical perspective, as well as its many social and political aspects, but the technology was not mentioned much at all. Loops and trunks were just "links," and the concept that humans can perform switching tasks just as well as a modern central office switch was covered.

This chapter examines the technologies behind the modern PSTN loops and trunks. Switches are mentioned as well, but a detailed discussion of the PSTN switching issues that ADSL is meant to address is reserved until the next chapter.

THE ANALOG PSTN

As was discussed in the previous chapter, the Public Switched Telephone Network (PSTN) in the United States was born around 1875 with the invention of the telephone. The network was "public" in that the network nodes (telephone switches, in this case) did not belong to any specific user or group of users (known as "subscribers" or "customers"); rather, all of the switches and network equipment (even the telephone handset itself, initially) belonged to the service provider. The essence of a public network of any kind is that any subscriber can reach any other subscriber. Now, users can block calls in voice networks, and users can be clustered into closed user groups and the like in data networks, but as long as the connectivity is allowed on a public network, notably the PSTN and the Internet, then this connectivity is at least possible. The network was "switched" in the sense that the network nodes were switches that established end-to-end connections on circuits between the source and destination telephone numbers—which are the network addresses of the PSTN, although few think of them that way. The circuits were dedicated to a given pair of subscribers for the duration of the call. This came to be known as *circuit switching*.

It took a while for the PSTN to take on the form shown in Figure 3-1, which is a slightly different representation than that shown in the previous chapter. By the end of the 1800s, however, the major pieces shown in the figure were in place in many metropolitan areas throughout the United States. With this architecture, subscriber telephones—which were analog, voice-only devices—fed central office switches over analog local loops, which were simply long twisted pairs of copper wire.

This is as good a time as any to go into more detail regarding just what is meant by the term *voice*. As was just mentioned, by 1900, voice-only devices such as telephones connected to central offices over analog local loops. The same is true today in most residential areas more than 100 years later. However, the "voice-only devices" in use today are not just telephones.

The PSTN was engineered first and foremost for voice, which has many implications. One is that much more bandwidth is available on a copper local loop than is needed for voice (about 1 MHz total versus the modest 4 KHz usually cited for voice). Another is that the easiest way to make something work over the PSTN is to make the devices look as

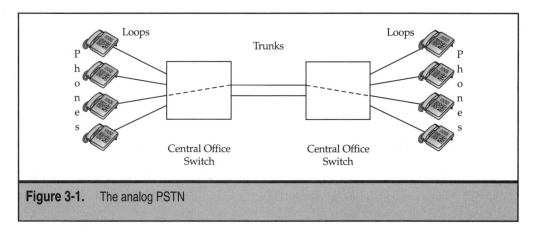

Figure 3-1.   The analog PSTN

much like telephones as possible to the voice network. So today, when people say "voice," they really mean four distinct things:

▼   Two people talking

■   Modems connecting computers

■   Fax machines

▲   Signaling for call control

Clearly, voice is not just the sound of human voices anymore. A few words about each of the items on the list is in order, because each of these is in the process of changing to adapt to the newer world of packet-switching networks designed primarily for data instead of voice, such as the Internet.

Voice in the form of two people talking is what the PSTN was engineered, built, and optimized for. But today, there is a movement to *packetized voice*, where digitized speech is compressed and placed inside packets, such as the same type of IP packets used to carry information over the Internet. Packetized voice over the Internet is often called *voice over IP* (VoIP), and in Chapter 11 there will be a detailed discussion of *voice over DSL* (VoDSL). Keep in mind that VoDSL does not mean sharing the local loop with digital information. It means packetizing digital voice for transport on the same portion of the local loop that DSL uses. This might sound confusing for now, but it will become clearer after the book covers the basics of DSL operation in more detail.

The use of modems to connect computers such as PCs has already been mentioned several times. Modems make digital information (data) "look like voice" as far as the PSTN is concerned. Modem calls just look like very long conversations, without many of the normal silences and gaps that humans require and use when they talk on the telephone. That is why these "never-ending telephone calls" have been a real problem for the telephone companies. But it makes more sense to put packet data on its own network, separated as much as possible form the traditional voice network. In a real sense, this item is what this book is all about.

Fax machines are really just simple modems. However, there are several differences between PC modem calls and fax calls on the PSTN. Fax machines are less concerned about errors during transmission, as long as the fax is still readable. Fax calls are shorter than modem connections, and the traffic is almost entirely in one direction, from sender to receiver. Lately, fax transmissions have also been digitized, packetized, and transported over the Internet and other types of data networks, which can be a real savings for international faxes.

Finally, the use of signaling for call control over the PSTN requires some explanation. Most people are familiar with the *touch tones* used by telephony devices to dial a number to request a connection. Basically all that can go through the analog local loop is sound, so these call control signals are just a special kind of sound. Telephony devices aren't the only items that can generate these sounds: whistles can do it (a cereal once inadvertently included a whistle of just the right pitch in the box), and so can people with high-pitched voices (when this results in a disconnect signal, it is called "talk off"). These call control signals are needed not only at the start and end of a call, but also in many cases during the call, as anyone with voice mail knows ("Press 1 for mailbox access..."). Naturally, these call control sounds are brief, intermittent, and very different again from speech or modems or fax machines. But they are still there and necessary on the analog local loop (once at the central office, the signal tones are almost always digitized and converted to packets as well).

So when people say loosely "the analog local loop was optimized for voice" they really mean "the analog local loop carries these four forms of traffic well."

These analog local loops, or access lines, were connected by "analog" switches, initially human beings whose job it was to say things like "Number, please" and connect the source to the destination. The analog switchboard was an electromechanical device firmly under human control. Naturally, as the network grew, not all telephones could be connected to one huge central office. For this purpose, the central office switches (technically, the switchboards) were connected with special analog circuits known as *trunks*. Trunks were also analog twisted-pair copper wires that connected not subscriber telephones, but switchboard operators; that way a call could be "switched" through the trunking network, usually without the caller even knowing what was going on, until the opposite party was contacted over their local loop, the operator said "Go ahead" and unplugged from the conversation (at least they were supposed to). This loop to trunk to loop switching process is shown in Figure 3-1.

So all of these central office switches were linked by trunks. The trunking network could be owned and operated by the local service provider for some calls, or the trunking network could be owned and operated by a "long distance" company, but they are all trunks. A trunk is just a common line that is shared by a number of users. Trunks are typically grouped together for use from a functional standpoint, but not necessarily a physical standpoint. A trunk may use the same physical facilities as lines, but the line coding and signaling (call control) requirements for a trunk differ a lot from those used in lines or loops. Trunks are also distinguished by a lack of user control over their use on a particular call.

Lines usually outnumber trunks, often only a couple of hundred trunks for up to 10,000 local loops. Because there were many more local calls within the central office serving area in those days, this ratio made sense. Not all switchboards in a given central office needed, or even had access to, the limited number of trunks available. The trunks typically ended up on the long distance switchboard under the control of the long distance operator (a prized position for a senior operator, because at the turn of the century there was lots of free time between such calls). On older phones from the early Twentieth Century, the "0" carried a "long distance" label to direct the call to one of these switchboards.

The whole point is that the network was analog from top to bottom, with analog local loops running through analog central offices linked by analog trunks. As noted earlier in this book, the term "analog" here refers to the signaling or coding method used on the loops and trunks and inside the switches. The term does not refer to the type of information content being conveyed, which could be analog itself (voice) or digital (PC files and e-mail).

DIGITIZING THE PSTN

Naturally, the PSTN service providers were quite concerned over the quality of the voice service they provided their customers. No real concern was paid to competition at that point because LECs were virtually guaranteed a life-long monopoly on local service with a given franchise area set up by the state. For the majority of people in the United States, the PSTN provider was an operating company under the AT&T or Bell System umbrella. The rest of the telephone companies were independents and operated in their own local franchise areas. The last independent company to have direct local competition with the Bell System was located in rural New Jersey and shut its doors in 1945. In short, there was little fear that the state government would revoke a local service franchise and give it to someone else.

Instead of competition, the major concern of the local telephone companies—BOCs and independents alike—was pleasing the state regulators with regard to service quality. There was one overriding reason for this concern. Subscribers who did not like the quality of their telephone service complained. Some grumbled a little, some griped all the time, but many took up the pen and wrote complaint letters to the state regulator, usually generically called the state *Public Utility Commission* (PUC), but the actual term used varied from state to state. When the state PUC periodically held public hearings about telephone service, the more disaffected tended to gather in groups and complain all at the same time.

The key to understanding the relationship between satisfying customers and satisfying state regulators is simple. In the vast majority of cases, all telephone service rate changes and increases had to be approved by the state regulator, typically after a round of public hearings. Naturally, the state PUC members tended to look unfavorably on rate increase requests when users were jamming the hearing room with signs and chants against the LEC whose rate increase was under discussion. The PUC commissioners, usually political appointees, were sensitive to situations that might become embarrassing to the administration that appointed them. Granting a rate increase to a telephone company whose service was being vilified by the public was a sure way to lose votes for all concerned. In fairness to the telephone companies, every franchise area seemed to harbor a small number of customers whose main goal in life seemed to be complaining about the quality of their telephone service. Some PUCs became fairly adept at filtering out this persistent background noise and concentrating on the major problem areas.

So the LECs got rate increases by pleasing the state regulators, who granted rate increases based on complaints remaining below a certain level for a set period of time (usually a year). Other factors were involved in this rate increase process, of course, but telephone companies with a stack of service complaints against them faced a tougher battle, without question. The bottom line was that the better the quality of service, the easier it was to get a fair hearing when a rate increase request was needed.

Voice quality depended on the quality of the electrical signal as it traveled through the PSTN. Analog systems have had one major problem when it comes to signal quality. Analog circuits are very noisy, and nearby thunderstorms and changes in electrical characteristics of the wire from rain or snow made the line hum, squeak, and pop. Shortly before World War II, Alec Reeve, a British engineer working for ITT in France, became so annoyed by the quality of analog voice signals that he decided to create a way to digitize the analog voice signal.

The digitization of analog voice involves a three-step process. First, the analog waves are *sampled* at a rate adequate to represent the analog waveform accurately. For analog voice, the international standard sampling rate is 8,000 times per second. Next, the samples are *quantized*, which means that they are represented by strings of 0s and 1s. For analog voice, the international quantization standard uses 8 bits per sample.

The choice of 8,000 samples per second was established by the voice passband bandwidth. The voice passband of 300 to 3,300 Hz was basically "rounded up" to 4,000 Hz. A rule known as the *Nyquist Theorem*—which states that digital sampling must take place at twice the highest frequency component to accurately reconstruct an analog signal—is then applied. Twice 4,000 Hz is 8,000 and so the sampling rate was set at 8,000 times per second. The 8 bits per sample is less rigidly maintained. In fact, the only real reason that 8 bits are still used is that most early computers and processor chips handled exactly 8 bits at a time instead of the modern standard of 32 or even 64 bits; it simply makes sense today to generate 8 bits for efficiency and ease of processing. Originally, the number of bits used to represent a sample was the subject of considerable debate. There was a lot of quantization noise with 7-bit samples, and 9 bits did not sound much better than 8 bits and was very wasteful of scarce resources on the older 8-bit processors. So 8-bit samples became the norm.

Finally, the bits are *coded* for transmission over a digital transmission link. Standard line codes come in many forms, but they all must operate at 64 Kbps (8,000 samples/second × 8 bits/sample = 64 Kbps) for digitized voice at the standard sampling rate. The three steps of sampling, quantizing, and coding are done in hardware and together make up a unit known as a *channel bank*. In the channel bank, the sampling and quantization is done by the codec, and the line coding is performed by the DSU/CSU.

The new quantized digital signals used on digital trunks were much less susceptible to noise and other impairments. The system was much too expensive, given the state of electronics in those days, and was a little like using a Rolls Royce as a New York City taxicab. World War II soon put an end to these experiments. The technical term for the process just described is *pulse code modulation,* or PCM. Sometimes, digitized voice is just called *PCM voice.*

It is important to realize that this digital invention in no way affected the information content sent over the line. This was a change from analog signaling on a line to digital signaling on a line, and it did not even matter whether the line in question was used as a loop or trunk. This technique did not change whether the information content being sent was analog or digital. The interface device might change, of course, from a telephone transducer to a codec for voice, and from a PC modem to a CSU/DSU for e-mail, but the information was the same.

After the war, the expanding world economy, pressure on the "telcos" to provide more and better service, and advances in electronics made the digitization of the PSTN not only feasible, but almost mandatory. Even the new electronic digital computers developed during the war were being researched and experimented with. The plan was to introduce the computer into service in the central office to switch the increased number of calls faster than humans, enabling the same number of trunks and lines to handle more calls. As it turned out, this ambitious plan took many years to happen.

Although originally developed for line side (local loop) use, digital transmission was perfected and extended to increase trunk capacity and alleviate the situation in metropolitan areas, where literally not another trunk cable could be run in a conduit. To understand why, a brief discussion of the multiplexing techniques used in analog and digital systems is necessary.

MULTIPLEXING AND TRUNKS

It is not in the least bit ironic that the telephone was invented as a result of the effort to multiplex, or combine, many telegraph messages over a single telegraph line. Multiplexing involves expanding the bandwidth available over a single medium—in this case, copper wire. Bell accidentally expanded the bandwidth so much that the wire could carry the major part of the voice bandwidth, the part containing about 80 percent of the power of the human voice. Figure 3-2 shows the bandwidths involved in human voice communication over the PSTN.

In the figure, the amplitude, or strength, of a signal is on the vertical axis, and the frequency is on the horizontal axis. Don't read too much into the actual shape of these curves—they are intended for comparison and informational purposes only.

The human ear can generally respond to frequencies from 20 Hz (cycles per second) to about 20,000 Hz, although the sensitivity to higher frequencies falls off with age. When it comes to speaking, human voice operates in the 100 to 10,000 range. Note that humans are capable of hearing more than the human voice can say, which is probably a result of needing to listen for the low rumblings of wild beasts and the high-pitched shrieks of many birds.

In any case, most of the power (a function of frequency and amplitude together) of the human voice occurs roughly between 300 and 3,300 Hz (this extends to 3,400 Hz in many places outside the United States). Any more, and telephone engineers found that transmitters and receivers became more expensive quickly. Any less, and people could not distinguish individual voices very well, so everyone tended to sound like they were speaking over an intercom at a fast food restaurant. This 300 to 3,300 Hz *passband* became

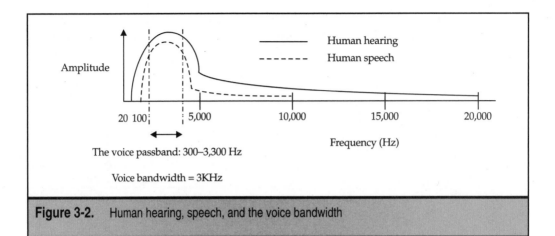

Figure 3-2. Human hearing, speech, and the voice bandwidth

the Bell System standard, and most other independents also adopted it (however, odd exceptions still exist here and there, where smaller passbands are used, typically at the high end (for example, 3,200 Hz)).

The 300 to 3,300 Hz passband meant that 3,000 Hz (3300–300), or 3 KHz of bandwidth, was needed to adequately transmit an analog human voice conversation, which was a lot more bandwidth than was ever needed to send the simple dots and dashes of Morse code, but that was the whole point. The 3 KHz bandwidth was easily provided on twisted-pair copper wire in the PSTN, which could carry frequencies up to and even above 1 MHz for considerable distances, as I have pointed out.

Note that there is nothing fixed by electricity or physical laws that limited the bandwidth on the copper wire to 3 KHz. That's just where the voice was, and this is one of the keys to understanding how ADSL works. Now suppose that a given transmitter and receiver can pass not just 3 KHz, but 36 KHz along two pairs of copper wire. Obviously, this wire could be used for more than one telephone conversation. These new transmitters and receivers would cost more than the ones that just sent and received 3 KHz, but maybe that was okay.

More expensive hardware is definitely okay on a trunk. Trunks are fewer in number and much longer than local loops in most cases, so it is financially sound to spend extra money to carry as many conversations on one or two pairs of copper wire as possible. As long as the money spent on transmitters, receivers, and multiplexing equipment is less than the money spent on new trunks, it would be a wise course of action. In addition, state regulators often mandate that the local service franchise holders show that they are operating the network in the most efficient and technically advanced way possible. Multiplexing voice conversations on trunks is all part of this process.

Note that multiplexing voice conversations on a local loop is also possible. However, with little financial or regulatory incentive to do so until the Web came along, coupled with the inability of LECs to control the wiring and CPE devices in the United States, multiplexing and increasing the bandwidth available on the local loop has been an enormously expensive and technically difficult thing to do (but it has been done in some places). Of course, ADSL changes all of the rules, so to speak.

It is desirable in parts of PSTN to carry more than one voice signal over the same transmission facility. The transmission facility may carry analog signals or digital signals. Both can be multiplexed, but it makes economic and technical sense to use different multiplexing techniques for analog and digital signals.

Analog signals typically use *frequency division multiplexing* (FDM). FDM gives a voice circuit "some of the bandwidth all of the time," which is done simply by dividing the passband of the transmission facility into separate frequency ranges, each corresponding to a voice channel. One voice circuit is carried in each channel, and usually one channel is provided in each direction, as is common for trunks. An FDM multiplexer device is installed on each end of the transmission facility to multiplex and demultiplex the various channels.

As an example, consider the twisted-pair copper wire with a bandwidth (technically, passband) of 48 KHz instead of 3 KHz. Suppose the entire passband was divided into 4 KHz channels. A voice conversation would comfortably fit inside each channel. For electrical reasons, the voice channels cannot sit "side by side," frequency-wise. The 4 KHz passband includes an adequate guardband to prevent crosstalk between the channels. Now the wire pair can carry not one, but 12 voice channels (48 ÷ 4 = 12). Another wire pair could carry the returning voice signals. The multiplexers on each end of the trunk would combine and split off the 12 voice circuits carried on the two pairs of wires. Note that if a voice channel is idle, a special "idle" signal is needed to indicate this.

FDM works best with analog signals. In the analog PSTN, FDM trunking networks were quite common. The Bell System developed many of these systems for their own use, and the technology became common even among the independents. The system just described, with 12 analog voice channels on two pairs of wires, corresponds in its basic idea (although not in detail) to something called *N-carrier*. The term *carrier* just meant that the voice circuits were multiplexed ("carried") and did not appear in their normal 300 to 3,300 Hz bandwidth.

The FDM methods used in the Bell System had their own terminology that showed both a firm grasp of simple words and a singular lack of imagination. In the FDM hierarchy, a *group* consists of 12 voice channels (as carried on N-carrier). A *supergroup* was made up of 5 groups for a total of 60 channels. Ten supergroups composed a *mastergroup* of 600 voice channels. Finally, a *jumbogroup* consisted of 6 mastergroups forming 3,600 voice channels. Frequently, the groupings were carried on coaxial cables that formed a family known as *E-carrier*. This should not be confused with the *digital* E-carrier groupings such as E1 and E3 that are part of the international digital multiplexing standards today.

All of these analog carrier groupings with FDM work just fine. As it turns out, however, if the signal being carried is digital and not analog (regardless of analog or digital information content), FDM becomes enormously inefficient. After the analog voice conversation was digitized, the rationale for FDM was weakened considerably.

Fortunately, another form of multiplexing was known and used for analog signals as well. *Time division multiplexing* (TDM) was quickly abandoned as a viable multiplexing technique for analog signals due to expense and annoying technical glitches. As opposed to the "some of the bandwidth all of the time" approach of FDM, TDM functions according to an *"all* of the bandwidth some of the time" approach.

With TDM, the transmission facilities bandwidth, perhaps 48 KHz as before, is divided into *time slots*, into which go not pieces of the analog signal (that was hard), but bits from a digital stream (that was easy). The multiplexer now loaded time slots with bits on the sending side and took them off on the receiving side. The "ownership" of bits was determined by position in the bit stream. In other words, time slot #1 always had voice channel #1's bits, time slot #2 always had voice channel #2's bits, and so on. Note that even if the voice channel is idle, the time slot must still be sent from sender to receiver. In this case, the bit pattern in the time slot says to the receiver, "ignore this, the channel is idle."

The differences between the analog FDM and the digital TDM approaches are shown in Figure 3-3. Note that the total bandwidth and numbers of channels in the figure are the same. Only the organization has changed. After the digitization of voice became possible and cost effective, it quickly became apparent that TDM was a better way to carry digital voice conversations. Of course, telephones and modems still generated only analog signals, and local loops still carried the 300 to 3,300 Hz passband only, but TDM was intended for the trunking network. TDM not only flourished there, it took off, thanks to a digital TDM technology known as *T-carrier*.

T-CARRIER

Although some think that digital signaling technologies for voice and data are relatively new, the T-carrier system has been around since the early 1960s. The voice quality advantage that digitization enjoyed has already been discussed. An added benefit turned out to be the increase of trunking capacity without the need to run new wires between switches. As the need for increased capacity on the voice network became evident in the 1950s and 1960s, this was a big bonus. The enormous expense of these early electronic devices, the

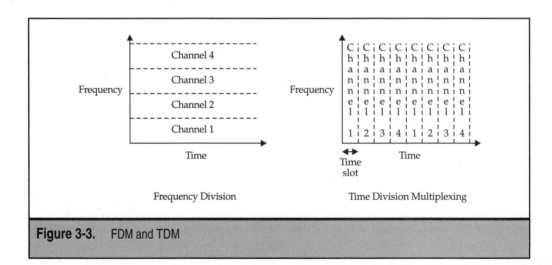

Figure 3-3.   FDM and TDM

codecs, and the CSU/DSU was offset by the savings enjoyed from fewer labor-intensive facilities installations. Thankfully, the price of electronics always goes down, while the population and wire costs always go up.

One of the most common ways to double analog trunk capacity on N-carrier links handling 12 simultaneous calls was to install the new digital T-carrier equipment on each end of the trunk. At its most basic level, this T1 could handle 24 voice channels on the same two pairs of wire. No new cable needed to be run between the central offices, and 24 simultaneous voice conversations could be carried. In another form, called T3, the T-carrier could handle 672 voice channels, or 28 T1s, on a single coaxial cable that might have been formerly used for an analog E-carrier.

By the mid-1970s, digital trunks were in use in many places for quality and increased capacity purposes, while electronic computer technology was beginning to be used for switching. In 1976, the first all digital local exchanges began to appear in small, rural central offices in the United States. However, even before that time, electronic computerization techniques had made advances in the telephone system possible on a large scale. The whole point was to complete calls faster and without the intervention of a human operator. By the late 1960s, subscribers in most urban and suburban areas could dial all local calls directly, thanks to electronic systems that handled all call routing.

Oddly, it was a while before digital trunks and digital switches were used together. The smaller, rural central offices employed digital switches first for two reasons: first, the switches were new and could only handle the smaller number of lines in rural areas; second, the human operators there were often underutilized given the smaller number of calls. On the other hand, large, urban areas got the new digital T-carrier trunks first. This was done mainly to alleviate conduit congestion and handle more calls on the same facilities.

Note that the analog local loop was not considered part of the problem(s) that digitization addressed, and so it was left alone. At the time, installing digital equipment in everyone's home would have been prohibitively expensive. Besides, such equipment would have been classified as *customer premises equipment* (CPE), and the telephone companies would have had as much control over its installation and use as they do with modems, which is to say, none at all.

INTEGRATED SERVICES DIGITAL NETWORK (ISDN)

This retention of the analog passband local loop had at least one consequence—the superior voice quality that digital transmission on the trunk provided was "masked" by the continuing presence of analog passband local loops. The new digital trunks did not dramatically improve voice quality, but at least they kept it from getting worse. Of course, there were still many analog trunks; because the voice quality "problem" was not the main reason for trunk and switch digitization in the first place, this approach still worked well.

It was only when long distance calling became common in the early 1960s (with the invention of area codes) that voice quality became a real issue. People objected to paying a lot more ($6–$12 a minute for a coast-to-coast call in 1961) for what was essentially the same voice service. AT&T Bell Labs responded by eventually proposing the complete and total end-to-end digitization of the telephone network. This plan formed the digital network portion of what eventually became known as *Integrated Services Digital Network* (ISDN). The whole goal of ISDN was to integrate all services reachable by a telephone connection (which included data by this time) and deliver them over an end-to-end digital network. By the mid-1980s, the digitization of switches in conjunction with digital trunks had begun. ISDN implementation basically boiled down to digitizing the analog local loops and the analog switch ports to which these loops were attached (if the switch was still an analog switch).

ISDN essentially digitizes the last segment of the PSTN, the local loop. No analog signals (this time this means both call control and encoding) at all are carried by an ISDN line. Although ISDN was a much more efficient platform for all kinds of digital information content, it required every telephone to either be converted to an ISDN digital device or use an expensive converter. Neither alternative was popular with service providers nor users.

The ISDN plan gave each wire pair running to a home the capability to carry not one but two simultaneous voice conversations, known as *bearer channels*, or B-channels. The voice calls were carried as 64 Kbps digital signals (128 Kbps together) along with a separate *digital signaling (call control) channel*, or D-channel, which ran at 16 Kbps. Some additional bits were added in the form of overhead to bring the total line rate up to 160 Kbps, of which 144 Kbps was represented by the B-channels and D-channel.

The D-channel could be used for packet data services when not needed for call control signaling. Some envisioned these services as being similar to what the Internet and Web provide today, but in those days this was just a dream. (In those days, 9,600 bps was seen as sufficient for almost any kind of data service.) The B-channels could be used for almost anything: voice, data, and even video. These B-channel circuits could operate in circuit mode (for voice), packet mode (for data), or today even frame mode (for frame relay "packet" services). ISDN included plans for new kinds of telephones that would have data displays and even video screens, as well as simple voice capabilities.

For its time, ISDN was an ambitious plan to funnel all forms of information down a simple pair of wires to everyone's home. Unfortunately, much of the best that ISDN had to offer is now available on the Internet and Web, and the funnel now ends not at the voice/video/data telephone, but at the voice/video/data PC.

ISDN became an international standard and could be applied not only by the Bell System, but also by the independents. A full discussion of why ISDN never took off as intended, in spite of years of publicity and deployment efforts, is well beyond the scope of this section. For now, it is enough to say that the digitization of the local loops and switches proved to be enormously expensive, even in the 1980s. This meant that ISDN services needed to be priced quite high in order to allow these expenses to be recouped in the amount of time established by state regulators. Also, not many people could figure out exactly what benefits ISDN conferred, at least until the Web came along. The combined result was that as late as 1996, there was absolutely no ISDN service at all available in Alaska, Montana, most of Nevada, or New Mexico.

ANALOG LOCAL LOOPS

Analog local loops have a number of characteristics that make any radical and comprehensive modifications (such as digitization for ISDN) difficult. Most of these characteristics are a consequence of actions that were applied strictly to improve analog passband performance and, therefore, become a hindrance in a digital environment.

Initial local loops were single wires (using something known as *ground return*) and only paired when a technician accidentally discovered that a *metallic return* improved voice quality dramatically by cutting down drastically on crosstalk. These long, parallel wires, however, suffered from signal loss (known as *attenuation*) because the wires acted as long, thin capacitors that tended to "store" the signal rather than allow it to flow freely. Twisting the pairs together to combat crosstalk was soon found to also help counter the attenuation effects slightly by adding an electrical characteristic known as *mutual inductance* to offset the capacitance of the wires. *Shielding* the wire by adding an outside metallic jacket or braid would have improved analog voice quality as well, but this approach was deemed too expensive and also tended to increase the attenuation.

The *unshielded twisted pair* (UTP) was invented very early in the history of the PSTN to minimize crosstalk. It also allowed the signals to flow more freely, but only marginally. After thousands of feet, the signal was just too weak and the voice was hopelessly muffled and tinny because only so much induction could be added by twisting the wires. Furthermore, the attenuation was much worse at higher frequencies on the loop, which is true of electrical signals when the electrical loads placed on them are said to be *reactive*.

After much trial and effort, the telephone companies eventually figured out that acceptable voice quality had the following practical limits. UTP analog local loop wire of 19, 22, and 24 gauge gave acceptable voice quality up to 18,000 feet (18 kft, about 5.5 km) from the central office. These *gauge numbers* are an *American Wire Gauge* (AWG) standard for the thickness of the wire; the smaller the number, the thicker the wire. (Legend has it that this odd usage related to the number of wires able to be placed in a hole of a standard size: the smaller the wire, the more could be placed through the hole.) The corresponding international wire gauges are 0.9, 0.63, and 0.5 mm. On thinner 26 gauge UTP (0.4 mm), only 15,000 feet (15 kft, about 4.5 km) of local loop gave acceptable voice quality. Nonetheless, there were many subscribers, especially in rural areas, who were more than 18 kft (about 3.4 miles, or 5.5 km) from the central office. How could the telephone companies reach these subscribers?

Adding induction would counteract the attenuation and signal loss, but only so much induction could be added by mutual twisting. The answer was to add extra inductance to the UTP. This process of extending the analog local loop with inductance is called loading. The electrical components used to produce these *loaded local loops* were known as *loading coils*.

WHY LOADING?

A more detailed look at loading requires at least a passing knowledge of electrical circuit terminology. Each of the terms used in this section will be given an analogy or example to help those unfamiliar with the terms (or those for whom formal education was longer ago than one would care to admit). This section just reviews some of the relationships between inductance, capacitance, and attenuation.

Power in a circuit is not solely dependent on either the voltage strength of a signal or the amount of electrical current flowing. Rather, power is the mathematical product of voltage and current together. Voltage is the electrical equivalent of pressure (as in a garden hose), and current is the electrical equivalent of the amount of water flowing through the hose. Thus, the total amount of water delivered to a barrel being filled by a hose is dependent on both the water pressure and the flow of water (that is, larger hoses deliver more water). More pressure through the same hose will fill the barrel faster, but so will a larger hose, even at the same pressure. This is true in all instances. A fire hose knocks you down not from flow by itself, nor the pressure, but from the combination.

In an electrical circuit, the capacitance changes the phase of the current relative to the voltage. This phase change means that the maximum voltage no longer occurs at the same time as the maximum current, which limits the power delivered over the path. Capacitance is the tendency for electricity to build up on a wire (or pair of wires) rather than flow freely. Interestingly enough, inductance (which is a measure of the "slipperiness" of the wire) changes the phase as well, but in the opposite direction from the phase shift due to capacitance effects.

This being the case, adding inductance to high capacitance circuits such as a long local loop can affect the power delivered to a receiver. Adding inductance in just the right amounts can bring the voltage and current back into phase again, but only in a certain bandwidth range. This boosts the received power, which is, of course, the whole idea.

Note that the additional inductance cannot be added haphazardly, but must be engineered carefully to affect only the bandwidth of interest. For the analog voice passband, this is the frequency range from about 300 to 3,300 Hz. Effectively, the adding of inductance "tunes" the circuit (in this case, the analog local loop) for voice transmission. The process of adding inductance to long local loops is done with loading coils.

LOADING AND ATTENUATION

Attenuation on analog local loop UTP limits acceptable voice quality to about 18,000 feet (about 5.5 km), in most cases. Attenuation means signal loss, and if the local loop is too long, the voice is hopelessly muffled. The attenuation is caused by capacitance effects between the wires. These effects may be countered by adding inductance, which was a side benefit of twisting the wires in the first place. However, only so much inductance can be added by twisting; loading coils must also be added to the local loop.

The effects of a loaded analog local loop on attenuation are dramatic. Figure 3-4 shows the effects of the three major families of loading. The attenuation is given in *decibels* (dB) per mile, a measure of signal strength. The horizontal axis shows the effects of the three major loading families when used on 22 AWG (0.63 mm international gauge) copper wire local loops. These are the "H," "D," and "B" loading architectures. The nonloaded curve shows that the signal loss is more severe as the frequency rises.

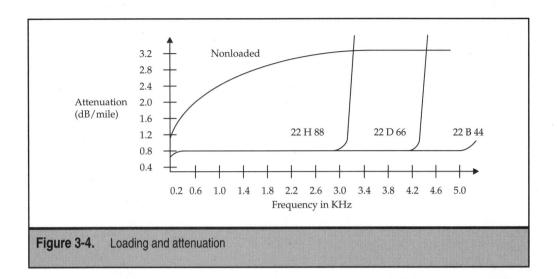

Figure 3-4. Loading and attenuation

The loaded analog local loops have much flatter signal loss profiles, which is the desired result and the key point to be made about loading effects. However, note that even with the common D and H loading schemes, attenuation rapidly increases above 3 or 4 KHz. That is okay for voice because most systems cut off local loop signals at about 3,300 Hz (the passband); but above 4 KHz or so, the attenuation of loaded loops actually exceeds the attenuation of a nonloaded loop of identical construction.

Loading posed a major problem for ISDN and for all other analog local loop digitization schemes. In order to deliver acceptable voice quality over analog local loops in excess of 18,000 feet (about 5.5 km), additional loading coils are usually necessary to add inductance to the loop. This inductance counteracts the effects of capacitance (which causes attenuation) over long distances.

Loading coils look like small iron "doughnuts," around which are wrapped each wire of the UTP loop. The inductance added is controlled by the size of the doughnut, the purity of the iron, and (most importantly) the number and spacing of the wire wrappings. The overall spacing between loading coils affects the distances at which the analog local loop operates as well. To avoid the haphazard deployment of loading coils, a series of standard "families" of loading coils and spacings evolved.

Figure 3-5 shows the general architecture of a loading system. Note that there is a central office end section, one or more spans between loading coils called the loading sections, and an outer end section. It is important to realize that the outer end section itself might extend thousands of feet. Also note that several phones might be serviced by the last loop, an arrangement usually known as a "party line." The interface between the outer end section and each individual phone in this arrangement is called a *bridged tap*. Bridged taps may still be present even when party line service is absent because the taps do not interfere with normal analog voice services. Bridged taps were used so that if a near subscriber

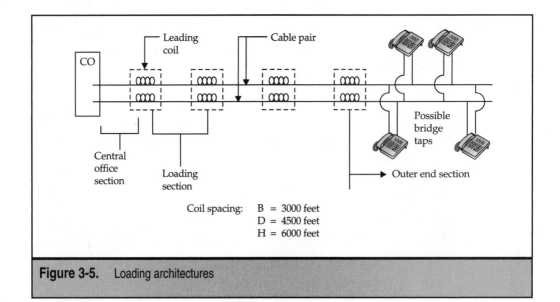

Figure 3-5. Loading architectures

terminated their telephone service (which happened frequently when service was expensive), that same pair could be used by a farther subscriber. This could not be done if the pairs were ever cut.

The loading "family" is distinguished by two related parameters: first, the spacing between the loading coils, and second, the number of millihenrys of inductance the coil adds to the loop at that point. Three loading schemes are common: B-44, D-66, and H-88. These are normally referred to as B, D, or H loading. A *millihenry* is a standard unit of inductance. For some strange reason, the standard units for capacitance and inductance in electrical standards were set much too large. Using a *henry* is like trying to buy a cup of coffee with a $1,000 bill. A millihenry is one-thousandth of a full henry and much more useful for common electrical situations. It is much easier to buy your coffee with a "milli-thousand-dollar bill"—a single dollar.

B-44 loading adds 44 millihenrys of inductance to the loop. The spacing is 3,000 feet (about 1 km). D-66 loading adds 66 millihenrys, so the spacing is longer at 4,500 feet (about 1.5 km). H-88 loading, the most common, adds 88 millihenrys and spaces coils 6,000 feet (about 2 km) apart.

One other important point is that the central office end section is always half of the normal spacing to the first loading coil. This compensates for loaded line to loaded line connections within the office itself. Thus, for H-88 loading, this central office end section span is only 3,000 feet (1 km).

OTHER ANALOG LOCAL LOOP FEATURES

Naturally, not all analog local loops suffered from excessive attenuation. In fact, the majority of PSTN subscribers, up to 80 percent by most accounts (although these figures are sometimes disputed), are located less than 18,000 feet (about 5.5 km) from the nearest central office.

The other subscribers are reached in a variety of ways. Loading, which adds extra inductance to counteract the effects of capacitance over distance, may extend the loop to up to 30,000 feet (about 5.7 miles or 9 km). Special *line extenders* (essentially amplifiers) may also be used to stretch the distance out to about 25,000 feet (about 4.7 miles or 7.5 km). Line extenders are typically nonstandard devices and so must be matched on the loop. This is not a problem as long as the same entity controls both devices (that is, one is not considered CPE). These local loop features are shown in Figure 3-6.

However, the figure also shows other analog local loop characteristics besides pure distance concerns. In many cases, the unshielded twisted-pair copper was "tapped" to service homes closer than the end of the wire, either to support party line arrangements (an older feature) or just because it was easier. In the latter case, if 100 pairs were run down a block and a new phone was installed, it was easier to "tap in" to a pair than to run new wire. Almost universally, the tapped wire was not cut because the new line service could be terminated in the future, with a new subscriber accommodated down the line at a later date. Not too long ago, when telephone service was much more expensive than it is today, people routinely cut off their telephone service if they were going away for a month or so. But there was no telling if the subscriber was ever coming back, so it made sense to have an option to reuse the pair with a bridged tap. These bridged taps posed no difficulty for analog voice service.

Finally, it was not uncommon to mix wire gauges on a UTP copper local loop. Going from 24 gauge to 26 gauge was most common, but other gauges were mixed as well. For analog voice this worked just fine, although sometimes faint "echoes" occurred because signals are always reflected, as well as transmitted, whenever the electrical characteristics of a wire path change, which is exactly what mixing gauges did.

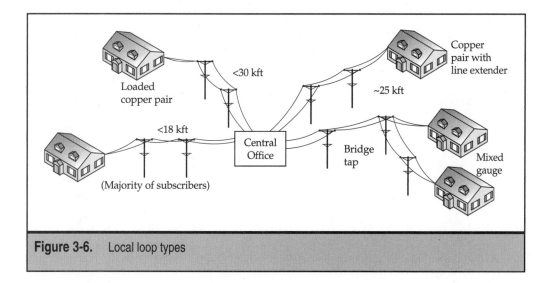

Figure 3-6.   Local loop types

When ISDN was proposed as a method to digitize the PSTN end-to-end, most of these analog local loop features caused concern. Loading coils, line extenders, bridged taps, and mixed gauges destroyed most digital signals. Bridged taps weakened and reflected them (the unterminated ends were long antennas), mixed gauges also reflected them, and loading coils and line extenders limited the available bandwidth because they were tuned for the analog voice passband of 300 to 3,300 Hz. Bridged taps act as "delay lines" and can extend anywhere from 1,000 to 5,000 feet (0.0 to 1.5 km) on their own. They also put an obvious "notch" in the line's attenuation at the frequency associated with the bridged taps' wavelength. This notch does not affect voice because voice is far down in the 4 KHz range of the local loop. A bridged tap would have to be far longer than practical to affect voice.

For the most part, the subscribers at the end of less than 18,000 feet (about 5.5 km) of analog local loop were pretty much okay, as long as there were no bridged taps or mixed gauges with which to contend. In fact, ISDN specifications called for runs of 24 AWG (0.5 mm) wire, with no bridged taps, with a maximum length of 18 kft (about 5.5 km).

ISDN, LOOPS, AND DAML

As if the different passband limited local loops variations were not enough to deal with, there is another local loop complication seen in the United States and around the world—*Digital Added Main Line* (DAML) local loops. In the United States, these lines are sometimes called to as *Digital Subscriber Single Carrier* (DSSC) links. Although not common, one form or another of DAML is used in about 10 percent of the United States. The reason is simple. One of the drawbacks that quickly became apparent in the PSTN was the need to add local loops to deliver second lines to users for Internet access or other services. Obviously, if some way could be found to offer second line service on a local loop engineered for a single analog conversation, the effort would be worthwhile. DAML allows the adding of this second telephone line. The service presented to the user is still two analog voice passbands. But this technique uses the ISDN DSL structure of two B-channels (64 Kbps each) to carry the PCM voice representing the analog voice over the local loop. The one D-channel (16 Kbps for signaling and packets) is not used. Of course, the new "2B+0" service would not run through an ISDN switch, and it effectively is still just two analog voice lines on one local loop. But because few people used ISDN to make telephone calls or had ISDN equipment in their homes, this was all right.

With DAML, a telephone company could add a second line to a home with existing analog service and provide the desired analog service on the new line. Think of DAML as "analog service packaged as ISDN without the ISDN switch." DAML uses readily available ISDN components on the local loop, but of course needs no ISDN switch upgrade. But because there is no ISDN switch at the end of a DAML link, the call control messages on the D channel are simply ignored. The service is still analog. DAML equipment is also used in Europe and other places around the world. Oddly, it is seldom considered or mentioned in surveys of local loop arrangements in the United States.

ANOTHER ARRANGEMENT: CSA

Loading systems are not the only way to extend the reach of the analog local loop. In addition, loading systems still had to be engineered and "balanced" for local conditions. Too much loading (technically, inductance) would make the loops "sing" or "hum" with annoying background noise; too little loading made the voice sound muffled and just too hard to hear; and of course, each loop had to have its own set of loading coils.

By the late 1960s and 1970s, digital technologies had matured enough to consider their use in the previously analog local loop arena. This brought the benefits of digital trunks to the local loops, namely better voice quality and higher call capacities. However, it was not a trunk anymore at all, but rather, a *carrier serving area* (CSA) that combined aspects of trunks and loops in one. The CSA architecture is shown in Figure 3-7.

The figure shows how this typically worked (there were many variations, so no generalization covers all instances). Usually, a T1 carrier was used to digitize 10 pairs of UTP from the central office to some point central to a cluster of subscribers; this central point, or "neighborhood," was the serving area itself. Normal 24 channel time division multiplexed digitized voice ran on eight of these pairs (two pairs were needed for each T1). The other two pairs were for network management or other purposes. Analog CSA technologies did exist, but digital T1 systems soon became more common. When T-carrier was used for this purpose, it was not called a bunch of T1s, but rather a *digital loop carrier* (DLC). In many cases, these systems were called *pairgain* systems because they "gained pairs" back for the telephone company to use for other purposes.

Now, local loops will not need loading coils unless the loops are longer than a certain distance. There could be mixed gauges, bridged taps, or both. It might sound a little like a free-for-all when it comes to the local loop, but there are actually two major architectures that are found in the local distribution plant that at least bring a little order into the local loop world. In order to better appreciate what CSA does for the local loop architecture, mentioning both is a good idea.

The first architecture, the *resistance design* rule, is often found on older local loops, and the second, *CSA guidelines,* is found mostly on local "cable plant" (the general term used to describe the collection of local access lines) installed since the mid-1980s. Although it is not completely true that all loops conform to one form or the other, there are relatively few loops that do not conform to either.

The resistance design rule is very old and was used right up until the mid-1980s when the CSA guidelines came in. The rule applies to straight copper pairs and limits the maximum loop DC resistance to 1,500 Ohms. This was the major consideration, because voice, ringing, and other supervisory voltages worked well as long as the 1,500 Ohms was respected. Beyond that, things were fairly relaxed. There was no real limit on the number and types of mixed gauges, nor the presence and length of bridged taps, other than the service worked and the 1,500 Ohms was not exceeded. The resistance design rule gives about a 9 dB maximum loss over the loop at 1,000 Hz, which was fine for voice services. Surveys have shown that the average loop has 600 Ohms DC resistance and a 4 dB loss at 1 KHz.

From the mid-1980s on, the CSA guidelines were used. These were designed primarily for DLC arrangements, but were soon also applied to straight copper runs for a variety

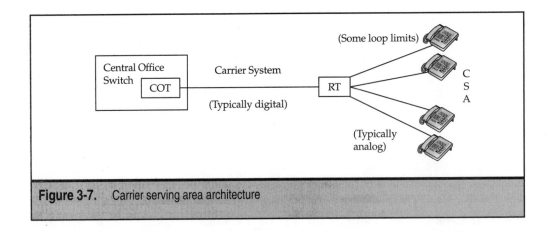

Figure 3-7. Carrier serving area architecture

of reasons. One of the major reasons was to avoid having one set of rules for straight copper runs and another set of rules altogether for DLCs.

CSA guidelines limit gauges and loop lengths (originally, at the end of the DLC). The customer end copper could be up to 9 kft (2.7 km) for 26 AWG (0.4 mm), or up to 12 kft (3.7 km) for 24 AWG (0.5 mm), but no more. In additional, there could be no loading coils. If there were bridged taps, a length of 2 kft (0.61 km) was established for the maximum individual bridged tap, and a limit of a total of 2.5 kft (0.76 km) for all bridged taps on the pair.

CSA allows for the use of only two gauges, with strict limits on the distance that each gauge could be used (a mathematical formula had to be applied).

Telcordia surveys have shown that some 60 percent of DLC loops meet CSA guidelines (they were only guidelines), and many single pairs meet the CSA guidelines as well.

There is a fundamental CSA guidelines design rule that is not emphasized in Figure 3-7. The loops must be less than 12,000 feet (about 3.7 km) with 24 AWG (0.5 mm in international gauge) pairs and less than 9,000 feet (about 2.7 km) with 26 AWG loops (0.4 mm).

To support the 96 voice channels (4×24), a device called a *central office terminal* (COT) was placed in the local exchange to interface with the 96 switch ports; and a device called the *remote terminal* (RT) was placed in the field, usually in an environmentally hardened casing. This arrangement was known as *SLC-96* within the AT&T Bell System—SLC stood for *subscriber loop carrier*, which was a specific type of DLC.

The advantage was that the 96 analog local loops now only had to run to the RT, and the same distance limits still applied at the end of the COT-RT link! So the SLC-96 (or whatever) DLC system could run 5 or even 10 miles to a housing development, and then the analog UTP local loops could extend another 12,000 feet (about 3.7 km) if need be. It was even possible to load these analog sections, but this was seldom done.

There are many CSA architectures that are variations on this simple theme. The general rule of CSAs was that they served customers more than 12,000 feet (about 3.7 km) from the central office. So in many places, there was an RT feed by the T1s in a cabinet right there at 12,000 feet.

The result was the modern local loop access plant shown in Figure 3-8.

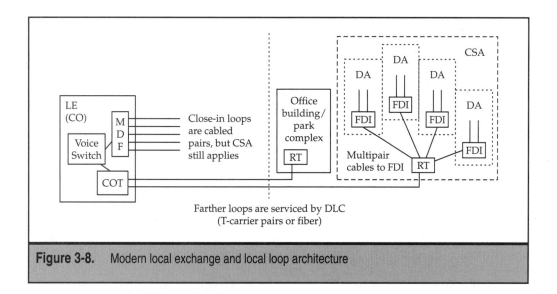

Figure 3-8.   Modern local exchange and local loop architecture

The figure shows how the CSA guidelines are applied in the modern local loop architecture. At the left, the local exchange (central office in the United States) exists mainly to provide customers and users with access to the voice switch. Ports on the voice switch may be accessed by individual pairs from the *main distribution frame* (MDF), also called the *wire center*. Although shown as a single line, there could be thousands of pairs leading to the switch from the MDF. All close-in, or near-by, subscribers are serviced by individual pairs through the MDF. The pairs could be cabled in 50-, 100-, or 200-pair cable binders, or even in higher pair counts.

Beyond the immediate area of the switch, things get interesting. Now the switch is accessed through a *central office terminal* (COT), which communicates with the RT over some loop carrier arrangement. Typically, this carrier is digital, and uses either T-carrier copper pairs or fiber optic cable for this main feeder section. The RT is usually located in some *controlled environment vault* (CEV) to protect and power the electronics, although the CEV can be as simple as a pale green box rising out of the ground. Usually these enclosures are jammed with equipment needed for the voice network, leaving little room to work with for DSL or other new technologies.

Each RT establishes a CSA, more or less by definition. If the RT is located in an office building or services a small office park, it is rare to refer to this as a CSA. This is just an RT, and that is all. Usually the term CSA is applied to residential arrangements, also shown in the figure. The same multipair cable binders used to service the close-in subscribers connect the RT to several *feeder distribution interface*s (FDIs). Each FDI establishes its own *distribution area* (DA), anywhere from 50 to 200 homes or sometimes more.

The whole architecture is well represented today, especially in "ex-urban" areas where strip malls and office parks alternate with housing developments and condominium complexes for mile after mile along the main roads. There is likely one or more RTs in each major location. The cables themselves can be above or below ground, depending on location or local rules. Some environments favor aerial cable (such as rock-hard ground coral) while others favor burial (such as dry, sandy soils).

THE TROUBLE WITH ANALOG LOCAL LOOPS

The progressive digitization of the PSTN to increase switching speeds and capacities proceeded in almost an "inside out" fashion. Initially, the central office switches and trunks were two targets of digitization efforts. The analog local loops, which needed neither more speed nor more capacity, were left alone.

Nevertheless, end-to-end digitization of the PSTN from phone to phone was attractive for many reasons. Once the whole network was digital, no conversion would be necessary, quality would improve, signaling (call control) would be more secure, and so forth. Even before ISDN was defined as the preferred method of digitizing the PSTN end-to-end, however, the problems inherent in analog local loop arrangements became obvious.

The main problem is that subscriber lines engineered for passband analog local loops in many cases limited the useful bandwidth available for digital signaling. This was not done on purpose, of course; it was just the end result of years of engineering for optimal passband analog voice signals.

Each of the analog local loop features, aside from basic, short-run local loops, turned out to be a concern. Loading coils cut off all the high frequency signal components required to transmit digital signals. Bridged taps cut the digital signal strength drastically because part of the signal went to each unterminated "branch," which reflected signals as well. Mixed wire gauges caused signal reflections that created echoes on the digital circuits. CSA terminal equipment limited the bandwidth available for any one loop, typically to 4 KHz analog or 64 Kbps digital, and this was true whether the outer loop was loaded or not.

All in all, these features of the analog local loop—developed over the years to solve problems of analog voice transmission—made the transition to digital local loops have a much higher price tag than anticipated.

REAL WORLD LOCAL LOOPS

It is all well and good to discuss the impairments of analog local loops in terms of bridged taps and mixed gauges in the abstract, but what do local loops look like in the real world? The statistics shown in the following list were gathered from a variety of sources, including the *Institute of Electrical and Electronics Engineers* (IEEE) and Telcordia.

▼ 200 million access lines in the U.S., 70 percent residential, 11 kft (about 3.3 km) average

■ Fifteen to 20 percent of all local loops have loading coils installed

■ Bridge taps and mixed gauges are almost universal

- ■ Typically 26 AWG (0.4 mm) out to 10 kft (about 3 km), then 24 AWG (0.5 mm) beyond (or even 19 AWG (0.9 mm))

- ■ Thirty percent are DLCs, usually 24 AWG (0.5 mm), but 9 kft (about 2.7 km) maximum

- ▲ Average loop has 22 splices, which pose corrosion and attenuation risks

It turns out that 200 million of the world's more than 700 million access lines (analog local loops for the most part, but a good 5 percent of the access lines in countries such as Germany are digital ISDN lines) are in the United States. The figures in the list apply to these 200 million United States local loops. About 70 percent of these run to homes, and the other 30 percent run to businesses. The average length of the local loop is about 11,000 feet (about 3.3 km), according to Telcordia. However, the larger number of shorter urban loops makes this number somewhat deceiving and lowers the impact of the much longer rural runs.

About 15 to 20 percent of these loops have loading coils, which means there are up to 40 million local loops beyond 18,000 feet (about 5.5 km). Moreover, bridged taps and mixed gauges, often presented as annoying anomalies, are just about universal (especially mixed gauges) because it was (and still is in many places) standard practice to run 26 AWG (0.4 mm) wire from the central office out to 10,000 feet (about 3 km). At that point, the wire was changed to 24 AWG (0.5 mm), or even 19 AWG (0.63 mm) in rural areas, which lessened attenuation effects. Bridged taps act as "delay lines" and can extend anywhere from 1,000 to 5,000 feet (about 0.3 to 1.5 km) on their own. They also put an obvious "notch" (more like a "ripple with notches") in the line's attenuation at the frequency associated with the bridged taps' wavelength (these are not a problem for voice). The bridged taps' wavelength determines the spacing between the notches.

Additionally, DLCs and RTs serve about 30 percent of this loop plant, or about 60 million homes and businesses. Almost 99 percent of all CSAs today are serviced by DLCs instead of analog pairgain systems. These DLCs usually have 24 AWG (0.5 mm) wire extending directly to the home, but they have a 9,000-foot (about 2.7 km) maximum in almost all cases.

The local loop has another impairment. Most outside plant cables, whether 200, 400, or even 3,600 pair wire, come in 500-foot reels, which means that they must be spliced every 500 feet (about 0.15 km); the average loop then has 22 splices (11,000 feet is the average length). Splices tend to form corrosion "collection points" and, if they are assembled improperly, can add a lot of attenuation themselves. In order to prevent corrosion, a small amount of electricity, called a *sealing current,* was applied to prevent oxide buildup (rust). Directly relevant to digital transmission, this sealing current could not be applied when the loop was used for purely digital signaling. However, DSL allows for the presence of the sealing current.

Figure 3-9 shows another perspective on the local loop issue in the United States. This figure is quite important when it comes to DSL. Here the 200 million access lines are first divided into those on DLCs and those that are "home runs" straight from subscriber to central office. Some 60 million users are limited to 64 Kbps until something is done about the digitization performed by the RT itself.

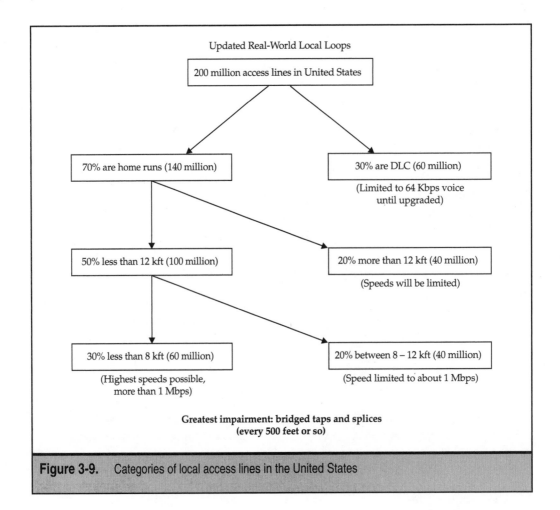

Updated Real-World Local Loops

200 million access lines in United States

70% are home runs (140 million)

30% are DLC (60 million)

(Limited to 64 Kbps voice
until upgraded)

50% less than 12 kft (100 million)

20% more than 12 kft (40 million)

(Speeds will be limited)

30% less than 8 kft (60 million)

20% between 8 – 12 kft (40 million)

(Highest speeds possible,
more than 1 Mbps)

(Speed limited to about 1 Mbps)

**Greatest impairment: bridged taps and splices
(every 500 feet or so)**

Figure 3-9.   Categories of local access lines in the United States

Of the 140 million home runs, 50 percent are less than 12,000 feet (about 3.6 km) from the CO (100 million access lines), leaving 20 percent of the total home run 70 percent more than 12,000 feet away (40 million). These 40 million users will have limited speeds, even with some form of DSL in most cases. Of the 100 million users less than 12,000 feet from the CO, 30 percent of the total are less than 8,000 feet (about 2.4 km) from the CO (60 million access lines), leaving 20 percent between 8,000 and 12,000 feet (some 40 million access lines). Now, the 60 million users on these shorter lines will get the highest DSL speeds possible, probably more than 1 Mbps. The other 40 million will get speeds up to about 1 Mbps.

TRUNK GROUPS AND THE LAW OF LARGE NUMBERS

Trunks in the PSTN consist of groups of voice circuits used between switching offices. Whenever a call is made that requires a trunk circuit, an idle trunk must be found in a group of a given size. If all the trunks in a group are occupied (busy), the call is blocked and must be redialed at some point in the future. This process is called *reorder* in the PSTN. Usually, the telephone company supplies a "fast busy" tone to the caller to inform them that the destination telephone may not be busy, but all the trunk groups that were tried during the switch routing process were occupied to a level that prohibits their use. Note that some trunks are always reserved for special uses and cannot be assigned for "regular" telephone calls.

Whenever things are grouped together for any purpose whatsoever, a principle known as the Law of Large Numbers takes over. Basically, this law states that large groups are more predictable than small groups when it comes to mathematical properties. To see why, consider the average height of people in groups of 100 rather than 10. There will be a lot of variation in the average height, determining 10 measurements at a time. One group may be all over 6 feet tall, while another is under 5 feet 6, and so on. But when taken in groups of 100 at time, one would expect much less variation in the average measurements from group to group, which should be closer to the average for the general population as a whole.

Telephony traffic is no exception to this rule. In the PSTN, service providers often use something called the *Erlang B traffic tables* to predict the probability that a call will be blocked. A level of service known as B.05 (Erlang B .05 blocking) will block only 5 percent of the calls attempting to use this group, or 1 call in 20. The B.1 level of service will block 10 percent of the calls attempted, or 1 in 10, and so on. These levels of service are built into tariffs and contracts and are closely watched by state regulators. Because larger trunk groups can support higher occupancies, the PSTN tends to aggregate traffic into large groups to achieve a given level of service more predictably.

So the PSTN is engineered for a given level of service when it comes to trunks. A level of service that is B.05 will block 1 call in every 20 (5 percent) attempting to use that trunk group. B.1 will block 10 percent, and so forth. Because the Law of Large Numbers applies, large trunk groups can be utilized closer to 100 percent at a given level of service, such as B.05.

For instance, trunk groups of 25 can be 75.9 percent occupied and block only 1 call in 20, whereas trunk groups of 200 can be 94.3 percent occupied. Larger trunk groups can be pushed closer to the edge, while smaller trunk groups need lots of slack.

However, this higher loading factor on larger trunk groups comes with a price. Because they operate closer to the edge, large trunk groups are more sensitive to traffic increases than small trunk groups. For example, a traffic increase of 14.2 percent will double the blocking to 1 call in 10 (B.1) on a trunk group of 25, but only a 7.9 percent increase will produce the same effect on a trunk group of 200.

In some cases, traffic increases have doubled the traffic loads on trunk groups. The level of service is lowered on a trunk group of 200 to B.5 service, which blocks 1 call out of every 2. Because these levels of service are built into tariffs and contracts and watched by state regulators, there are real consequences to these degraded service levels.

The impact is on the users as well as the service providers. Right after the 1993 California earthquake, for example, PSTN traffic increased dramatically. In larger cities, the added traffic pressures on the large trunk groups quickly overloaded the trunking network. Yet in smaller towns with modest trunk groupings, service degraded much less under the same increases.

The higher efficiencies of large trunk groups have resulted in large tandem switching offices (recall that tandems switch trunk to trunk) across the country. When one of these offices burned in Hinsdale, Illinois, the large trunks could not handle the traffic increases on alternate routes and quickly blocked many of the calls attempted on them.

INCREASING LOADS AND TRUNKS

In the PSTN, a trunk is occupied for the duration of a call. That is, the connection is mapped onto a trunk circuit as long as the connection between two endpoints exists. Traffic engineers use complex formulas and tables to determine the number of trunks needed to support a given number of local loops (that is, access lines) at the desired level of service (usually drawn as the Erlang B tables). This level of service may be mandated by the state or written into service agreement contracts or state documents known as tariffs. Either way, the service provider faces real consequences if the service level dips below the engineered level.

The problem is that the tables were created for relatively short voice telephone calls. The characteristic holding time (the duration of a call) for a voice telephone call is only three or four minutes (some studies place this as high as six minutes) and usage is relatively constant. This makes sense only for a human voice conversation. One person talks while the other listens, and when they are done speaking to each other, they hang up. A change in these underlying assumptions of trunk traffic engineering, in terms of traffic patterns or holding times, could degrade service. Even a modest 10 percent increase can degrade trunk group service from blocking 5 calls in 100 to a possibly unacceptable 10 calls in 100.

This is precisely what is happening to the PSTN today. Families with PCs (some 17 million or more), as well as *small office/home office* (SOHO) users, have been busily ordering second lines to enable their PCs to attach to the Internet and Web. In most cases this involves a dial-up modem connection to an Internet service provider with a router (a sort of "connectionless switch") connected to the Internet.

However, data connections differ from voice connections in two significant ways. First, the holding times are much longer, typically five times longer than voice calls. The modest modem speeds available on analog local loops virtually guarantees longer holding times for the smallest data transfers today. Secondly, the modems send bursty data packets over the voice circuits. It is not unusual for data connections to send a flurry of packets initially and then sit idle for many minutes on end until another "burst" of packets occurs. Both points are important and affect the efficiency of the trunking network. The basic problem is that voice circuits have limited bandwidth, and when trunk circuits on the circuit-switched PSTN are used for bursty packet transmission to and from the packet-switched Internet and Web, a lot of otherwise useful bandwidth is wasted.

Figure 3-10 illustrates this. As more local loops are used for data connections, trunk occupancy from PC users increases to the point where voice users cannot reliably make telephone calls at all. Service levels might degrade to the point where the telephone company must make ad hoc and expensive additions to add voice circuit capacity to their trunking

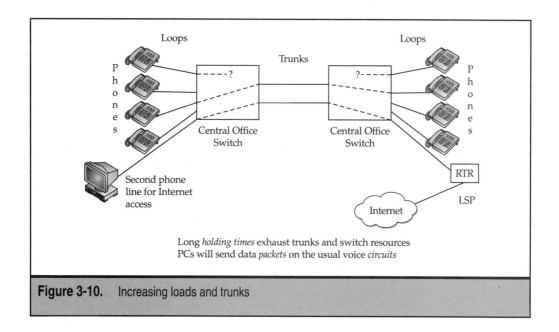

Loops   Loops

P   Trunks   P
h   h
o   o
n   n
e   e
s   s

Central Office   Central Office
Switch   Switch

Second phone   RTR
line for Internet   LSP
access   Internet

Long *holding times* exhaust trunks and switch resources
PCs will send data *packets* on the usual voice *circuits*

Figure 3-10.   Increasing loads and trunks

network. This is clearly not an ideal situation. Of course, this situation is exactly the one that ADSL is meant to address.

Actually, ADSL cannot totally rectify the situation in and of itself. But ADSL can be combined with a data "overlay" network that gets the data traffic off of the voice switches and network and more directly onto networks that were made for this purpose, such as the Internet. The next chapter shows why this is so.

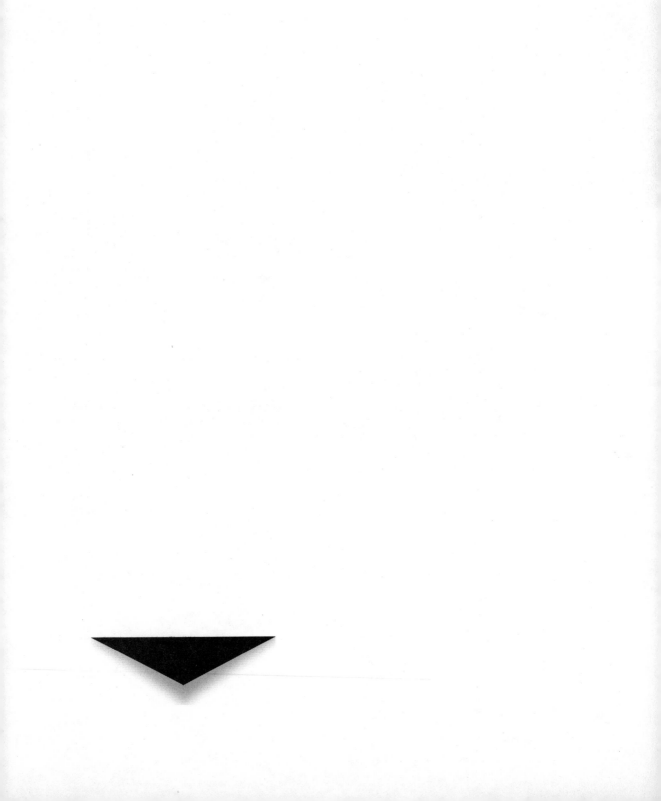

CHAPTER 4

Packet Switching and Circuit Switching

T his chapter examines one of the important issues concerning both the PSTN and the Internet/Web today. The issue is simply that the PSTN was designed, engineered, deployed, and optimized not only to be a voice network, but also to be a circuit switched network, and the Internet/Web was designed, engineered, deployed, and optimized to be a packet switched network. The differences, as has been pointed out earlier, are basically how the bandwidth on the trunks and lines of the network are handled and how the network nodes handle the flow of traffic from input port to output port.

A circuit switched network works best when the end devices or users have, for the most part, a constant interaction for the duration of the call. A circuit switched network dedicates "all of the bandwidth all of the time" to a particular call. If anyone makes a telephone call on the PSTN, the trunks (in most cases) will reserve 64 Kbps in each direction for the bits representing the voices. This 64 Kbps is usually coding "silence" in at least one direction because most people listen politely while the other is talking. This 64 Kbps cannot be used by anyone else, not even the telephone service provider, while the call is in progress. The two 64 Kbps channels that represent the voice circuit "belonged" to the end users for the duration of the call.

Newer voice technologies not only digitize voice, but also packetize the voice. This is done by taking digitally encoded voice and placing it inside a packet with a header, just as is done with data on the Internet. Once the voice is packetized, it also makes sense to eliminate the silences during a conversation (why packetize nothing?). Packetized voice will be dealt with more fully later in this book. For now, it is enough to note that 64 Kbps digitized voice is not the same as packetized voice.

As long as the holding times were only a few minutes, this dedicated bandwidth, circuit-switched technique was fine. The trunking and switching network could still handle plenty of calls one after the other. Sometimes, with metered service, this generated revenue because more calls completed meant more money. Other times, with flat rate service, the benefit was less direct, but real nonetheless. More calls completed meant more satisfied customers, who generated fewer complaints to the state regulators who looked more kindly on LEC requests for flat rate service increases, which meant more money.

In certain social situations today, putting someone on hold is considered a particularly obvious and not very inventive form of insult, but the hold button has changed the way that people use telephones. In turn, the familiar hold button on the telephone handset, which did not exist on residential telephones until relatively recently in the history of the PSTN, changed voice circuit holding times somewhat. Instead of saying, "I'll find that information and call you back," people said "Let me put you on hold for a minute." The minutes sometimes seemed like hours, but that was a different effect.

Consider two people on the telephone discussing plans to meet somewhere at a certain time the following week. Travel plans must be coordinated and checked, perhaps with spouses who are present, but not nearby. No problem—the hold button enables one of the people to check and then pick up again. But travel plans may affect the time initially proposed. Hold again, pick up again. But the new time may trigger a better travel arrangement idea, leading to a repetition of the cycle. In all, several minutes of talk may be punctuated by longer periods of silence. Even with metered service, because the rates typically drop after three minutes or so, it may make more sense to keep one conversation going than to redial again and again.

In this case, the pattern is not connect-talk-disconnect. The pattern is connect-talk-hold-talk-hold-talk, and so on. While the call is on hold, no meaningful bits are being sent across the 64 Kbps trunk channels nor through the circuit switches, but neither the trunk channels nor the switch ports can be used to send any other meaningful information through the PSTN. The circuit is sacred in that regard. Once again, nothing is wrong or bad with the PSTN functioning in this manner. This was the most efficient way to design, engineer, deploy, and optimize the PSTN if traffic is not bursty or intermittent, as those terms were defined previously.

But wait. The whole talk-hold-talk-hold pattern appears to be quite bursty and intermittent. In fact, it is. The question is whether enough people employ their telephone in this fashion to impact the circuit switching aspects on the PSTN. Before 1994, people did not impact the performance of the PSTN appreciably, no matter how often they hit the hold button, but the Web changed that in 1994. Data sent to and from client and server PCs is nothing if not bursty—and usually extremely bursty. This similarity is shown in Figure 4-1.

Another aspect of circuits is that they connect only one thing to another, which is the essence of the PSTN: it exists to connect the device at the end of one access line to the device at the end of the other access line. It matters little whether the devices are telephones, modems, or fax machines, as long as they are compatible. One major aspect of the Internet and Web, however, and probably the one that fueled such explosive growth in the first place, is that the Internet exists to connect the client device at the end of one access line to *everything*. As more and more people use PSTN access lines for Internet access and not telephone calls, perhaps circuit switched networks like the PSTN are not the best way to handle this traffic. Perhaps there is a better way, such as ADSL. In fact, ADSL and related DSLs can be better for all forms of packetized information, from data to voice and audio to video.

On the Internet, information is sent and received in the form of *packets*. Packets are variable-length units that have maximum and minimum sizes. On the Internet, packets conform to the rules set out in the Internet Protocol Suite set of standards, which was formerly and still is usually called TCP/IP. On the Internet and Web, all packets conform to the *Internet Protocol* (IP) rules. The information contained within packets can be almost anything at all. The contents of a packet on the Internet and Web today could be a piece of a file being transferred, an e-mail message, packetized voice (known as *Voice Over IP*, or VoIP), or even packetized video.

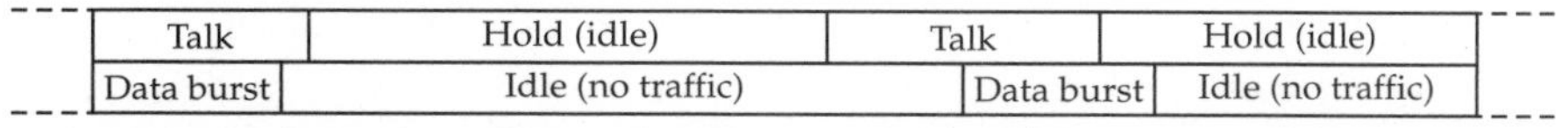

Talk	Hold (idle)	Talk	Hold (idle)
Data burst	Idle (no traffic)	Data burst	Idle (no traffic)

Figure 4-1.   The hold button and bursty data

In these last two cases, the term *packetized* means that voice and video have traditionally been handled by circuit switched networks (such as the PSTN and cable TV networks) rather than packet switched networks. Once packetized, with state-of-the-art digitization and compression techniques, this type of voice and video more closely resembles data than anything else. Packet switching also is better for bursty data.

The following plays fast and loose with "Internet" history, which was then just the ARPANET. But the intent here is to be informative about packet switching, not to be precise as to the date and times that one particular event or name change occurred. In any case, the Internet was planned as a packet switched network from the start for a very good reason: money. Packets were always used as basic information units on the Internet, but to connect Internet nodes (then called "gateways," not routers), long links were needed. Because these links connected Internet nodes, it made sense to call them trunks. From where did the Internet get trunks? From the PSTN, of course, in the form of point-to-point leased private lines (in some instances, they were dial-ups, but still circuits).

Of course, when a trunk connected Internet Node A with Internet Node B, that was all it connected. True, a dial-up could also connect to Internet Node C later, but not at the same time that Node A was connected to Node B. It was the familiar "telephone mesh" problem all over again. The solution this time was for the gateways to route packets from one link (or trunk) onto another. So far, this was nothing more than the old central office switch trick applied to the Internet. When anyone on the early network wanted to send a file somewhere, a circuit was needed there. If someone else using the computer wanted to send e-mail at the same time, and maybe even to the same place, they were out of luck—the circuit was busy. Circuits reserve all of the bandwidth all of the time for one thing. Naturally, these long trunks were paid for by the mile, but it was very expensive to require one circuit for each potential user. Of course, the applications that people were running remained as bursty and intermittent as ever. A lot of expensive bandwidth was tied up on these circuits while people examined and scratched their heads over experiment results and the like.

To save money and make more efficient use of these long and expensive trunks, the Internet (ARPANET) people made one big leap beyond basic switching to bring packet switching to the Internet. In this way, individual packets could be switched from trunk to trunk, not based on what circuit the packet represented, but for which end application the packet was carrying information (which is how a single link on a packet switched network can connect one end device to everything). The packet switch, now called a router on the Internet, could send a packet literally anywhere on the network, based on the individual address carried in the packet header.

So instead of a client PC needing a separate circuit for sending e-mail to an e-mail server, another for transferring files to another PC, and another for accessing a Web site, only one link was needed for all these activities. The packets just had different addresses in them that allowed the routers as packet switch network nodes to distribute them properly. This link could still be a circuit but make the distinction between this link and a circuit because packets still must flow on something. This "something" may even be a PSTN dial-up circuit, but there are other forms of links that might be used. In fact, many of these alternate forms are detailed in Chapter 5.

This individual address information attached to every packet gives another perspective for distinguishing circuit switching from packet switching. Circuit switches, such as PSTN local exchanges, switch the entire bandwidth of the circuits (all the bandwidth all the time) from one place to another. Packet switches, such as Internet routers, switch the individual packets that form the content of the bandwidth. If a PSTN circuit switch was a packet switch for voice, it would be possible to establish one call to the switch itself and then say something like, "This next sentence is for Person A," and then, "This next sentence is for Person B," and so on, all on one telephone call. The sentences would be voice "packets," and the "next sentence" statements would contain the "addressing information" for the "packet." The key is that each packet is individually routed to its destination over links shared by many users, not links with dedicated bandwidth for one pair of users.

In short, packets can be mixed to a whole host of destinations on the same link, which may still be a PSTN circuit, dialed or leased. In ISDN, which shares many features in common with the Internet today, this link for ISDN services would be the *digital subscriber line* (DSL) B-channel, which operates in either *circuit mode* or *packet mode*, but not both at the same time.

The current discussion has become quite complex. The whole point is that sending packets on circuits, especially passband-limited voice circuits, may not be the wisest thing to do. It is time to look at a real example of how the transfer of Internet packets on the circuits of the PSTN impacts the performance of the PSTN.

CIRCUIT USE AND PACKET USE

Figure 4-2 shows the basic differences between circuit switched networks (the PSTN) and packet switched networks (the Internet and Web). There are other differences, of course, but the figure shows the major issue discussed in this book.

In a circuit network, a local exchange has trunks to a tandem switch, which in turn connects to an *interexchange carrier,* or IXC, for long distance calling, and the ISP, for Internet access. In the simple network shown in the figure, only two trunks lead from the central office to the tandem switch, and there are only four devices, but the effect is the same, regardless of absolute numbers. Usually, 10,000 potential users are served by as few as 600 trunks. Note that the trunk itself, which may have a lot of bandwidth in and of itself, is always divided into 64 Kbps bandwidth circuits (naturally). For example, a T-3 has 45 Mbps in bandwidth, but has only 672 channels of 64 Kbps each. No user can ever get a circuit with more than 64 Kbps.

The point of the figure is that if the two residential users have both telephones and PCs, they may want to make both calls (circuits) and access the Web (packets). However, the packet equipment (routers) is in the ISP office. The only way to get there is through a circuit (local loop to trunks). Because circuits are assigned for the duration of the call, it is easy to see that any two connections, voice or data, from telephone or PC, will exhaust the trunking capacity between central office and tandem switch. If two PCs are currently accessing the ISP router ports, no long distance calls can be made. As more and more homes obtain second local loops for PC access, it is easy to see that the number of trunks must increase, as well as the number of switch ports to service these extra access lines and trunks.

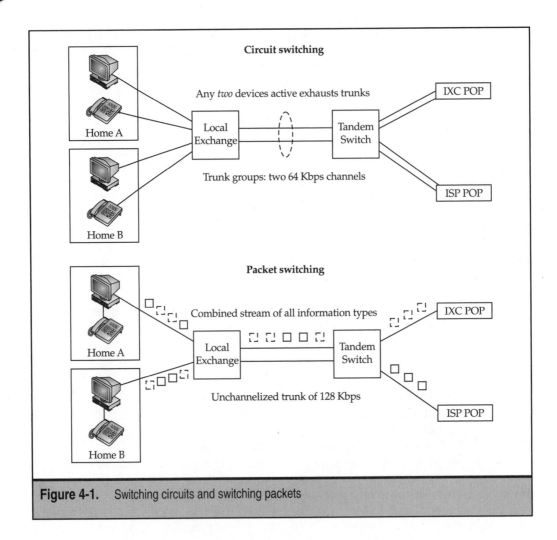

Figure 4-1.   Switching circuits and switching packets

   The figure also shows the result when local loops, trunks, and switches are "packetized." When packets are used for data and compressed digital voice, it makes no sense to have circuits anywhere at all. Now a combined digital stream of voice and data (and perhaps video!) packets can easily share the unchannelized loops and trunks to the IEC or ISP. No dedicated bandwidth exists to eat up the PSTN resources.

   The figure shows the effects of hooking both packetized voice equipment and a PC to a single "packetized" local loop. A voice call no longer consumes all of the bandwidth on the loop. Voice compression and silence suppression (the twin features of voice packetization techniques) leave plenty of bandwidth for data packets (or any other type of packets, such as video). The figure codes the voice packets with a distinctive border and shades the packets from one home. Note that the packets share the loop and trunk bandwidth, but still arrive at the IXC or ISP switch correctly. The local exchange and tandem switches must now switch on content, not just bandwidth. Of course, the issue of whether this is still the PSTN or not is open to debate.

The "packetization" of local loops and trunks has many technical solutions. ISDN was and is a possible solution, but only when used as a packet mode connection, not in pure circuit mode. B-channel circuits are still 64 Kbps circuits. ADSL, or any DSL technology, can be used, and there are other alternatives, such as ATM. The big questions for today are: who gets to do it, and who pays for it?

THE LOCAL EXCHANGE SWITCH
AND THE INTERNET ROUTER

Much has been said so far about a PSTN switch and an Internet router being just different types of network nodes, but to look at them, one would never guess that they had much in common at all. The PSTN switch is housed in a huge building called the local exchange, or central office. The Internet router is housed in an ISP office which may be—believe it or not—someone's bedroom. How can two things be at once so different and so alike?

Part of the answer is that the PSTN switching office is the culmination of over 100 years of backward compatibility and technological evolution. The Internet dates back to 1969, but it has only been a force to be reckoned with regarding computers, networks, and society since about 1995. Another part of the answer is that the PSTN switch might service 10,000 local loops and 1,000 or so trunks. ISPs that are housed in bedrooms rarely handle more than 30 lines. If an ISP does handle 10,000 lines, it is often as big as a local exchange building (well, maybe not quite so big).

Now is the time to look inside the local exchange switching office and the ISP site to see just what is in there that is generating so much excitement today.

The typical structure of a PSTN local exchange is shown in Figure 4-3.

In the figure, access lines (local loops) and trunks (to other switches) enter the local exchange or central office below ground level. It does not matter whether the access lines run on utility poles or underground for most of their length, the entrance facility to the local exchange is below ground, with rare exceptions. This sub-basement is known as the cable vault, and technicians sometimes dread having to work there, depending on the lighting and drainage conditions. In some major metropolitan areas, there may be 40,000 lines and trunks all snaking their way from underground conduits into the cable vault.

The access lines and trunks make their way through the cable vault to the *main distribution frame* (MDF), which is the facility commonly called a *wire center* by the LECs (to the IXCs, it usually means the actual building where the trunks to the LEC terminate). The MDF is essentially one huge patch panel. Every trunk and access line appears on the MDF. There may be other frames in the local exchange just for this group of access lines or trunks, but everything sooner or later ends up on the MDF. Each cable on the MDF is identified by a vertical and horizontal number assigned to it. V=365 and H=4533 is the way a subscriber's access line or trunk to an IXC POP might be identified within the local exchange itself.

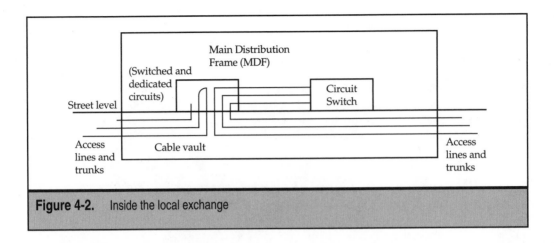

Figure 4-2.   Inside the local exchange

For switched services, the access lines and trunks must be patched through the MDF to a circuit switch port. Naturally, these switch sizes and capacities range from small in rural areas (a few hundred ports) to gigantic in major metropolitan areas (up to 40,000 ports). Some are modular and digital; others are barely upgradable at all and still analog; some use computers only to control the switching process; and others are nothing more than big computers in their own right. Because there are some 25,000 local exchange switches in the United States today, some installed last week and some installed in 1932, it is hard to make valid generalizations about the switch itself without some qualifying statements.

Generally, information makes its way through the local exchange or central office in two ways. Circuits may be switched through the local switch, directed by the telephone number dialed by the originating telephone equipment. Alternatively, circuits may be "nailed up" as leased private lines that always lead from point A to point B. These were called "nailed up" because early implementation permanently connected the wires forming the circuit, which made modifications difficult and awkward. No call control is needed on the private line circuits, so a line appearing on the main distribution frame may go directly into the PSTN switch or into another piece of equipment known as a *digital access cross connect* (DACS or sometimes just DCS). The DACS equipment got around the need for physically "nailed up" private lines by allowing the input and output connections to be made in a software configurable "mini-switch," but calling a DACS a true switch is probably a gross overstatement of its capabilities. Whether routed through the DACS for private line services or through the PSTN switch for switched services, the line eventually made its way into the trunk transport system, unless the other end of the circuit was served by the same local exchange. This detail of DACS/PSTN switch alternatives is shown in Figure 4-4. In either case, the switching is always port to port, bandwidth to bandwidth, and is transparent to any of the bits or packets traveling on the circuit. Note that the number of links leading between the components is not intended to imply relative numbers, but to show connectivity.

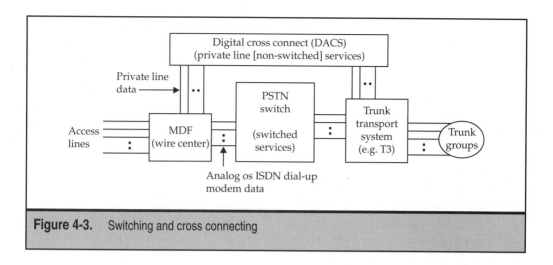

Figure 4-3. Switching and cross connecting

Another point is that the DACS equipment operates on digital access lines, not analog local loops. That is, if an analog access line was to be converted to private line use and it was necessary to put this private line through a DACS, the access line had to be digitized first. There were a number of ways of doing this, which are beyond the scope of this discussion. The emphasis here is on switched services as opposed to private line services.

The point is that the circuit switch is needed in all cases for switched services. Switched services are those where the endpoint is determined by some call control procedures (signaling), such as a person dialing a call to another telephone or a PC speed-dialing an ISP POP. Obviously, the switch capacity can be exceeded in terms of the number of ports currently supported, especially as more and more people order second lines to attach to the Internet. The only thing to do is to install more switch ports, which is sometimes easier said than done. Only so many switch ports can be added before a total overhaul or upgrade to the entire switch is needed. This can cost anywhere from a few hundred thousand to a million dollars.

So every circuit that needs to be switched must have an input and output port on the circuit switch. Whenever a home PC user dials in to their ISP, the access line must be switched to a trunk leading to the ISP POP. Note that once this is done, the PSTN switch adds absolutely nothing to the circuit. The packets are invisible to the circuit switch.

Switched services are not the only type of services that the local exchange can offer, however. These are all *dedicated services,* such as those represented by leased private lines. Because the MDF is basically one huge patch panel, it is easy to provide these dedicated circuits. All that needs to be done is to "patch" the vertical and horizontal position of one access line on the MDF to another vertical and horizontal position on the MDF. No switch port is needed because no switching needs to be done on the circuit. In the PC user to ISP world, this is the equivalent of using the second telephone line as a dedicated connection to the ISP POP. However, the user could never use this line for anything other than ISP access, and the ISP POP presently cannot support any other user on this router port. Dedicated services are point to point, which saves having to run the circuits through the PSTN switch.

A local exchange has many other features, of course. The emphasis here has been on the handling of access lines and trunks and the role of the MDF and circuit switch.

The typical structure of an ISP office is a little tougher to categorize. Thousands of large and small ISPs are still in business today, and some have as few as 100 customers, whereas others have millions of customers. Of course, after the number of customers grows past a few thousand, it is nearly impossible to service them all from a single ISP site. Nevertheless, small or large, most ISP offices have many features in common. The typical structure of a medium-sized ISP is shown in Figure 4-5.

Figure 4-5 has two distinct sets of outside lines. The first set consists of the trunks that lead to the PSTN switch ports, which is how the ISP handles users dialing in from home with their PC modems. It is not unusual for the ratio of users to trunks to be on the order of 10 to 1. In other words, if an ISP has 1,000 customers, the number of trunks could be as small as 100, which means that only 10 percent of the customers could ever be attached to the ISP at one time. All the others would get a busy signal. This practice started before the Web became popular in 1994 and simply reflected the fact that not everybody wanted to be on the Internet at the same time. If customers complained loudly enough, the ISP could add trunks, but this added to the ISP's monthly expenses without necessarily increasing revenues, which were usually billed at a flat rate. ISPs tended to resist adding trunks until absolutely necessary. Today, even a 3:1 user-to-trunk ratio might not prevent user dissatisfaction and complaints. The lines terminate with a series of modems in a *terminal server*,

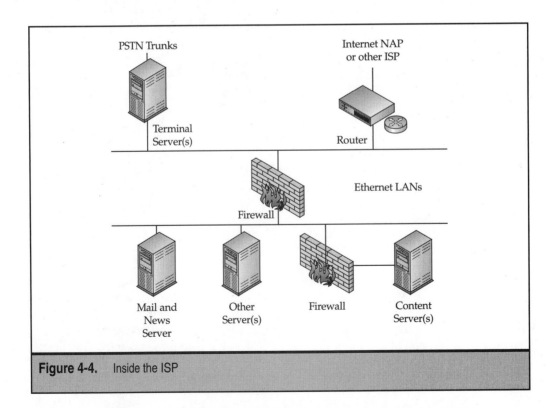

Figure 4-4.   Inside the ISP

a term that reflects older technology when this device handled calls from dumb terminals and not PCs, but the term seems to have stuck.

The second set of outside lines is usually just a single link. This leads to the Internet itself. The link may lead to a major Internet NAP link or to another ISP's router (this is especially true of small ISPs). Note that only one link is needed because the router is essentially a packet switch and not a circuit switch. A T1, if used as this Internet link, can be used not as 24 channels of 64 Kbps, but rather as one unchannelized link running at a full 1.5 Mbps.

User traffic need not be isolated by circuits because each user's traffic is distinguished by the addressing information in the packet—an important point. It means that if there is only one active user, that user's packet gets to use all 1.5 Mbps for packets, instead of just a 64 Kbps channel. Even if 25 users are active, they will gracefully share the T1 bandwidth, although, in this case, things would be slower than having a 64 Kbps channel of their own. But statistically, with bursty traffic, this should not happen often. With circuits, supporting the twenty-fifth user would be absolutely impossible.

The ISP exists to shuttle packets to and from users dialed in on the trunks, or with dedicated leased-line access links, and the Internet itself. In between, there is usually an Ethernet LAN running at 10 or 100 Mbps. A special *firewall* device separates the terminal server and router from the rest of the ISP equipment to prevent unauthorized users from gaining access to the site, through either the Internet or one of the dialup ports.

Behind the main firewall is another Ethernet LAN. Authorized ISP customers can access a variety of servers here, at least one of which will be the e-mail and Internet newsgroup server. This server stores e-mail sent to the ISP's customers until the user logs on and retrieves the messages. Newsgroups are a special application of e-mail where like-minded users can exchange e-mail messages among themselves. Typically, users will belong to or monitor several newsgroups.

Other, more specialized servers may exist, depending on the size and focus of the ISP. Some ISPs try to attract scientists, others are geared toward commercial artists, and so forth. These servers contain the specialized services that the ISP is trying to promote. Today, many ISPs that are experimenting with voice and video over the Internet employ these servers here as well.

Behind another firewall lies the heart of the ISP itself. This, the *content server*, may also be comprised of more than one server. The content server stores the ISPs *home page*, the Web page that users first see when they dial in to the ISP. Users may store their own pages on these servers, although many ISPs limit the size of user pages. Note that nothing stops a user from having a Web site on their home PC other than the fact that others on the Internet can only access the page(s) when the user is dialed in and attached to the ISP.

Smaller ISPs may skimp on some of the firewalls and number of servers. Even a large ISP may have several sites around the country that appear the same as this medium-sized ISP's office, so the figure is a good representation. The actual equipment is very small. There are seldom more than seven separate "boxes" that need to be housed, each about the size of a PC. A modest office in a small building will do. Of course, the ISP can handle only a few hundred simultaneous users with this arrangement, whereas the LEC switch might handle several thousand, but no ISP needs to worry about switches installed in 1932 nor access lines installed in 1919.

Please note that many LECs and IXCs are also ISPs today. In that case, the ISP equipment is usually located in a separate space, and in the case of a regulated LEC, the ISP line of business is carefully isolated from the regulated activities of the telephony aspect of the business.

INTERNET SERVICE PROVIDERS (ISPS)

As millions of U.S. households and every organization worth its salt hooks up to the Internet and World Wide Web, more and more Internet Service Providers are appearing all over the United States to satisfy this hunger for Web access. In most cases, these ISPs supply Internet access on switched local loops and trunks, and then onto leased trunks from carriers to ISP *points of presence* (POP); the ISPs are generating more and more traffic on the public switched telephone network.

However, ISPs are quick to claim that they really generate no traffic at all on the PSTN! The ISP is just a way to connect clients (user PCs in most cases) to servers (Web sites in most cases) on the Internet. Technically, the users and servers generate the traffic. All the ISP does is deliver it. The point is that the traffic the ISP adds to the PSTN would not be there without the ISP.

Unfortunately, the PSTN is engineered for three-minute voice calls. Actually, for engineering purposes, call lengths can vary from three to seven minutes, but the three-minute residential call (that's when metered rates went down, right at the end of most calls) is enshrined in the hallowed halls of PSTN traffic engineering guidelines everywhere. This has been the guideline and general rule of thumb for close to 100 years.

The problem is that Internet access and Web browsing sessions last much longer. One oft-repeated tale speaks of a two week call that was manually terminated by the LEC, only to get a call five minutes later: "What happened to my Internet access?" Flat rate monthly charges from most ISPs (as low as $15 in some cases) give many people little incentive to ever hang up at all! Why not just attach to the Internet the minute the PC comes up in the morning? Then it's there if you need it.

As a result, other calls may be blocked on the PSTN. An access line ties up at least two switch ports (the loop and trunk) and an outbound trunk to the ISP itself (usually a 64 Kbps T-carrier channel). Tales have gone around of 911 calls being blocked (no switch capacity or outbound trunk) due to increased Internet usage in a neighborhood. Now, these 911 calls usually have priority in the trunk group, but only so many trunks are set aside for priority purposes. The rest can be claimed by any call. The problem is that when 100 drivers see an accident, or 100 civic-minded citizens spot a fire, there are instantly 100 calls to 911. The reserved trunks are overwhelmed, and if the "regular" trunks are all full, the 911 call is blocked. If one of these 100 callers is trying to report a robbery or a fire, they are out of luck.

In one sense, the problem is obvious. Using a series of switched 64 Kbps circuits (with their dedicated bandwidth all the time) for a long-holding-time data call is an inefficient use of PSTN resources.

To make matters worse, from the carrier perspective, the ISPs are often exempt from *access charges,* which are paid to the LECs at each end of the call by the IXC for completing the local calls to and from an IXC POP. This exemption was put in to reflect the fact that the ISPs actually add value to the basic telephone service. Also, the federal government explicitly wanted to encourage data services by keeping financial burdens to a minimum when the whole system was set up for divestiture in 1984. All an ISP needs to do is lease a trunk (usually at what are known as wholesale, *reseller rates*) from switch to ISP POP, and then on to the Internet network access point (or other ISP), and they are in business. This flat rate monthly cost makes it easy for ISPs to offer flat rate monthly service to their customers. The only initial concern is whether the number of customers for a given ISP can cover the fixed costs of the ISP.

THE "PACKETS ON CIRCUITS" LOAD

In a very real sense, all packets flow on circuits anyway. Sometimes, the packets are said to flow on "virtual circuits" when the packet flow follows a *connection,* as defined by international packet switching standards. The Internet protocol suite even defines a concept known as a *flow,* which for all intents and purposes is nothing more than a virtual circuit on the Internet.

So although technicians talk loosely of "packets on circuits," the real issue is more one of "PSTN circuit switching versus Internet packet switching," as was covered in detail previously.

In any case, when bursty packets from long, but intermittently used, Internet connections are used across dedicated circuits built for short voice connections, the impact on PSTN resources can be considerable. This impact is shown in Table 4-1.

The table shows the results of an extensive study done by Bell Atlantic (now part of Verizon) in the summer of 1996. It compares the traffic load from ISPs' access (measured in traditional voice traffic units, hundreds of call seconds, or hundred call seconds per hour) to the traffic load from traditional telephony use. Although the study is somewhat old, it shows the impact of the Internet and ISPs on the telephone network at the time.

Customer Type	Average Busy Hour CCS	Busy Hour
ISPs on 1MB lines	26 CCS	11:00 P.M.
ISPs on ISDN (PRN)	28 CCS	10:00 P.M.
Business with MLHG	12 CCS	5:00 P.M.
Local exchange average	3 CSS	4:00 P.M.

Table 4-1.   Impacts of Packets on Circuits

This only happened once in a given area, so it is not even possible to update the study in that sense.

The first two lines show ISPs' load over a 1-MB line (one line, measured business) and through an ISDN *Primary Rate Interface*, or PRI (circuit mode, not packet mode). The last two lines compare a business with a *multiline hunt group* (MLHG), common for businesses and the average for the entire local exchange, with voice residential users factored in. The units are in *hundreds of call seconds per hour* (CCS), a telephone traffic engineering concept. The precise definition of a CCS is not necessary here. Once utilization hits 36 CCS, which equals 100 percent utilization, no more calls can be made.

CCS stands for centum call seconds, where "centum" means one hundred. So 36 CCS is 3,600 call seconds, the number of seconds in a hour. This is also called an *erlang*, after a German voice engineer. By definition, a simple voice circuit, trunk or not, can carry at most one erlang of traffic and no more. There must be enough trunks so that busy days and busy hours can pass without restricting the number of calls unrealistically. If the CCSs for the whole central office are added up for a peak period, then this information can be used to determine the number of trunk circuits required.

For example, 9 trunks can carry 9 erlangs of traffic from all of the local loops. If each call is assumed to last 3 minutes on average, then the office can handle $(9 \times 60) \div 3 = 180$ calls per hour. But if the average length (holding time) of a call rises to 20 minutes or more due to ISP dial in sessions, the office can only handle $(9 \times 60) \div 20 = 27$ calls per hour at best. All other things being equal, there could now be 150 blocked calls per hour—calls that were completed before the ISPs came to town.

The impact of the ISPs on voice traffic is amazing. The ISP use of 26–28 CCS approaches the maximum on the circuits (that is, 36 CCS). Voice use peaks at a modest 12 CCS and averages only 3 CCS. Note also that the busy hours (the study calls them "peak hours") have shifted from late afternoon (voice) to late evening (data).

This timing affects the trunking and overall capacity of the central office, which is engineered, built, and provisioned around 3–12 CCS with a late afternoon peak. The LEC claimed that in order to prevent voice call blocking (like that of 911), an additional $17.8 million had to be spent to increase trunking capacity (that is, 44 percent trunk growth for ISPs versus 9 percent "normal" voice growth). Dial tone delays grew as well, impacting state service requirements, and added personnel were needed to service additional trunks.

The plea of local exchange carriers at this time is clear. Why should ISPs be immune to access charges when it is the ISPs who are killing the PSTN? Why should ISPs be treated as "special cases" with reduced interconnection charges (wholesale), when clearly they are special cases, but in another sense entirely?

Of course, the ISPs and others concerned about inexpensive Internet access have not taken all of this lying down. The Bell Atlantic study, in particular, was dismissed at the time by consulting firms such as Puttre, Incorporated, as "silly" and criticized for using obscure and "obsolete" units, such as CCS, to make their point. This criticism goes to show how little understanding, sympathy, and respect many of the technical consulting firms and Internet Service Providers have for the basic engineering concepts employed in the PSTN.

FLAT RATE SERVICE

The assessment of rates charged to subscribers by telephone companies for their services fall into two broad categories: metered service and flat rate service. With *metered service,* charges accumulate based on how much of the service is used. With *flat rate service,* a fixed monthly charge is assessed, regardless of how much of the service is used. In the telephony world, metered charges are usually assessed by connection time (the duration of the call); however, this requires that all calls be monitored by the service provider for duration, which is a very complex process and one that tempts users to cheat.

To understand why, consider someone who rents a log splitter for a day. The rental agency could assess charges on a metered basis—the cost would increase for every log split. How does one determine the exact number of logs, short of looking over the person's shoulder all day? Cheating by undercounting the logs actually split would yield the benefit of lower cost to the user. If the rental amount charged is a flat rate for the day, however, the renter does not have to watch the user and no one cares if the device is used or not.

In the United States, a number of regulatory decisions have resulted in LECs offering flat rate local service. Customers definitely prefer this billing method. Most local telephone companies have adopted flat rates for their basic telephone services. The ISPs have followed right along because flat rate users are easier to bill, track, and administer. Revenues are stable and not as dependent on fickle user patterns. In any case, few ISPs have the software or hardware needed to meter customer usage.

Most calls to an ISP are just flat rate local calls through the LEC switching and trunking network. The ISP charge is flat rate as well, generally $20 a month or so for unlimited access. The cost is the same whether the ISP session lasts three minutes, three hours, or three days!

Also, some home office users and telecommuters have begun hosting their own Web sites on their home PCs. After all, the days are past when a powerful, $20,000 SUN SPARCstation was needed to do so—any $1,000 PC will do today.

The question is, with flat rate local service, why should someone ever hang up? Why not just connect to the ISP and stay there forever?

TYPICAL RESIDENTIAL NETWORK USAGE

The limitations imposed by the continued presence of the analog local loop and its restricted bandwidth and, to some extent, the trunking network and its persistent circuit connections, are most apparent when one considers what people do at home today. Nowadays, networks and service providers do not merely provide simple voice connectivity; they can supply a wide range of services, from home shopping to telecommuting, as well as Internet access.

Table 4-2 shows how all these new services differ in terms of bandwidth and holding time. Note that most of them are *asymmetric* in nature, which means that the characteristics in terms of bandwidth and holding time differ in the upstream (out of the home) and downstream (into the home) directions. Sometimes the differences are drastic (broadcast-quality

Type of Service	Minimum Bandwidth (Minutes)	Holding Time (Minutes)	Downstream Minimum	Upstream Holding Time Bandwidth
Video on demand	3.0 Mbps	110	64 Kbps	0.1
Teleshopping	384 Kbps	7	64 Kbps	0.7
Broadcast TV	3.0 Mbps	120	64 Kbps	0.1
Near VOD	3.0 Mbps	110	64 Kbps	0.1
Delayed broadcast	3.0 Mbps	30	64 Kbps	0.1
Video games	384 Kbps	60	64 Kbps	60
Telecommuting	384 Kbps	60	384 Kbps	60
Web audio/video	3 Mbps	20	128 Kbps	20

Source: ANSI/TIA-1558B (1995)

Table 4-2. Typical Residential Network Usage

digital TV, for example). All the differences reflect the traffic patterns of client to server (upstream: not much) and server to client (downstream: a lot). The low holding times upstream for interactive services reflect the brief messages needed for start, stop, reset, and so forth.

Some of the services may be unfamiliar. Near Video on Demand means that the movie is not started at the instant a user requests it, but within a specific time interval, usually the nearest 15 minutes. Delayed broadcast services can start a TV show (for example) that is normally broadcast at 9:00 PM on a Wednesday and show it at 10:00 PM on a Friday.

Naturally, the more bandwidth the better. These figures are minimum requirements, but improvements in technology are driving these bandwidth needs down all the time. It should further be noted that there are more than 17 million home PC users.

CURRENT ISP TRAFFIC WOES

The rise of flat rate ISP service, coupled with flat rate telephone company local service, has caused real troubles for ISPs, as well as for the telephone companies. The links to the ISPs suffer from occupancy overload effects. The equipment at the ISP must still answer the call and connect the user to the Internet. More processing is required to validate the user login request.

Heavy loads all conspire to defeat these simple steps at one point or another. A study done by Inverse Networks in March of 1997 showed that users failed to get onto America Online a whopping 60 percent of the time between 6 P.M. and 12 A.M. in March of 1997. The problems were either no answer (no port process available to handle the request), line busy, connect fail (the overloaded equipment drops the user), or login fail (the ISP database process is too busy to handle one more request for validation).

Even a respected ISP like Sprint had an 11 percent failure rate, while the top-rated ISPs in the survey, IBM and CompuServe (now CSi and part of America Online), had a 5 percent failure rate. Now, these figures have aged somewhat, but it was considerations like this that led directly to the rise of ADSL. This is a key point: a lot of DSL technologies being deployed today are based on studies done five or more years ago. This lag might be regrettable, but it is almost inevitable when it comes to new technology.

The line busy failures are the direct result of trunk pressures either within the PSTN or to the ISP. Limited speed, analog local loops feeding trunking networks chopped up into voice bandwidth circuits lead directly to longer holding times and higher trunk group occupancy. It is immediately apparent from the survey that if something is done to increase the bandwidth available to ISP users and move them off of the voice circuit switch and onto a packet switch, the absence of the "line busy" problem will improve service dramatically.

By the way, CompuServe is one of the few ISPs that did not offer flat rate service at the time the survey was done! They had no busy lines due to users sitting on trunks forever.

SWITCH BLOCKING

The pressures brought on by the rise of Internet packet usage on telephone circuits extend not only from the analog local loop (in terms of speed) and the trunking network (in terms of occupancy or holding time), but also to the very heart of the PSTN itself—the central office switch.

Switch blocking occurs when there is no path available from an input port to an output port. The call request to set up a connection through this switch from loop to trunk (or loop to loop, or trunk to trunk) is now blocked. After all, switches can only handle so many circuits and connections at one time. Even the most powerful processor has limits in terms of memory and data bus capacity. Furthermore, the vast majority of central office switches cannot be replaced fast enough to keep up with the rapid advances in computing, making these switch computers less powerful than some desktop devices.

The problem becomes most apparent when a significant number of connections are used for data packets rather than voice. The long hold time associated with data usage can overload switch resources, causing blockage to occur in the switch, even if adequate trunk resources are available.

Under extreme conditions, essential calls to services such as 911 or fire departments may experience dial tone delays. Not only is the impact on the service provider negative in this case (in terms of state regulation or contracts), there is obviously a negative social impact as well ("I have cheap Internet access, but my house just burned down").

IS THIS THE NETWORK NEEDED TODAY?

Many people, including high government officials, computer hardware and software company executives, and computer network visionaries, once had a glorious dream of an up-and-coming information superhighway. This superhighway would link homes and businesses through high-speed digital access lines (local loops) to state-of-the-art switches and routers linked together with higher speed digital trunks (nonchannelized to reflect increasing packet use and ease the restrictions and inefficiencies caused by trunks installed for voice circuit bandwidths). People would have fast and easy access to all types of information and even entertainment—much like the Internet today.

Unfortunately, for many people, this dream seems more like a nightmare. The lack of access charges assessed to ISPs for linking to the local network encourages them to set up shop almost anywhere and charge unbelievably low, flat monthly rates for unlimited Internet access. The flat rate service encourages long (or seemingly never-ending) "phone calls" to these ISPs. Furthermore, the lack of adequate bandwidth on the analog local loop only makes the problem worse. Building more circuits in the form of trunks may turn out to be self-defeating as these facilities quickly are pressed into service for ISP access. Things were so bad that in late 1996, the New York Times reported that a central office in California had 2,000 trunks (lines) running to a single ISP.

Segregating data on the local loop will help. With separate bandwidth on the local loop available for data packets, data could be routed directly to a packet switch network.

THE TROUBLE WITH LOOPS TODAY

The primary trouble with local loops in today's environment of long data "calls" to the Internet and Web, as well as a need to provide enhanced services (video, for example), is that the local loop firmly remains analog, with few exceptions.

The passband (sometimes called "narrowband") analog local loop was not considered to be an issue when digital switches and trunks were introduced, and so were not addressed by this initial wave of digitization. When digitization arrived with ISDN, several stubborn characteristics of analog local loops confounded attempts at quick and cost-effective digitization.

Loading coils, which add inductance to counteract the attenuation caused by capacitance over long distances, are a big obstacle. Each coil must be painstakingly removed to digitize the loop; and even then, the possible presence of bridged taps and mixed UTP wire gauges only makes the situation worse.

ISDN deployment methods did not resolve these situations cost effectively for many reasons. A major issue was that ISDN, as an international standard, was a "package deal" that required not just the digitization of the local loop, but major changes to the central office switches that cost about $500,000. This cost left little to gracefully digitize the analog loops.

The relatively low-speed modems that analog local loops support encourage longer holding times for data sessions. This contributes to trunk group "exhaustion" and lower service levels for one and all.

Finally, the continued use of analog modems means that data traffic goes through the switch the same as voice traffic (that is, the data "sounds" just like voice to the switch), but the use of circuits for Internet packet usage merely ties up switch resources and leads to blockage all around.

Why should the PSTN worry about rebuilding the voice network for packets? Because packets will not go away, and, in fact, there will be more packets on the PSTN, Not fewer. Even voice will be packetized as the new international standards for silence suppression and compression become common in the next few years. In late 1997, Reed Hundt, then chair of the *Federal Communications Commission* (FCC) said during a speech that "packet-switched networks will soon carry most of the country's bits, and that will change the economics, the structure, and just about everything else about the telecommunications industry. We are not sure how it will happen or how long it will take, but all indications are that it will happen."

What technologies exist both to increase the bandwidths available on local loops and relieve some the pressure on congested trunks and circuit switches? ADSL is one of them, but there are a host of others, most of which are introduced in Chapter 5.

CHAPTER 5

Possible Solutions

Many solutions are possible for the congestion caused by the limited bandwidth and modern data traffic patterns on analog local loops. In fact, the issues extend to the trunking network and even the central office switch itself. One of the key solutions is ADSL, of course, but there are others. This chapter reviews them all and concludes that the family of technologies that includes ADSL and related DSL technologies might offer a number of distinct advantages over the others.

For a while, many thought that the congestion problems caused by people spending a lot of time of the Internet and Web could be solved by giving them a faster way to obtain information. So instead of dialing in at 33.6 Kbps or below, depending on the line conditions at the time, give them a new kind of modem that ran at 56 Kbps (at least in one direction). If Internet sessions have long holding times because of slow dial-up links, faster links will lead to shorter holding times. At least that's what was supposed to happen at one point.

But it now appears that what people need is not only faster Internet access, but faster Internet access *all the time*. A faster modem is all well and good for intermittent Internet access. But if the Internet and Web are indispensable for people just living their lives (as now is the case), this is not really a solution at all. Anything that still makes the Internet run through a telephone central office while people are still trying to make "normal" telephone calls is not going to be effective.

These new modems allow for connections at 56 Kbps (these often operate more in the 40 Kbps range). This speed was formerly thought of as impossible, given analog local loop passband bandwidth. The catch is that one end must be digital (usually the remote end!) and that the 56 Kbps is only one-way (that is, "downstream"). For example, a user with a 56 Kbps modem must connect to an ISP that has leased a digital line for user dial-in access in order to get speeds between 33 Kbps and 56 Kbps downstream to the user. This is because it is the digital conversion from analog signals that adds noise to the link. So only one analog to digital conversion can take place for 56 Kbps modems to work above 33 Kbps.

In addition, marginally faster downloading of Web pages has not reduced holding time. Users simply look at more pages. Hold time, not data rate, is still the real culprit in this speed range.

Cable TV companies have been playing around with "cable modems" for a while now. These devices essentially bypass the local loop altogether and use the CATV network (usually itself analog) to provide high-speed Internet connections directly to a router or terminal server.

In the wireless arena, services such as *multichannel multipoint distribution services* (MMDS) and *local multipoint distribution services* (LMDS) have attracted attention initially as forms of wireless cable TV, but now as much for their data service possibilities as anything else. Satellite service providers have also considered data and interactive video services, notably DirecPC and newer systems, such as Teledesic, which have their own limitations.

Newer cellular phones can be used to access the Internet and Web, and an even newer technology called Bluetooth is coming to PDAs and related devices. These wireless solutions might solve not only Internet access issues, but mobile Internet access issues as well.

The technology that has generated the most amount of interest in terms of solving the local loop issues (as well as most of the trunk/switch issues) once and for all is known as *Asymmetric Digital Subscriber Line* (ADSL). ADSL is a particular flavor of a whole family of technologies known as DSL technologies.

The technologies introduced as potential solutions to the local loop, switch, and trunk capacity problems are shown in Table 5-1. The rest of this chapter surveys them briefly and points out some of their advantages and drawbacks.

OVERVIEW OF 56K MODEMS

Because 56K modems have been dismissed as a potential solution to the Internet access problem, you might wonder why I consider them at all. But 56K modems are still worth looking at for a couple of reasons. First, many DSLs still use the same fundamental analog line code methods used in analog modems. And second, 56K modems have so pushed technology and engineering to the limit that many people see 56K modems as the culmination of many years of effort to push data over noisy 4 KHz voice channels. But this section is brief.

The technology used in analog modems has increased in power and decreased in price dramatically over the past few years. Analog modems quickly went from 9600 bps to 14.4 Kbps to 28.8 Kbps to 33.6 Kbps in a short time. However, things tended to stall as theoretical limits and physical impairments seemed to establish 33.6 Kbps as the upper speed limit for full-duplex operation on the 3 KHz (or so) analog local loop bandwidth.

This is a good place to briefly review the history of modems in general and try to appreciate just how difficult it is to create modem devices that operate at higher speeds. Because many modem concepts are still used in DSL, this will not be a waste of time.

As I mentioned previously, the term "modem" is a sort of acronym and abbreviation that stands for *MOdulation/DEModulation*. A modem enables two devices generating

Possible Solutions	Comment
56 Kbps modems	No breakthrough technology, but good use of what is there.
Cable modems	Plenty of bandwidth downstream on cable TV network.
MMDS	Sometimes called "wireless cable TV."
LMDS	Sometimes called "cellular cable TV," but much more.
Satellite systems	DirecPC is a good example.
ADSL	One of the DSL technologies.

Table 5-1.   Possible Solutions

digital information (such as computers, but not *always* computers) to communicate by using the public switched telephone network. The PSTN is optimized and limited to the frequencies used for human voice and can carry only sounds. Modems need to translate the computer's digital information format into a series of sounds that can be transported over the phone lines. When the sounds arrive at the destination, they are demodulated and turned back into digital information for the receiving computer or digital device.

All modern modems also use some form of compression and error correction. Compression algorithms enable the modem throughput speed to be enhanced from two to four times over normal transmission rates. Error correction examines the incoming digital data stream for absence of errors and requests retransmission of a frame when it detects a problem.

Modems in the early days of telecommunications networks—the 1950s—were all proprietary. They used simple techniques to operate at 300 to 2400 bps. These simple modems either used or were built on technology borrowed from radio frequency techniques developed during World War II and then applied to wireline telecommunications. Mostly, these early modems connected simple teletype machines. Teletypes were devices with keyboards for typing and rolls of paper for printing on the other end of the line. Some would even hesitate to classify these as true modems, because modems as people think of them today did not appear until the early 1960s.

Prior to 1969, all modems had to be made by AT&T and installed by the Bell System. Customers played no role at all in the modem selection or installation process. The modem was as much a part of the PSTN as the telephone itself in those days. Naturally, this practice limited the spread of modems, as absolute control over technology usually does.

The rise of the modern market for analog modems can be directly traced to July 1968. In a landmark ruling (part of the famous Carterfone decision), the FCC decided that "the provisions prohibiting the use of customer-provided interconnecting devices were unreasonable." On January 1, 1969, AT&T had to revise their tariffs to allow the attachment of customer-provided devices (such as modems, answering machines, and so forth) to the public switched network. This was subject to three important conditions that AT&T imposed:

▼ The customer-provided equipment was restricted to certain output power and energy levels, which guaranteed that the customer device would not interfere with or harm the telephone network.

■ The interconnection to the public switched network had to be made through a telephone-company-provided protective device, which in those days was sometimes referred to as a *data access arrangement* (DAA).

▲ All network control supervision and signaling, such as dialing, busy signals, and so on, had to be performed with telephone company equipment at the interconnection point.

The biggest problem turned out to be the protection device, which had to be purchased from the Bell System, was expensive, and was often not available. By 1976, the

FCC recommended a plan in which the current protective devices would be phased out in favor of a type of registration plan. Registration would allow for the direct PSTN electrical connection of equipment that had been inspected, certified, and registered by an independent agency, such as the FCC, as technically safe for use on the PSTN.

In 1948, Claude Shannon, a Bell System scientist, wrote a paper that established for the first time firm theoretical limits on the speed that a modem could operate. The paper considered the top speed for a power limited and bandwidth channel hampered by noise. In other words, the typical analog telephone channel. The paper did not explain just how to reach this upper speed limit; it simply stated that this channel capacity limit could be approached by using the proper techniques.

Approaching this upper limit did not become an issue for years, but as more customers started buying and using modems, speed (and reliability) became important issues. Each individual modem vendor tried to get as close to "Shannon's limit," as expressed by "Shannon's Law," as they could.

By the 1960s, networks had become important enough to try to increase modem speeds. In the 1964 session of the ITU-T (then the CCITT), the first modem Recommendation, known as V.21 (1964), established an international standard for a 200 bps modem (which is now 300 bps). Oddly, V.21 is still used in very modern modem "handshakes," which establish basic compatibility between modems. Significant advances in line coding techniques came in 1968 and 1984 (V.22bis).

Also in 1984, a major advancement in modem technology and standards came in Recommendation V.32 with the addition of echo cancellation and trellis coding. Trellis codes were a major breakthrough because they paved the way for providing some degree of *forward error correction* (FEC) techniques to modems. Trellis codes arrange line code signaling states not haphazardly, but according to a complex mathematical plan. So only certain transitions from one state to a limited number of other states are allowed. If one bit is errored, the receiver is often able to tell that the received bits only fit one allowed state, and so the error can be corrected. The state changes were often arranged in a ladder-shaped figure looking much like a garden trellis. Shortly thereafter, ITU Recommendation V.32bis improved on this use of trellis coding and increased the data rates to 14.4 Kbps.

These breakthroughs only made technicians eager to achieve higher operational modem speeds. Work on the V.34 standard turned serious in 1989 and 1990. Acknowledging improvements in basic telephone network equipment such as switches, the initial V.34 goal of 19.2 Kbps was moved to 24.0 Kbps and then on up to 28.8 Kbps. The latest version of V.34, from 1996, supports modem speeds up to 33.6 Kbps. These latest modems achieve 10 bits per Hertz of bandwidth, which closely approaches the theoretical limits established by Shannon.

It came as somewhat of a surprise when several modem makers, most notably US Robotics, announced plans to market a modem that would operate at speeds of 56 Kbps (usually called "56K modems" and not "56 Kbps modems," for some reason). This became the V.90 modem standard and is the most commonly used modem type for dial-up Internet access today.

Now, the "normal" T-carrier digitization technique for voice quantizes and encodes speech into 8-bit *pulse code modulation* (PCM) "words" 8,000 times per second, giving 64 Kbps operation (8 bits × 8,000/second). 56K modems use a similar trick, but because these modems operate over nonconditioned pairs (conditioning tunes the electrical characteristics to support 64 Kbps), not all PCM words could be used, so the modem makers dropped a bit, leaving 128 signal levels instead of a the full 256 in 64 Kbps operation. Technically, they use only 7 of the 8 companding regions for PCM voice. *Companding* is part of the voice digitization quantization process. This translates to 56 Kbps operation (7 bits × 8,000/second).

Now, if it were this simple, surely such a step would have been taken before. As it turns out, it is not simple at all, and 56K modems are not just a trivial upgrade to devices at each end of a link (although this upgrade is still a requirement). 56K modems require one end of the link to be essentially a DS-0 running with 56/64 Kbps digital line signaling and an internal digital trunking and switching network. Fortunately, these conditions are quite common.

Furthermore, there can be only one digital-to-analog conversion along the path from end to end. Usually, this is at the interface to the analog local loop in the downstream direction to the home. In the downstream direction, this digital-to-analog conversion is not susceptible to quantization noise and makes 56K operation in this downstream direction possible. The operation in this direction is straightforward: the 128 PCM word generates a tone (just as in voice PCM) that can be recognized at the far end so that the 128 PCM word can be reconstructed at the 56K modem attached to the user's PC.

Upstream is another matter. Here, analog-to-digital conversion noise (quantization noise) can be fairly severe, and so 56K modems are limited to "regular" 33.6 Kbps operation in this upstream direction (where analog-to-digital conversion is necessary to enter the digital trunking/switching network).

Oddly, 56K modems are a good example of "asymmetric" operation, where the speed in one direction is not equal to the speed in the opposite direction. The general idea of 56K modem operation is shown in Figure 5-1.

As mentioned, in a very real sense 56K modems are at the evolutionary end of the traditional modem using only the single 4 KHz voice channel available on the local loop. But maybe not.

The greatest advantage of the 56K modem over almost every other technology in this chapter is that the 56K modem is a *switched* solution. That is, any ISP can be reached over the PSTN simply by having the 56K modem dial the proper number. Almost all of the other technologies in this chapter are *dedicated* solutions, meaning they are used to reach one ISP and one ISP alone. When the Internet access path is taken out of the local telephone switch, the call control signaling used to make a switched connection is lost too. If a user wants to reach many ISPs, or just more than one, all the time, no other solution is available except the 56K modem.

And plans are even in the works to make the 56K modem even faster. But how can this be when Shannon's Law sets a maximum for analog modem operation over a noisy

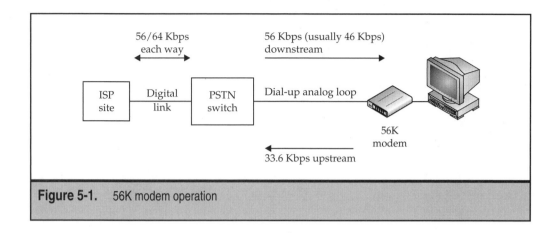

Figure 5-1.   56K modem operation

voice channel of a given bandwidth? The key here is *noise*. Shannon's Law limits modem speed not only due to bandwidth, but due to noise. If the bandwidth of the voice channel cannot be expanded, lowering the noise will also enable faster operation.

The usual figure cited for the *signal-to-noise ratio* (SNR) on a twisted-pair copper local loop is 1,000 to 1. So if the signal level is 1,000, the noise level is 1. But if the noise level is reduced so that the SNR is not 1,000, but 10,000 or even 100,000 to 1, the voice channel analog modem might have some life left in it after all. And the increased use of fiber in the local loop has been reducing voice channel noise for years, so some further speed increases are possible even now.

OVERVIEW OF CABLE MODEMS

To the providers of cable TV service, the battle over the best way to increase the 4 KHz bandwidth on the analog local loop to residential dwellings seems silly. After all, the cable TV networks have delivered hundreds of MHz for years, and some systems now operate at 1 GHz (1,000 MHz). Of course, these signals were almost universally one-way analog video, but this is no longer true.

In many cable TV networks, cable modems are used to convert digital data signals for transmission over the coaxial cable. Cable modems are true modems in that they convert digital information content input and output for transmission as analog signals over an analog path. Cable modems carry data packets on *radio frequency* (RF) cable channels, the same as analog video. It should be noted that even cable TV systems with fiber installed (that is, *hybrid fiber/coax* [HFC] systems) still operate, almost universally, in an analog fashion. Only in these newer HFC systems can two-way communication take place easily. Only HFC systems easily support traditional two-way, analog voice services over the cable TV systems.

However, the addition of fiber optic cable to coaxial cable TV networks has been going on for a long time, and not just because of two-way services. These reasons include the following:

▼ There are fewer amplifiers in the HFC networks, so there was much less additive noise to affect service quality.

■ The few fiber optic amplifiers consume much less power individually and collectively.

■ There are more "trunk" fibers, so a single amplifier failure is much less disruptive.

■ The few fiber amplifiers are easier to push upstream signals through, and upstream "noise" at the head end was less.

▲ Two-way services enjoy much higher upstream capacity.

So HFCs were not built with just cable modems in mind. But HFCs certainly help cable modem operation.

Cable modems take advantage of the enormous bandwidths that cable TV networks enjoy. Even a single "Internet channel" can run at 6 Mbps or better. Moreover, all cable TV systems have upstream channels reserved for interactive services. The FCC set these aside roughly in the 5 to 50 MHz range for interactive services years ago. A lot of cable TV networks cannot use these upstream channels, however, because few devices in the home could generate upstream signals other than noise. So anything flowing upstream to the cable TV head end was interpreted as noise and suppressed. In other words, cable amplifiers are usually one-way only (to the home) and interpret anything in the opposite direction (from the home) as noise to be ignored or even eliminated. This made sense given the one-way, broadcast nature of current cable TV networks.

Also, the upstream channels must be shared by potentially large numbers of users. There are only a few useful upstream channels per fiber feeder in an HFC, usually between four and eight. But claims that these upstream channels would become hopelessly congested from sharing them among hundreds of Internet users have generally been unfounded. The network slows as users and use grows, but much of this can be attributed to portions of the network other than the access link.

The use of cable modems neatly sidesteps all of the issues regarding analog local loops, trunking capacity, and switching resources. Cable modem data traffic can be fed directly into an Internet router at the cable TV head end. If voice channels and services are provided, this is the only traffic actually fed into the PSTN. In fact, the voice switch may even be in the cable TV head end site itself, especially if the telco is also the cable TV company. The general idea behind cable modem operation is shown in Figure 5-2.

The nice thing about cable modems is that they can integrate both TV and PC functions into either device. Both exist as *Internet TV* (a TV with PC capabilities) or *PC TV* (a PC with a cable TV board and coaxial cable connector). Generally, PCs with cable TV video boards can watch 16 channels at once (but only listen to one), and these can even be

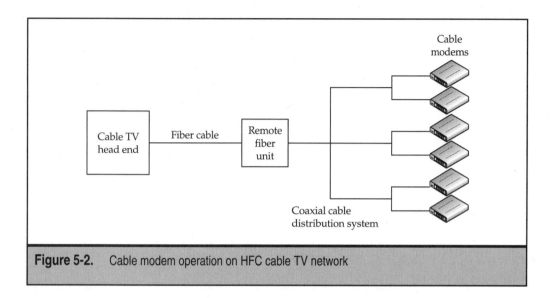

Figure 5-2. Cable modem operation on HFC cable TV network

iconized. The progression to allowing data communication seems a natural one. In fact, cable modems actually encourage this type of "convergence."

Cable modems are an important enough technology and a serious enough competitor to DSL technologies to spend more time on cable modems than any of the other technologies in this chapter. Cable modems once enjoyed a commanding lead on DSL in terms of deployment, but the gap has closed, and it appears that DSL will ultimately become the access technology of choice in many areas.

Early cable modem systems used a simple Ethernet frame without even basic LAN *media access control* (MAC) methods. Contention was a problem, especially with the shared upstream channels, which were few in number compared to the potential number of downstream channels that could be used for data instead of video signals. These early upstream channels were easily overloaded, and the whole architecture was unsuited for deployment with the anticipated high *buy rate* (the percentage of customers taking the service when offered) for the Internet access.

So modern cable modem systems are distinguished by the addition of an *intelligent controller* to prevent contention and sharing from "freezing out" some users. The intelligent controller adds a *scheduler* to the normal MAC processor and WAN interface to the Internet. With a few enhancements, the scheduler can even guarantee *quality of service* (QOS) parameters such as bandwidth and delay, making the system much more suitable for voice over IP (VoIP) and eventually video over IP.

The overall architecture of a modern cable modem cable TV system is shown in Figure 5-3. The upstream and downstream channels are logically separated by frequency and might also be separated onto upstream and downstream fibers in an HFC. Two-way operation is also possible in all-coaxial cable systems, but most cable modem services are offered over HFC networks.

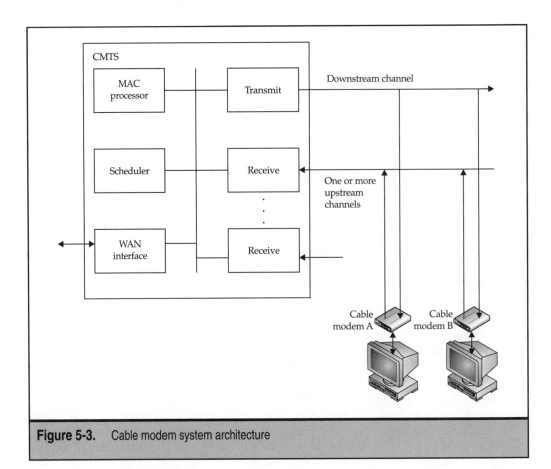

Figure 5-3. Cable modem system architecture

By the late 1990s, cable TV operators had determined that certain characteristics were necessary to deploy a successful *cable modem termination system* (CMTS) at the head end to regulate traffic to the cable modems at the customer premises. These features were as follows:

▼ *Cable modems had to share a single 6 MHz downstream channel.* The speed needed to make sure delays were not too high was in the 30 to 40 Mbps range, depending on which line code method was used. Any channel between 54 and 860 MHz could be used, and there could be no interference with adjacent video signals.

■ *There had to be one or more shared upstream channels using an analog line code.* The range was to be from 5 to 42 MHz. Upstream channels could be from 180 KHz to the full 6 MHz (actually 6.4 MHz) for flexibility, because most cable modem traffic flows in the downstream direction (assuming that there are no servers on the customer premises).

▲ *The user interface had to be Ethernet on the premises and at the CMTS to connect to the Internet.*

There were additional features concerned with privacy, QOS, voice support, management, and so on. The cable operators worked with many standards groups to come up with CMTS and cable modem products to fulfill these characteristics. The IEEE 802.14 cable TV Working Group (known as "IP over cable TV") got involved, as did the ATM Forum, and their work involved using ATM on a cable TV network (an idea that seems almost quaint today). A group called the *European Cable Modem Consortium*, and another European group known as the *Digital Audio Video Council* (DAVIC) industrial consortium contributed as well. In the United States, most of the work was done by a group of vendors forming the *multimedia cable-network system* (MCNS) group. Their work was known as *DOCSIS* (Data Over Cable Systems Interoperability Specification), and now the group is simply known by their work: DOCSIS. DOCSIS now adds a *Baseline Privacy Interface* (BPI) which supplies a lot of security to cable modem implementations.

Although DOCSIS was originally intended for use in the United States, today there is also *EuroDOCSIS*. DOCSIS now guarantees data rates, even with shared channels. Typical guaranteed rates are 512 Kbps downstream and 128 Kbps upstream, which compare very favorably with DSL. What DOCSIS does is to try to guarantee interoperability by establishing standards for carrying IP packets over an HFC cable TV network, while at the same time allowing and acknowledging the differences in detail in many of these HFC systems. The general idea of DOCSIS is shown in Figure 5-4.

DOCSIS is not the only way to do this, of course. DAVIC is still around, and it has combined with the *Digital Video Broadcasting* (DVB) group to explore standards for cable modems, known as *DVB-RCCL* (DVB Return Channels for Cable and LMDS). *Local Multipoint Distribution System* (LMDS) is a type of "cellular cable TV" or "wireless fiber" that delivers services wirelessly. Today, there are significant differences between the way that DVB/DAVIC and MCNS/DOCSIS cable modems operate, mostly at the lowest, physical layer. Standards will help, and DOCSIS has been accepted by the ITU as J.112. DVB still has wide support in Europe, where it is seen as better for European cable TV systems and more of a "home grown" solution. The DVB 2.0 specification is now a

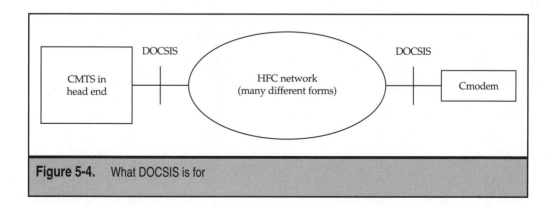

Figure 5-4.   What DOCSIS is for

European Telecommunications Standards Institute (ETSI) specification, ETS 300800. This now has also become DAVIC 1.5, so all three (DVB 2.0, ETS 300800, and DAVIC 1.5) are essentially the same.

The main thing that DOCSIS adds to a cable TV system is the intelligent controller, which creates a standardized CMTS. The scheduler divides the total available bandwidth on each link into numbered *minislots* that can be used through contention or by a type of reservation called a *direct grant*. Minislots are 2^k times 6.25 microseconds in length, and the allowed values of k are 0 to 7. The minislots therefore vary in duration from 6.25 microseconds to 800 microseconds (128×6.25), depending on the needs of the cable TV operator. Bigger slots allow for more bits per grant to be sent, but increases delays for other users waiting for a grant. Grants assign numbered slots to the cable modem and also the number of slots the cable modem can use before stopping the send process.

The cable modem has a smaller version of the CMTS scheduler. This is the trickier part of the operation, because upstream bandwidth is typically less than downstream. The bandwidth is usually either 27 Mbps or 39 Mbps but shared among many users. DOCSIS also adds security to cable modems through the BPI. The BPI adds encryption for the users' traffic, some features to prevent neighbors from browsing each other's computers, and so on. Newer cable modems have as many security features as a personal firewall or a decent router.

There is much more to cable modems and systems than this brief section can investigate. A longer treatment would consider the impact that the newer *digital TV* (DTV) standard will have on boosting the popularity of cable modems even higher. But I've covered enough to give you an appreciation that cable modems are the most mature and serious competition to DSL.

Pros and Cons of Cable Modems

New technologies are invented all the time. Some flourish (VHS and cassette tapes, for example) whereas others fade away to occupy niches or become footnotes (BetaMax and 8-track). All technologies are born with a set of advantages and disadvantages that follow them throughout their technical lifetimes. Sometimes these seem obvious, but they usually become apparent only in retrospect. Also, there is no official list of pros and cons, agreed upon by everyone, that characterizes a particular technology. Therefore, what follows is, of necessity, at best subjective and at worst all wrong.

This section covers the pros and cons of cable modems that are listed in Table 5-2.

On the plus side, cable modems take advantage of existing cable TV networks. No brand-new infrastructure is necessary, although changes to the cable TV network could be inevitable.

In addition, the radio frequency components required for cable modem operation are now inexpensive and plentiful. Many chipsets (the components of circuit boards that perform the communications tasks) are available at competitive prices and interoperability is quite high.

It should not be overlooked that in many ways, cable modems fulfill the early promise of interactive cable TV systems and services. Instead of building the functions into the set-top box, however, the PC has taken its place.

Pros	Cons
Runs on existing cable TV networks	May require extensive premises and drop cable rewiring
RF components relatively inexpensive and plentiful	Operates in upstream "garbage" bandwidths
Fulfills early promises for interactive cable TV systems	Shared upstream channels liable to easily overload

Table 5-2.   Pros and Cons of Cable Modems

On the minus side, some premises rewiring may be needed to reach the area where the PC is located. However, where new premises cabling is needed, it is either coaxial cable to the cable modem or UTP from the cable modem to the PC, both of which are inexpensive and well understood.

Cable modems may also require reinstallation of 90 percent of all drop cables that deliver services to the premises. Drop cables are often hastily and poorly installed, and lead to what is known as ingress noise. Another major problem is homeowner-installed splitters that hook up several TVs to one cable TV coaxial cable feed into the home. These also are often poorly installed.

A more serious drawback is that the upstream 5 to 50 MHz bandwidth picks up a lot of "garbage" and noise from other home devices. Freezers, refrigerators, and a host of other devices radiate signals in this bandwidth, which is concentrated as it makes its way upstream through the branching bus network to the head end. In self defense, many cable TV networks filter out this "noise." In cases where only one-way downstream operation of the cable modem is possible, a normal PSTN modem must be used for upstream communication.

Finally, because the upstream channels must be shared by many users (perhaps thousands), a real danger of traffic overloads exist as all of these data packets converge on the single head end site, in spite of all efforts to make this a non-issue. Current practice supports between 100 and 200 or so customers on a 6 MHz channel, but this might not always be possible as the customer base increases. This has already been identified as a potential trouble spot by the cable industry, and various techniques for dealing with this issue have been proposed, from DOCSIS to more advanced techniques. For instance, DOCSIS 1.1 includes methods that will make it possible not only for cable TV operators to guarantee bandwidth on the shared channels, but also make it possible for the operators to charge more for higher service levels.

In fairness, it should be noted that almost all forms of high-speed access services can be prone to shared bandwidth bottlenecks. Later DSL discussions point out shared bandwidth bottlenecks with the same effect as with cable modems, just further from the user than the access link itself. Much debate revolves around whether remedying these bandwidth bottlenecks is easier in cable modems or in DSL. The intent here is not to consider this issue in favor of one side or the other, but to point out that solutions to the problem exist in both cable modem and DSL technologies.

As an aside, cable modems have been somewhat plagued lately, especially in the United States, by the issue of *open access*. Open access just means that the user of the service has a choice of service providers, just as a person making a long-distance call can direct the call to any one of a number of long-distance service providers over the same local loop. Anyone with a dial-up modem can reach many ISPs. But people with cable modems currently have their choice of ISP restricted to exactly one: the one cable ISP with which the cable TV network operator has a exclusive contract. In North America, 80 percent of cable modems are using @Home or Road Runner exclusively, and both are now controlled by AT&T. Only two other cable modem ISPs—High Speed Access Corp and ISP Channel—service the rest of the cable TV networks that do not provide their own cable modem ISP service. The major self-contained cable TV operators that do not require a separate cable ISP are CableVision and Adelphia in the United States, and Videotron in Canada.

DSL users often have a choice between at least two, and often more, ISPs to provide their Internet access. The LEC, incumbent or competitive, provides the DSL link in all cases, of course. DSL promoters are fond of pointing out the lack of open access competition in cable modems to regulators. In reply, the cable modem ISPs usually point out that open access is a specifically telephony issue and does not apply to cable TV. This is not the place to debate these issues, and both sides have some points of merit as well as some degree of distortion. Open access for cable modems is a tricky issue that does not appear to have an easy or quick resolution.

Of course, it never pays to be dogmatic about technology. The best strategy for users seems to be "never pass up a cable modem service to wait for DSL, and never pass up DSL to wait for cable modems." If given the choice between the two, pick the cheapest unless you have hard evidence that one is much faster than the other.

MULTICHANNEL, MULTIPOINT DISTRIBUTION SYSTEM

Multichannel, Multipoint Distribution Systems (MMDS) were first proposed as a form of "wireless cable TV." It seems odd to position a service in this way, because the original free broadcast TV stations are essentially "wireless cable TV." But MMDS differs from broadcast TV as much as cable TV differs from broadcast TV. And MMDS is worthy of discussion because the MMDS service has been repositioned away from just being a TV and movie delivery platform and is now positioned as a very-high-speed adjunct system for cable TV or as a wireless broadband data service all on it own.

MMDS merges what are essentially TV channels from the *Instructional Television Fixed Service* (ITFS) and *Multipoint Distribution Service* (MDS). Some other channels were added so that MMDS has 33 analog channels, each 6 MHz wide, just as in broadcast and cable TV networks. The MMDS channels that came from ITFS have upstream bandwidth defined for user-to-network communications. ITFS was intended to be an experimental network for schools, but few schools ever used the ITFS channels. However, if there is a school near an MMDS network using one of the ITFS channels, the corresponding MMDS channels must be shut down and cannot be used.

The channels that make up MMDS in the United States are

▼ 2 MDS channels between 2,150 and 2,162 MHz

■ 16 ITFS channels between 2,500 and 2,596 MHz

■ 8 MMDS channels between 2,596 and 2,644 MHz

▲ 4 more ITFS channel interleaved with 3 more MMDS channels between 2,644 and 2,686 MHz

Outside of the United States, MMDS runs with similar channel structures between 2 and 3 GHz.

For a while, MMDS was seen by the telephone companies in the United States as a way to compete with the cable TV companies for video services. Several trial MMDS networks were built and piloted by the telephone companies, but were eventually abandoned. At MMDS frequencies, an unobstructed line-of-sight is needed from user (home antenna) to central tower. A pine tree, for example, could be a serious obstacle to the signal. In hilly and heavily wooded areas, adequate *coverage* (the number of potential customers for the service) was hard to guarantee, especially as trees grew and buildings were erected. In most cases, only a small percentage of the people living within range of the signal could actually get the services. And 33 channels, once adequate for a TV-based service, were soon not enough to compete with newer cable TV architectures.

Outside of the United States, MMDS has been more successful. MMDS made sense for island nations and nations without a well-developed history of cable installation. This was especially true in nations with heavy rainfall or long rainy seasons. The 2.15–2.16 GHz band and the 2.5–2.686 GHz band that make up MMDS are not as susceptible to rain fade as other bands.

Lately, MMDS has been redefined as a high-speed service for IP packets and Internet traffic. On order to help in this process, the bands that make up MMDS have been reallocated for two-way communication. However, the channels are not paired in any way. Moreover, there are two competing technologies for how to use the MMDS bands: *frequency division duplexing* (FDD) and *time division duplexing* (TDD). FDD uses two different frequencies for downstream and upstream traffic, whereas TDD uses the same frequency, but at different times to "bounce" traffic back and forth between end points.

The basic MMDS architecture has a sectored (divided) cell site that has multiple "subscriber terminals" associated with each channel in every sector. A tricky issue is coordinating the MMDS service channels if there are pre-existing ITFS and MDS operators in the area. The older systems are broadcast-only, so in some cases there might not be any upstream channels available for the MMDS operator to use.

MMDS can deliver up to 30 Mbps of data over a channel. Depending on the frequencies used, coverage can be to users 10 miles (about 16 km), 20 miles (32 km), or even

35 miles (56 km) away from the central tower. The high bandwidth available has made MMDS attractive for small- and medium-sized businesses. Few homes require 30 Mbps at this time. Although MMDS has upstream channels available, most MMDS service providers use what is called *telco return* for traffic sent from user to network. That is, the user must still dial-in to an ISP (the MMDS service provider in most cases) at 56 Kbps or less to send anything at all *to* the Internet. Only downstream traffic *from* the Internet can flow at 30 Mbps. The basic organization of an MMDS network used for Internet access is shown in Figure 5-5.

The "MMDS modem" functions in almost exactly the same way as the cable modem. The upstream modem banks do not have to be owned and operated by the MMDS service provider. In fact, many ISPs could be customers of the MMDS service provider and the MMDS service provider need not be an ISP at all.

Each 6 MHz channel can support up to about 9,000 subscribers. Naturally, this all depends on user traffic patterns in the same way that cable modems depend on the same thing. When there are many low-traffic users, the number of users per channel can even be increased by using directional transmitters and more advanced equipment.

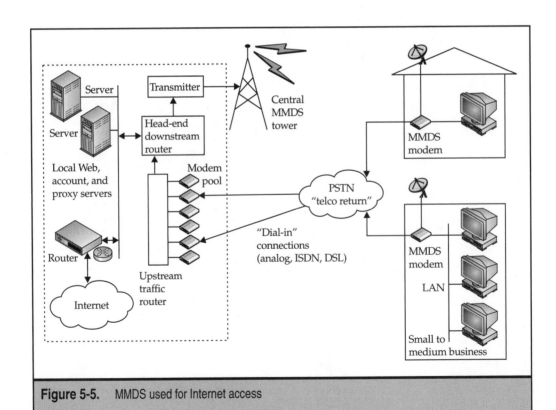

Figure 5-5. MMDS used for Internet access

An ideal ISP location is not often the ideal MMDS head-end location, so the ISP and MMDS locations are split about 50 percent of the time, regardless of the relationship between ISP and MMDS operator. For areas located in the "shadow" of the central transmitter, a low power broadband "booster" can be used when adequate line of sight is not possible. This type of arrangement is shown in Figure 5-6.

MMDS receivers must usually be installed with a signal level near the high end of the range, because signals tend to fade over time and not increase. Antennas can be mounted on the roof, the side of a building, or even in a window, as long as the line of sight is adequate. The antennas vary widely in appearance, from traditional parabolas to foot-square flat arrays to dome-shaped designs.

The use of wireless returns for MMDS systems is relatively new. Current designs use the 2.4 GHz band upstream (the same ISM band used for wireless LANs). If the Internet service provider's office is closer than the downstream tower, this point of presence (POP) can be the target of the upstream transmissions, again with line of sight a concern. However, this requires separate sending and receiving equipment, each with its own alignment. But the upstream power requirement is considerably less.

Upstream speeds vary from 256 Kbps to 5.12 Mbps per user. Lower speeds support more users, of course. If two 6 MHz channels are combined into a 12 MHz band and split into 200 KHz "subchannels," the data rate of each subchannel would be around 256 Kbps to 320 Kbps. These 60 user channels could support about 3,000 customers for data services. A full 6 MHz channel can deliver 30 Mbps, or 27 Mbps with forward error correction.

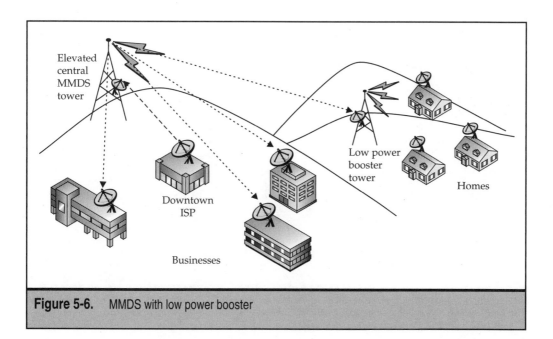

Figure 5-6.   MMDS with low power booster

All in all, MMDS is a technology in transition. Faced with difficulties in the video services arena, the success of MMDS in the wireless Internet services arena is far from assured. Compared to other wireless systems, the basic MMDS channel speed of 30 Mbps is appreciable, but service providers have been cutting this speed drastically in the interest of supporting more users per channel and per MMDS system.

New technologies like MMDS struggle with even simple issues. Assume that MMDS is the perfect technology to solve all access woes. This still does not answer basic users concerns with regard to cost, availability, applicability for business use, and so on. Indeed, no book could explore these issues and hope to be accurate. It is worth pointing out, however, that the "new technology" of cellular telephony, first available in 1984, is still not useable in many remote parts of the United States.

Pros and Cons of MMDS

Some of the pros and cons of MMDS are listed in Table 5-3 and then discussed.

A distinct advantage of MMDS is that the future could belong to wireless. The acceptance of cordless and cellular phones provides ample precedent. Restricting people to certain locations has long been considered a punishment of sorts. Wireless shows great promise in removing these restrictions.

Also, wireless bandwidth and speeds have been growing in leaps and bounds. On most MMDS channels, 54 Mbps downstream is achievable with state-of-the-art equipment (this equipment is admittedly expensive, however).

A further plus is that the FCC has begun issuing two-way licenses, which is necessary because the FCC is firmly in charge of which signals and strengths can be transmitted legally. Obviously, two-way wireless transmission requires a sending unit in everyone's home, though the FCC seems willing to grant such licenses on a routine basis now.

However, an important issue that remains is whether users will embrace what is, at heart, simply another TV package. The history of such "me too" services has been spotty. Without a clearly new service to offer or drastic price advantage, MMDS offers little to entice customers in huge numbers. Also, early MMDS trials have been disappointing in

Pros	Cons
The future could belong to wireless.	Will users embrace another TV package?
Bandwidth is becoming plentiful (54 Mbps possible)	Early trials have been disappointing.
Two-way licenses from FCC are beginning.	Yet another totally new technology and system.

Table 5-3. Pros and Cons of MMDS

the United States. Signal strength has been erratic, even in the most carefully engineered systems.

Although not listed in the table as a separate issue, at MMDS wavelengths, a pine tree can be like a stone wall to the signals, so line-of-sight from the receiver dish to transmitter is necessary. But this limits coverage, especially in tree-rich areas such as the Northeastern United States.

Finally, MMDS utilizes totally new technologies and systems. They might take a while to mature, and a lot of money will be needed to deploy them. In comparison, consider that after starting out in 1984, cellular phone service is still not universal in the United States.

Recently, a lot of interest in the wireless broadband industry has shifted from MMDS to a related technology, LMDS.

LOCAL MULTIPOINT DISTRIBUTION SERVICE (LMDS)

The LMDS is also a form of wireless broadband technology. If MMDS is a kind of "wireless cable TV," LMDS might be characterized as a type of "cellular cable TV." That is, the larger, centralized transmitter towers characteristic of MMDS give way to smaller, distributed transmitter towers in LMDS.

In fact, LMDS is often seen without a transmitter tower at all. LMDS equipment has become inexpensive enough to deploy on a point-to-point basis between buildings in a metropolitan area, where it is often called "wireless fiber" or just "wireless broadband." City governments in the United States have pioneered this use of LMDS, and LMDS is often the most cost effective and quickest way to provide broadband speeds between two buildings located in congested downtown areas.

LMDS can be used as a video delivery system, and in fact the first LMDS test sites did offer exactly that type of service. But today, LMDS is seen as a type of wireless access technology, especially for Internet access. LMDS is ideal for point-to-multipoint configurations where a central site connects multiple buildings. Traffic flow is determined by a system of queuing and multiplexing.

Although LMDS operates in the same microwave frequency range as point-to-point microwave systems, the small size and power of LMDS devices allow many more subscribers (really Mbps per square mile or kilometer) than any point-to-point microwave system. LMDS increases capacity by using a frequency reuse pattern similar to that used in cellular telephony, but based on signal polarization. But there is no handoff in LMDS, of course. LMDS can even be used as a perfectly adequate *wireless local loop* (WLL) technology (so can MMDS).

The small coverage area of an LMDS transmitter is sometimes a blessing. Service providers can start small and build up the network over time. And the equipment needed is quite inexpensive, especially compared to the cost of burying fiber.

LMDS networks consist of a physical transport network and a set of services delivered on this physical platform. The physical transport can handle both circuit-switched and packet-switched traffic. A series of base stations provide wireless access to users and provide the links to the central LMDS service provider office. Overall, the LMDS network is similar in layout to the cellular or PCS telephony network. Of course, there are no handoffs, as mentioned already, and different frequencies and speeds are used on the channels. This overall architecture for LMDS is shown in Figure 5-7.

The base station is usually configured as a hub with four sectors. There are multiple users assigned to each sector, as with MMDS. The number of channels and the overall frequency plan for the network vary, depending2 on the type and number of users, the geographical area, and the capacity required.

A nice feature of LMDS is the ability to *oversubscribe*. The idea behind oversubscription (also called *overbooking*) is that bandwidth available can be assigned to many more customers than it should be able to support, because rarely are all of the customers using the bandwidth at the same time. For example, 12 Mbps might be assigned to a sector on an LMDS base station, but the users in that sector are paying for 24 Mbps in aggregate. Quality of service concerns might arise, but oversubscribing is at least possible with LMDS.

The concept of LMDS overbooking is similar to the idea of engineering the voice network to handle only 1,000 calls from 10,000 users, but so is the risk. When usage patterns

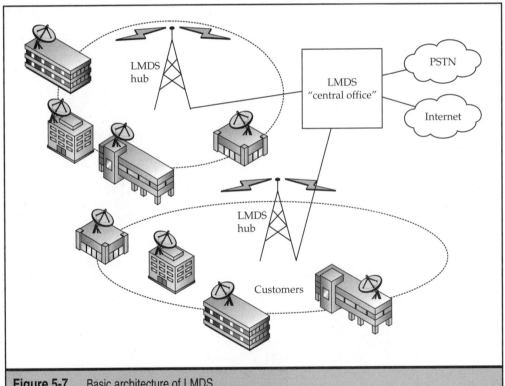

Figure 5-7.   Basic architecture of LMDS

change to add 1,000 online users, the network does not work so well. The risk is that usage patterns might change yet again (broadcast TV over the Internet?) and LMDS might also struggle.

LMDS networks do not have to cover an entire geographical area. Services can be provided only where there are customers. LMDS is at its best when the LMDS channels serve a whole building. All of the users in the building share the channel without the need for each one to have a separate transmitter and receiver. Usually the LMDS equipment is located on the roof, and the individual customer sites are connected with fiber optic cable.

When it comes to services, the following types are being delivered over LMDS today:

▼ T1/E1 circuit replacement (1.5 or 2.0 Mbps)

■ Fractional T1/E1 (up to 1.5 or 2.0 Mbps in 256 Kbps increments)

■ LAN connectivity (typically 10 Mbps Ethernet speeds)

■ Frame Relay/ATM network access (usually at T1/E1 speeds)

■ Voice and video services (such as teleconferencing)

■ Internet connectivity

■ Web services such as Web site hosting

■ Voice and faxing over IP

▲ ISDN (64 Kbps and 1.5 or 2.0 Mbps)

There are multiple ways that an LMDS system can be deployed. Channels can be shared with TDD, FDD, or TDM. Various modulation methods can be used.

In the United States and Canada, LMDS uses the 24 GHz, 28 GHz, and 39 GHz bands. Spectrum is assigned in the 24 GHz band in pairs of 50 MHz blocks, so licenses can get a total of 100 MHz (sometimes as much as 400 MHz). Subchannels have 10 MHz block allocations. The channels are paired for communication in each direction and "blocked" so that there is sure to be competition between two service providers in each area.

The 28 GHz band also follows the cellular practice of A and B blocks to ensure competition, and the subchannel structure is up to the service provider. The 39 GHz band has 50 MHz channel blocks, again paired for competition, and some 175 licenses will be given in each of 14 different 100 MHz blocks.

LMDS can also be used to extend the reach of a traditional HFC cable TV network. This use of LMDS is shown in Figure 5-8.

LMDS is not only for use in the United States. Outside of the United States (and Canada), LMDS is known as *Fixed Wireless Point to Multipoint* (FWPMP). The equipment and architectures are nearly identical, but the frequency allocations are more suitable for the European environment.

Oversubscription is possible in FWPMP, as it is with LMDS. There are WLL and cable TV extension applications. However, more restrictions are generally imposed on the FWPMP license than in the United States. For example, there is a minimum coverage requirement and a time line for universal services to be deployed (five years).

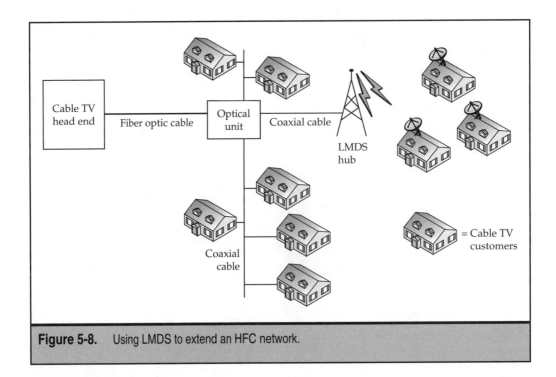

Figure 5-8.   Using LMDS to extend an HFC network.

FWPMP operates in the 26 and 28 GHz bands (actually, 24.5–26.5 GHz and 27.5–29.5 GHz). The frequency blocks are multiples of 7 MHz, and channel allocations are usually in increments of 28 MHz. There is also a 38 GHz band, each having a channel block of 50 MHz and four channels 12.5 MHz wide, duplexed as expected.

FWPMP can also operate in the 3 GHz and 10 GHz bands, but these channels do not have the bandwidth that the higher frequencies offer.

Although private LMDS links can operate at SONET/SDH link speeds of around 155 Mbps, most LMDS and FWPMP equipment is designed to run at the basic T1 rate (1.5 Mbps) and E1 (2.0 Mbps) rate. Multiple links can and are provided to a single customer, however, and a large customer might have LMDS or FWPMP running at an aggregate speed of 4.5 Mbps or 6.0 Mbps.

It should be noted that MMDS and LMDS are not the only systems that can be used for high-speed wireless Internet access. For instance, in the United States the *Unlicensed National Information Infrastructure* (UNII) can be used by anyone without paying for an expensive license. The UNII bands are at 2.4 GHz and 5.8 GHz. Speeds usually range from 128 Kbps to 2 Mbps. The biggest problem with UNII-based equipment is interference. When two vendors have UNII customers close to each other, the band becomes saturated, both systems suffer, the problem is really no one's fault, and little can be done to improve things.

Pros and Cons of LMDS

Table 5-4 shows some of the pros and cons of LMDS for high-speed digital services.

Most of the advantages and disadvantages of LMDS closely mirror MMDS and need not be repeated. The wireless advantage, channels' capability of up to 56 Mbps operations, and two-way licenses are the same. A very good sign for LMDS is that a large amount of interest exists, although much of it has been in Canada.

If TV services form the heart of LMDS, however, the issue is still whether users will embrace what is, at heart, just another TV package. LMDS in the form of wireless broadband seems destined to form a new LAN interconnection package for urban businesses. Video services may be an afterthought. Also, both MMDS and LMDS lack an easy way to add more capacity. This is not so much of a problem in one-way broadcast systems where everyone shares outbound channels, but for inbound traffic, MMDS systems especially might be overwhelmed with no easy way to increase the licensed bandwidth to add capacity. A similar concern is present in the cellular telephone network.

With LMDS, the smaller transmitter size and coverage area (cell) can keep deployment costs manageable, especially in dense urban areas. However, this same small cell size could be a problem in suburban and rural areas. Suburban residents are notorious for opposing "unsightly" towers, and rural areas would need a lot of two-mile-across transmitters for any kind of penetration. Is it really feasible to give every farm its own LMDS tower?

Finally, like MMDS, LMDS is a totally new technology and system. The same comments made at the end of the MMDS section are relevant here. New technologies like LMDS struggle with even the simple issues of cost, availability, applicability for business use, and so on, even if LMDS is the perfect technology to solve all access woes.

Pros	Cons
The future could belong to wireless.	Will users embrace another TV package?
Bandwidth is becoming plentiful (54 Mbps possible).	Small coverage area could be a problem in suburban and rural areas.
Two-way licenses from FCC are beginning.	Yet another totally new technology and system.
Plenty of interest and manageable costs.	

Table 5-4.   Pros and Cons of LMDS

OVERVIEW OF SATELLITE SYSTEMS

Satellite systems form a special case of wireless technologies used for broadband access. Satellites have been used for voice communications for many years, although in countries with highly developed communications infrastructures, the high delays are often unacceptable. This is also true for data services in some cases, but data satellite services such as DirecPC have been popular in spite of the delay.

The data service delay issue in satellite systems has been handled in the past in two ways. First, only the *downstream* (Internet-to-user) traffic has typically been sent over the satellite link. The upstream (user-to-Internet) traffic is usually handled over a traditional, dial-up link to the satellite ISP. This works well because the user device is usually a client PC, and the overwhelming proportion of the traffic flows not from client to server, but from server downstream to the client. Second, there is a lot of bandwidth on the downstream portion of the link, usually 400 Kbps of more. So even though it might take a few seconds for the Web page to start to appear, once it does the whole page is essentially there all as a whole (most Web pages are only about 50 to 100 kilobytes (KB) in size). There are exceptions, of course, and it all depends on how many independent connections have to be set up between client and server, but the overall effect is much better than with dial-up alone.

However, some satellite service providers do regularly employ some form of upstream satellite communications. And other service providers have begun to experiment with satellite systems that do not follow the normal geosynchronous orbits traditionally used by communications satellites. The geosynchronous orbits are the cause for the high satellite delays. *Low earth orbit* (LEO) and *middle earth orbit* (MEO) satellites will have much lower delays and make two-way communication with low-powered transmitters possible.

These systems have been promoted since the late 1990s, and yet always seem to be a few years away. Rather than reproduce the system plans and schedules that always seem to slip further and further into the future, the interested reader is referred to the Internet for sources of information, especially regarding the Microsoft-backed system known as Teledesic.

Direct broadcast satellite (DBS) systems already deliver high-quality TV pictures to many homes. Most deliver digital TV pictures (although to the same old analog TV for now) by using *Motion Pictures Experts Group II* (MPEG II) encoding. Audio (sometimes a kind of "jukebox" service) is generally Dolby Surround Sound delivered with MPEG I Layer II audio encoding. This is CD-quality stereo sound, not the tinny voice service or TV audio that most are used to.

Driven by equipment vendors and satellite system operators faced with delay problems for two-way voice and data services, DBS systems have found a ready market for their one-way, all-digital TV services. Most use some form of MPEG compression and carry multiple channels to provide a "near" video-on-demand service (that is, movies begin every 15 minutes). DBS systems have been hampered by problems carrying local broadcast channels, delays due to compression processing (sports events are delayed about one second), and lack of two-way operation (although some subscriber-to-satellite systems do exist). Most DBS systems rely on a telephone modem connection for upstream communication.

From time to time, even more exotic plans for sky-based systems are proposed. For instance, a company recently proposed flying blimps over metropolitan areas with wireless equipment. Others have turned to tethered hot air balloons over Los Angeles and other cities. Blimps are one thing, but last time tethered balloons were tried in the 1970s, they had a disturbing habit of shearing the wings off aircraft that wandered into the guy wires. Such failures will not stop some from trying.

Pros and Cons of Satellite Systems

Some of the pros and cons of using satellite systems for high-speed digital services are listed in Table 5-5 and discussed here.

The biggest advantage, especially for geosynchronous earth orbit systems, is the wide ground coverage. Satellites of any type maximize the area that a given transmitter can reach for service coverage—an ideal situation (although some fading occurs at the edges of the area).

Along the same lines, satellites give absolute terrain independence, and function in valleys, mountains, and at sea with equal ease. The services that satellites offer are totally insensitive to distance as well. So, a satellite serving sites thousands of miles apart is just as efficient as a satellite serving sites in the same neighborhood, as long as the sites are not in these fringe areas.

Another nice touch is that the satellite systems leverage the combined use of digital voice, video, and data technologies. After all, the very first satellites sent and received data; voice service was not added until years later.

The last advantage may seem silly, but the satellite people have played this card to maximum advantage for years. Where do you think the cable TV signal comes from? From satellites, of course! All that satellite systems do, the sales pitch goes, is eliminate the cable TV company in the middle.

Pros	Cons
Maximum potential service area coverage.	Solar effects and outages.
Terrain independence, distance insensitivity, many others.	Capacity and licensing could be an issue.
Where do you think the cable TV signal comes from?	Crowding and sharing may become a real problem.

Table 5-5.   Pros and Cons of Satellite Systems

On the other hand, satellites suffer from a number of serious and annoying technical problems. Satellite services may be eclipsed by planes and even lower orbit satellites. Solar flares may adversely affect signals, and they must routinely be shut down for a while when the sun passes behind them on its annual journey between the solstices. Some solar effects cause the atmosphere to literally "swell up," and low-orbiting satellites must be constantly moved to avoid the cumulative effects of atmospheric drag. The lifetime of a satellite is also determined by the amount of fuel it carries. Small gravitational effects must be compensated for by periodic rocket firings. When the fuel is exhausted, an otherwise functional satellite is pretty much useless.

The delays due to distance and digital compression, complex signal handoffs needed with LEO systems, and a need for true two-way service could become issues quickly. There are already some satellite transmitters designed for home use, but they are presently very costly.

In addition, naturally, the more two-way there is, the more important the issues of capacity and licensing become. The FCC rules the airwaves, but a number of agencies, including NASA and the ITU-R, decide what goes into orbit and where.

A final drawback is that crowding of prime orbital positions may become a problem. The most desirable GEO positions have been occupied for years, and the same may happen with LEO orbits, although not for years to come. And, of course, the sharing of limited bandwidths among truly huge numbers of users could prove to be a problem.

BLUETOOTH

No discussion of wireless technologies today would be complete without a short mention of Bluetooth. Harald Bluetooth was king of Denmark from about 935 to 985, and his name has been taken for *Bluetooth*, a new wireless communications technology. Bluetooth is a wireless technology that is neither WAN nor LAN, but something entirely new, really: the *personal area network* (PAN). PANs are intended to network the new breed of *personal digital assistants* (PDAs) and other small, hand-held digital devices such as portable game-playing units.

But Bluetooth does not really belong in this chapter. Bluetooth is not a serious contender in and of itself for wireless Internet access. This section explains why, and should be read as a balance for some of the more exaggerated claims for Bluetooth in terms of speed and distance. Bluetooth is mentioned again with regard to high-speed Internet access later in this book in the more conventional role as a *premises distribution network* (PDN) for users with more than one PC needing Internet access.

Handheld PDAs and PCs hold a lot of information in common. Schedules, address books, meeting notes, e-mail, and so on can dwell as easily on a PC as on a PDA. The problem is *synchronization*: taking information that has been updated on the PC or PDA and making sure it is the same on both units. Today, this synchronization of information is achieved mainly by a short cable running to the serial port on the back of the PC.

Bluetooth is designed to be more or less automatic. Walk into a room, and the Bluetooth devices automatically find each other. Whenever Bluetooth devices are brought into close proximity, they start networking—it's what they do. No need to turn Bluetooth on, even if the unit itself is powered off. Airlines have had an issue with this feature of Bluetooth, of course, and early prototypes had to have their batteries removed to eliminate any chance of the units interfering with aircraft navigational equipment on landing and takeoff.

Bluetooth, which usually runs at under 1 Mbps, just does not have the bandwidth needed for more than occasional use. But Bluetooth is a good, low-power standard for both voice and data. Bluetooth-equipped pagers and cell phones can automatically switch to vibrate mode in restaurants or theaters or turn the phones off upon entering hospitals.

Bluetooth operates in the 2.4 GHz ISM (industrial, scientific, and medical) frequency range. Bluetooth devices have three power classes: 100 milliwatt, 2.5 milliwatt, and 1 milliwatt. The 100 milliwatt device range is 100 meters (a little more than 300 feet), the 2.5 milliwatt range is 10 meters (about 30 feet), and the 1 milliwatt range is only 10 centimeters (only a few inches). As with all wireless technologies, the ranges are approximate and can vary quite a bit depending on the local environment. Outdoors, some 2.5 milliwatt Bluetooth devices have been demonstrated to function over 100 meters.

Bluetooth has two possible network topologies: *piconet* and *scatternet*. Bluetooth has both a point-to-point and point-to-multipoint mode of operation. When using a point-to-multipoint connection, the connection channel is shared among the Bluetooth units. These units form the piconet.

When two or more Bluetooth devices come into range, they automatically form a piconet. One unit becomes the *master*, while the other unit becomes the *slave* (the choice of term is unfortunate, but defined in the standard). The master unit controls all of the traffic on the piconet, and both units will shift, or *hop*, among the same set of frequencies at the same time. The master gives all the slaves in a piconet its own clock-device ID and sets other parameters. Up to seven Bluetooth devices can be active on a piconet.

Bluetooth scatternets are formed when multiple masters exist within range of each other. In other words, a Bluetooth scatternet consists of multiple, overlapping piconets. A master unit might also be a slave on another piconet.

But whether piconet or scatternet, Bluetooth just does not have the range nor the speed to allow the use of Bluetooth for any form of serious Internet access.

SWITCH-BASED SOLUTIONS

Clearly, a lot of the congestion problems of switches and trunks in the PSTN could be relieved by identifying long-holding–time, digital information streams and redirecting this traffic directly onto a dedicated network built for this purpose, such as the Internet. The question is now one of finding the best or most cost-effective solution for doing so. Solutions such as cable modems or wireless systems avoid the PSTN switch and voice trunking network entirely for this traffic by shifting it off the local loop altogether. Perhaps this traffic could still be identified as it arrives from the local loop and redirected before entering the full voice switch and trunking network. This is sort of a "make a left at the switch" approach. Several vendors have developed equipment to do this. A potential problem is that many of the solutions are tied to a particular vendor's switch or architecture. In other words, these are hardly standards-based solutions at the present time. Five major vendors, which are outlined in the following paragraphs, are using this approach.

Lucent Technologies' *Access Interface Unit* (AIU) and Access Gateway are software-based methods used to identify long-holding-time traffic, usually at the ISP-attached central office. The software enables a service provider to move this traffic onto a more cost-effective part of the digital switch or to the AIU itself. The AIU provides a "virtually nonblocking" (that is, it rarely blocks) path through the switch and offers high capacities. The AIU works only with Lucent (AT&T) digital switches, but the Gateway boxes can re-route traffic arriving on the voice network for transmission over a dedicated data network.

Premisys' Interlude operates in two modes: for directly dialed or ISDN traffic, or in a remote hub configuration. In direct dial, the box just shunts aside data or ISDN traffic from the central office switch. As a remote hub, the Interlude gathers traffic, and directs voice to the PSTN and data somewhere else (usually the Internet).

DSC Communications' Intelligent Internet Solution combines hardware and software methods already on the market. They use the signaling network of the PSTN (based on a protocol called "SS7") to redirect online service and ISP traffic away from the PSTN switch. Other signaling software detects these data connections and hands them off directly to a digital cross-connect (from DSC, of course) to make a connection to the data service provider.

NORTEL's Internet Thruway product is naturally based on NORTEL equipment. A NORTEL AccessNode, a digital loop carrier (DLC) device, looks at the dialed number and routes the call to the voice or data network as appropriate. Voice goes through the DLC to the switch, but data goes over a separate trunk to a NORTEL Rapport dial-up switch and from there to a packet-based data network such as the Internet. This package is very popular, and NORTEL plans higher speed support, "always-on" connections, and data-over-voice support.

Finally, Telco Systems' Intelligent Call Routing employs an "access server" that intercepts incoming calls from sites with ISP/PSTN service and sends them to the CO switch or onto a packet network as appropriate.

ADSL AND OTHER TECHNOLOGIES

All DSL technologies, of which ADSL is an important member, represent just one effort to solve the problem of broadband residential access for advanced services. Other methods exist, including direct broadcast satellite, MMDS/LMDS wireless CATV systems, cable modems for cable TV systems, ISDN (the original DSL, with two B-channels running at 64 Kbps each and a D-channel running at 16 Kbps), and even digital data services (DDS), which are leased lines running at 64 Kbps. The only technology that is not really new is the 56K modem, for reasons already discussed. All the others are compared according to the four criteria shown in Table 5-6.

Technology	High Initial Deployment Cost	Low-Speed Return	Major Wired Infrastructure Change	No Analog Voice Support
DBS	X	X		X
MMDS		X (typical)		X
LMDS	X (but will diminish)			
Cable modems on HFC	X		X	
Cable modems		X		X
ISDN	X		X	
ADSL				

Table 5-6. DSL and Other Technologies

The table compares these other technologies to the DSL family. The table contains four criteria, explained here:

▼ **High Initial Deployment Cost** Some of the other technologies will require the investment of huge sums of money just to get off the ground. DSLs are incremental costs—that is, analog local loops can be converted to DSL almost on a home-by-home basis. (Of course, the services also have to be put in place, but these can be done incrementally as well.)

■ **Low-Speed Return** In this context, low-speed does not mean asymmetric, but rather that some of the other technologies are inherently one-way at present, and so must use a "normal" analog modem for the upstream path. Ironically, maybe DSL will help these technologies in this area! The slowest DSL runs at 64 Kbps upstream, and many will operate at much higher speeds.

■ **Major Wired Infrastructure Change** Some technologies, although not requiring an entirely new infrastructure to be put into place before services can be offered to even one customer, will still require extensive changes to the way such systems currently operate. With DSL, the intent is to maximize reuse of the existing infrastructure.

▲ **No Analog Voice Support** Some of the other technologies make no provision for older analog phones. A word of caution is in order here: although ISDN supports analog telephones, a conversion device (a terminal adapter, or TA) is needed. Also, some DSL schemes (such as HDSL) make no real provision for analog phone support at all.

Note that the ADSL row in the table is blank. ADSL, and indeed almost all other DSLs, suffer from none of these limitations. All in all, DSL offers more advantages and fewer disadvantages than any other technology. The time has come to see how.

CHAPTER 6

Introducing the DSL Family

ADSL is more than just a technology that allows for broadband access from the residence or small office to a network service provider, ISP, or not. ADSL is one of a number of access technologies that can be used to convert the access line into a high-speed digital link and to avoid overloading the circuit-switched PSTN. These technologies form a family loosely called digital subscriber line (DSL) technologies, or sometimes *x-type digital subscriber line* (xDSL) technologies, where the *x* stands for one of several letters of the alphabet. It is important to note that some of these technologies are based on modems. That is, some of the DSL family use analog signaling methods to transport analog or digital information content across the access line or local loop; they have much in common with other modem technologies, of course. Other members of the DSL family use true CSU/DSU arrangements. These members use true digital signaling to transport digital information content (seldom analog information) across the access line or local loop. They have much in common with T-carrier arrangements.

This chapter details the operation of each of the DSLs and provides additional information about them, as well. The focus shall be on ADSL and its close relative, RADSL (pronounced RAD-sil). The common "thread" among them is that they are all based on existing pairs of copper wires installed as local loops, as opposed to most of the alternative solutions examined so far, which rely on totally new networks and technology infrastructures for the most part. Once the frequencies above 4,000 Hz can be used, higher speeds can be attained on the local loop. The DSL family is a set of copper-based solutions.

COPPER-BASED SOLUTIONS

There exist many possible solutions to the problems of overloading the *public switched telephone network* (PSTN) voice network with packetized data and interactive broadband services. Some involve building entirely new systems based on wireless and satellite networks, and there is certainly nothing wrong with this. However, it may be better to start with something that already exists and builds upon or improves the operational capabilities of the copper-based analog local loop. The only solution based on using the copper local loop not considered here is 56K modems because these do nothing to avoid problems with the circuit switches in the PSTN. Indeed, they may make matters worse by encouraging even longer holding times on voice switches.

This approach is not meant to detract from cable modem or other solutions. It is simply a more practical approach. People are comfortable with using a telephone line to access the Internet, and many are replacing their dial-up modem arrangements with some form of DSL. DSL technologies do not require a complete rebuilding of an existing infrastructure, nor do they require a completely new infrastructure that can essentially be used only for Internet access. Many cable TV systems do offer voice telephony services, but these require even more changes to the basic cable TV infrastructure than cable modems do. Few people using cable modems throw out their telephone service and rely on the cable TV company entirely. If the telephone access line is there, why not use it to it maximum capability? So a realistic and cost-effective solution using the available access line infrastructure would be the following:

1. Maximize reuse of existing analog local loops.

2. Include some provision for backward compatibility with existing voice telephony equipment (that is, the analog handset).

For the time being, it seems that only copper-based solutions satisfy these two criteria.

DSL's full name, Digital Subscriber Line, began with the *Integrated Services Digital Network* (ISDN), which was created to foster the total digitization of the PSTN end-to-end, from user device (handset, PC, and so on) to user device. DSL is basically an ISDN term. ISDN was the first DSL service, and its position as the first of the DSLs should never be forgotten or minimized. Many of the advanced features of ADSL became possible only through the experience gathered with ISDN DSL methods.

For residential services, the ISDN DSL takes the form of the *Basic Rate Interface* (BRI). The BRI operates at 144 Kbps full-duplex, organized into 2 Bearer (B-) channels running at 64 Kbps and one D-channel for signaling and data running at 16 Kbps. The two B-channels may be bonded to yield 128 Kbps in most circumstances, although not universally. The ISDN switch (called the local exchange [LE] in ISDN) must allow this bonding to occur through the assignment and use of *Service Provider Identification numbers* (SPIDs).

Newer DSL technologies are more interesting and promising. As was stated previously, this is sometimes seen as xDSL, where the *x* represents any one of a number of letter designations, but this book just uses the term DSL, which is the current practice. Some DSL technologies are sometimes called *duplex,* in the sense that the speeds are identical in both directions. Note that this use of the term *duplex* differs from the usual sense in the United States of "both directions." When applied to DSL it means "both directions *at the same speed,*" and so contrasts with "asymmetrical." In spite of this use of the *duplex* term, DSL speeds are far more commonly described as simply "symmetric" (same speed in both directions) or "asymmetric" (different speeds in each direction). For this reason, the duplex terminology will never be used in a symmetry context in this book.

However, many broadband residential services are distinctly asymmetrical, such as video-on-demand or Internet Web access. That is, the amount of traffic sent upstream from a home or client PC is much less than the traffic sent downstream to a home or from a server Web site (there are exceptions such as FTP, but this general statement is still true in the case of the Web). When the client is in the home or office, and the server is at a site connected by a leased private line to the Internet, it makes more sense to allow for higher speeds downstream (into the home) than upstream (out of the home). Indeed, many versions of DSL (ADSL, RADSL, and VDSL, for example) are inherently asymmetrical.

High bit-rate DSL (HDSL) and HDSL2 (a more up-to-date version of HDSL) is a duplex technology (again, in the sense of "symmetric speeds"). The speed upstream and downstream is either 1.5 Mbps in the United States or 2.0 Mbps in most other areas around the world. This aligns HDSL with the existing T-carrier DS-1 speed in the United States and the existing E-carrier E1 speed elsewhere. In fact, HDSL and HDSL2 are intended for transporting a DS-1 or E1 over copper lines and are most often deployed as a more cost effective way to deploy DS-1 or E1 services. The customer still sees and buys a DS-1 or E1, but it is provisioned as HDSL or HDSL2 within the network. As just a "newer and better" DS-1 or E1, HDSL and HDSL2 are most often used for the same purposes as traditional private lines, namely carrier services for feeder plants such as DLC pairgain systems, or to the customer (as a DS-1 or E1) for LAN interconnection or leased-line WAN or Internet access.

Next to be considered is *Symmetric* (sometimes seen as "single-line") *DSL* (SDSL). For a while, SDSL seemed to be a promising variation on HDSL, intended for all that HDSL could do, and more. For instance, SDSL was capable of both 1.5 Mbps and 2.0 Mbps operation in both directions, and often over very long distances. However, it now appears that SDSL will be seen as a forerunner of HDSL2, which preserves many of the best features of SDSL and eliminates some of the drawbacks that SDSL suffered.

Some DSLs are asymmetrical in nature. The speeds upstream (always defined as from the customer site to the service provider) are typically much less than the speeds downstream (always defined as from the service provider site to the customer). Given the extremely asymmetric nature of most client-server interactions, however, especially on the Web, this should not be a drawback in most cases. A possible exception is when a home PC user or SOHO business wishes to run a Web server in their home or home office. In this case, naturally, it is desirable to have at least symmetric (that is, duplex) speeds. Perhaps HDSL or HDSL2 would be more suitable for such home Web site arrangements.

Asymmetric DSL (ADSL) and its close relative *Rate-Adaptive* DSL (RADSL) do not differ much, if at all, in terms of speed and distances. In fact, because most ADSLs are now rate-adaptive, it makes less and less sense to distinguish ADSL and RADSL, but this distinction is still made here for historical and educational reasons. Both function between 1.5 Mbps to about 6–8 Mbps downstream and 16 Kbps to about 640 Kbps upstream, but these are only common figures. Both have a variety of uses, all centered on interactive multimedia applications. For simple Internet or Web access, either ADSL or RADSL will do just fine. And even for video-on-demand services, of simplex (one-way "broadcast" quality) video TV services, either ADSL or RADSL is more than adequate at higher downstream speeds. For remote LAN access for telecommuters, ADSL and RADSL will be a key service, as well. Of course, both ADSL and RADSL allow for the continued use of existing analog telephones, which is a real plus.

ISDN DSL (IDSL) sounds odd because ISDN already employs a DSL, but this combination actually makes sense. IDSL supports the ISDN 2B+D BRI structure running at 144 Kbps in both directions. The problem in part is that the ISDN BRI is used in large measure for fast Internet and Web access. ISDL gets the ISDN BRI off of the circuit switch when the line is used simply for Internet and Web access, alleviating a great deal of the switch congestion and allowing more "real" ISDN users to be supported through the switch. The D-channel in the BRI can no longer be used to set up ISDN voice connections on the PSTN, but if the BRI is used exclusively for ISP access, this is not a major consideration anyway.

Finally, *Very high data rate* DSL (VDSL) is at once promising and ambitious. Usually considered asymmetric, the VDSL specification calls for an optional symmetric configuration. VDSL speeds cannot be achieved wholly on the longest lengths of local loop copper and must employ fiber DLCs for at least half of the access line distance to the switching office. Speeds are an amazing 13 Mbps to 52 Mbps downstream, and, a not inconsiderable 1.5 Mbps to 6.0 Mbps upstream. The applications supported include all that ADSL/RADSL is intended for, plus *high definition TV* (HDTV) digital television services. VDSL is usually seen as the ultimate evolutionary goal of all asymmetrical DSL technologies.

Please note that supported DSL speeds vary in both directions, depending on the physical characteristics of the analog local loops on which they are deployed. Although one home may be able to use ADSL at 1.5 Mbps downstream, another home located nearby may enjoy only 768 Kbps transmission speeds downstream. In most cases, though, speeds greatly exceed those available with "regular" modems or ISDN BRI.

Another point is that one should treat all DSL speeds and distance limits with caution. In most cases, these are merely design parameters and not set in concrete. Vendors are always pushing the DSL envelope of speed and distances. But it is always true that higher speeds must be balanced by lower distances.

THE DSL FAMILY IN DETAIL

The major characteristics of the current DSL family of technologies are shown in Table 6-1. The emphasis here is more on technical operation than applications. This table lists mainly the forms of DSL that have been the subjects of international standardization efforts. Other forms of DSL exist, such as *Multispeed* DSL (MDSL), championed by one particular vendor or another, but these forms are dealt with shortly. The order is roughly by age of the technology. That is, HDSL essentially came first, and VDSL is the newest, with the exception of G.lite, for reasons discussed later.

Name	Meaning	Typical Data Rate	Mode	Comment
HDSL	High bit-rate DSL	1.544 or 2.048 Mbps	Symmetric	Uses two pairs (only DSL that does)
HDSL2 (SHDSL)	High bit-rate DSL #2, or single-pair HDSL	1.544 or 2.048 Mbps	Symmetric	Uses only one pair for same speeds
SDSL	Symmetric DSL	768 Kbps to 2.0 Mbps	Symmetric	One pair, as far as 24 kft (about 7 km)
ADSL/RADSL	Asymmetric DSL, Rate-Adaptive DSL	1 to 6–8 Mbps 16 to 640 Kbps	Downstream Upstream	18 kft (5.49 km) maximum

Table 6-1. The DSL Family in Particular

Name	Meaning	Typical Data Rate	Mode	Comment
G.lite	"G" is ITU series, "lite" ADSL form	Up to 1.1 Mbps Up to 128 Kbps	Downstream Upstream	Simpler version of ADSL
IDSL	ISDN DSL	Same as ISDN BRI	Symmetric	Called "BRI without the switch"
VDSL	Very high data rate DSL	13 to 52 Mbps 1.5 to 6.0 Mbps	Downstream Upstream	Short copper fiber and ATM needed

Table 6-1. The DSL Family in Particular *(continued)*

Remember to treat all DSL speeds and distance limits in this table with caution; consider them design parameters, not hard and fast rules. Vendors are always pushing the DSL envelope, so in many cases today, it is possible to see VDSL products without a need for fiber feeders, full 1.5 Mbps ADSL all the way out to 18 thousand feet (kft), and SDSL with analog voice support (just to mention a few of the variations).

The DSL family members listed in Table 6-1 are explained in more detail here:

▼ **HDSL—High bit-rate DSL**   As was mentioned previously, HDSL runs at 1.544 Mbps (T1 speeds) in the United States (really throughout North America) and at 2.048 Mbps (E1 speeds) almost everywhere else. Both speeds are symmetric (the same in both directions). The original HDSL at 1.544 Mbps used two-wire pairs and extended to 12,000 feet (about 3.66 km). It was primarily intended as a more efficient way to link the DLC to the local exchange switch. HDSL at 2.048 Mbps needed three-wire pairs for the same distance (but no longer). HDSL is the only DSL that still requires two pairs of twisted-pair copper wires to achieve high speeds.

■ **HDSL2—High bit-rate DSL version 2**   The version of HDSL, known as SHDSL to the ITU—and therefore almost everywhere outside of North America—preserves all of the features of HDSL and at the same time requires only one pair of copper wires. This is a great improvement, because the need for two wire pairs greatly limits deployment possibilities. HDSL2 (SHDSL) is expected to replace HDSL sooner or later.

■ **SDSL—Symmetric (or single pair) DSL**   If the goal of DSL technology is to reuse analog local loops, perhaps it would be better to squeeze as much speed as possible out of a single wire pair, which is what SDSL does. SDSL uses only one wire pair and can be extended on some loops as far as 24,000 feet (about 7 km), at least in some vendor's versions. Usually, however, the speeds are the same as with HDSL or even below. SDSL is sometimes provisioned at 768 Kbps using a single-pair version HDSL. Because SDSL cannot run freely when closely packed with other wire pairs in a cable bundle—causing interference with other services—SDSL has remained on the fringes of the DSL world, except in Europe. As time goes on, it seems likely that HDSL2 will do all that SDSL can do and with less interference with other services nearby. HDSL2 is expected to replace SDSL in many cases.

- **ADSL—Asymmetric DSL**   SDSL uses only one wire pair, but the need to limit the power that SDSL generated (to minimize interference with other services in the same copper wire bundle) also limited distance. ADSL acknowledges the asymmetrical nature of many broadband services and at the same time extends the reach to 18,000 feet (about 3.4 miles or about 5.5 km).

- **RADSL—Rate-Adaptive DSL**   Although treated separately here, RADSL and ADSL are essentially one and the same today. Typically, it is assumed when equipment is installed that some minimum criteria for line conditions are met to allow for operation at a given speed. At least this has been true of former digital technologies, such as T-carrier or ISDN. However, what if line conditions vary or operational speeds make equipment sensitive to small environmental changes? RADSL, which is an inherent property of ADSL using Discrete Multitone (DMT) line coding, can actually adapt to changing line conditions and adjust speeds each way to maximize the speed on each individual line. Usually, this is performed only when the line first becomes active, but in principal the speeds can vary even while the line is in operation.

- **G.lite**   Although closely related to ADSL and RADSL, G.lite is sufficiently different to rate an entry of its own. The *G* is from the ITU series of specifications that govern line technologies, and the *lite* refers to the fact that G.lite does not have all of the features of full ADSL/RADSL. Also known as ADSL lite or Consumer DSL (CDSL), it is generally more modest in terms of speed and distances compared to ADSL/RADSL, but it has a unique advantage. With G.lite, no remote devices—known as splitters—are needed on the customer premises. The function of the splitter is to allow existing analog voice telephone and others types of equipment, such as fax machines, to continue to operate as before. The splitter needed with ADSL/RADSL is discussed more fully later in this chapter. G.lite, however, needs no remote splitter and its associated wiring. G.lite uses a small electronic filter instead and is essentially a subset of full ADSL/RADSL.

- **IDSL—ISDN DSL**   This technique takes the normal 2B+D channels of the ISDN BRI, which runs at 144 Kbps (two 64 Kbps B-channels and one 16 Kbps D-channel), and runs the BRI not into the ISDN voice switch, but into the DSL equipment in the local exchange. IDSL also runs on one pair of wires and extends up to 18 kft, exactly the same as an ISDN DSL. Some versions of IDSL allow for the use of the full 144 Kbps (2B+D), whereas others allow only 128 Kbps (2B) operation.

▲   **VDSL—Very high data rate DSL**   The newest member of the family, VDSL is seen by some as the "ultimate goal" of DSL technology. Speeds are the highest possible, but only over 1,000 to 4,500 feet (less than a mile, or about 1.4 km) of twisted-pair copper wire. This is not a problem for VDSL. VDSL expects to pick up a fiber feeder at this 1,000 to 4,500 foot point, and it is also intended to carry ATM cells, not as an option, but as a recommendation. I explore this aspect of VDSL more fully in Chapter 14.

ANOTHER VIEW OF THE DSL FAMILY

It is possible to visualize the relationships between DSLs in another way. This visualization would separate symmetrical DSLs from asymmetrical DSLs and also allow room to add the nonstandard and vendor-specific flavors of DSL to the visualization. This type of DSL family tree is shown in Figure 6-1.

The family tree branches into symmetrical (same speed defined upstream and downstream) and asymmetrical (higher speeds defined downstream than upstream) families. Note that even the asymmetrical DSLs all have symmetrical options, meaning that they can all be configured if desired to run at the same speed upstream and downstream. However, this symmetry is achieved by limiting the downstream speed to equal the upstream speed, so this is not often done.

Each of the major families is further divided into members that are firmly based on international or at least national standards and those that are based simply on what engineers and vendors can make possible. This is important because only with those DSLs firmly based on international or national standards can any form of interoperability between the devices from different vendors be even remotely possible (even with

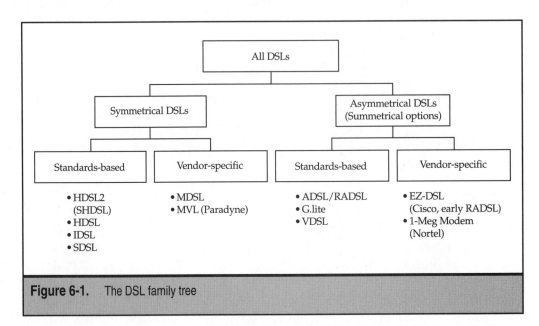

Figure 6-1.   The DSL family tree

standards, interoperability is rarer than it should be). But there is no real hope of interoperability without standards in the vast majority of cases.

The symmetrical DSLs have an odd mixture of standards-based and vendor-specific members. Only HDSL2 (SHDSL) is intended to be standardized enough to allow interoperations between different vendors' implementations. HDSL itself is the subject of many standards, but not enough were spelled out in the text to allow for easy interoperability of different vendors' HDSL equipment. IDSL is based on standards, but not DSL standards. IDSL is based on ISDN standards, and it still suffers from interoperability woes as a result of differing vendors interpreting the ISDN specifications for their type of IDSL. SDSL has also been the subject of many specifications, but not enough to achieve more than a happy circumstance of interoperability.

Finally, the purely vendor-specific forms of symmetrical DSL are listed. MDSL is a form of DSL running from a DLC arrangement that allows for multiples of 64 Kbps service, usually up to 768 Kbps, to be provisioned for one customer. Multiple Virtual Line (MVL) was the invention of Paradyne (now part of Texas Pacific Group), and this technology extends not only to the customer premises, but right to the user device. Both MDSL and MVL are sometimes the subject of a wave of interest, but the emphasis today is firmly on the side of standards-based solutions.

On the asymmetrical side of the DSL family tree, the situation is a little more clear-cut. ADSL/RADSL, G.lite, and VDSL are all the subjects of intensive and extensive standardization efforts. Interoperability is important because the DSL equipment used on the customer premises must be compatible with the DSL equipment used in the local exchange. Most regulators want to give consumers a choice of equipment to purchase, and in open markets there can be no control of buying decisions, by definition. In many DSL cases, interoperability remains a dream, but one worth pursuing.

The vendor-specific asymmetrical DSLs are of more interest here, because these forms aren't mentioned again in this book. Cisco, the major router vendor, purchased a company that made early ADSL equipment and incorporated it into the Cisco product line as EZ-DSL. EZ-DSL was an early form of RADSL, available while most other forms of ADSL required technicians to fix the speeds by hand. Nortel has the 1-Meg Modem, an early form of G.lite designed specifically for use with Nortel central office switches. Both Cisco and Nortel have announced their intention to support the standard forms of RADSL and G.lite respectively, but have been unclear on exactly when and how this transition will take place.

There is another important aspect of the DSL family that should be examined here. Later on in this book, mention will be made of voice over DSL (VoDSL). VoDSL is not a flavor of DSL, it is just a feature of some DSLs that allows the DSL to carry digitized and packetized voice over the DSL link. Some versions of DSL cannot support analog voice directly, so they must add voice support in the form of VoDSL if they wish to carry voice packets along with data packets. (The actual situation with VoDSL is somewhat more complicated, but this simplification will do for now.)

The issue of voice support revolves around the fact that not all DSLs use digital line codes. The sounds very strange at first. How can it be a *digital* subscriber line if the line code is analog and a modem is required? Not many sources are clear on this, and the DSL terminology is too entrenched to change now.

Some DSLs use digital line codes and some use analog line codes. *Analog line codes* require devices that are true modems (digital information to and from analog line codes), and *digital line codes* require devices that are DSU/CSUs (digital information to and from a digital line code). It is only the analog line codes that can support analog voice at the same time that the DSL link carries digital information. The engineering reasons are somewhat complex. Simply note that an access line can run an analog line code or a digital line code, but not both at the same time. An access line can be either analog (voice okay) or digital (VoDSL required), but not at the same time.

If a DSL cannot support analog voice directly, special equipment is required to allow a customer to use their old telephones on the new DSL line. This adds expense and complexity to the DSL link. VoDSL and derived voice will be explored more fully in Chapter 11.

But which DSL is which? Table 6-2 shows whether digital or analog line coding is used with the DSL and so whether the interface device is a modem or a DSU/CSU. The lone coding method is even given, but these will be discussed in Chapter 11. For now, the possibility of voice support and how this voice must be achieved for each of the DSL is shown. Again, the emphasis is on standard forms of DSL.

When a DSL supports voice other than in its analog form, this is called *derived voice*. VoDSL is a form of derived voice, but it is not the only form of derived voice possible for use on a DSL link. SDSL is special because there are many forms of SDSL, and it all depends on the line code that the vendor of the SDSL equipment has chosen to implement. Most newer SDSLs use PAM and require derived voice. There is some question about VDSL because the final line codes for use with VDSL have not been finalized yet. It appears that VDSL might have options for both analog and digital line codes. Early implementations of VDSL have used DMT, however, making analog voice support relatively easy.

After this investigation into the relationship between the various DSLs, the time has come to look at each of the major DSLs in a little more detail.

DSL	Analog/Digital Line Code	Line Code Type Used	Device Is a:	Type of Voice Support
HDSL/HDSL2	Digital	PAM	DSU/CSU	Derived voice
SDSL	Analog/Digital	PAM, CAP, and DMT	DSU/CSU or modem	Analog or derived
IDSL	Digital	2B1Q (4-PAM)	DSU/CSU	Derived voice
ADSL/RADSL	Analog	DMT	Modem	Analog
G.lite	Analog	DMT (subset)	Modem	Analog
VDSL	Analog/Digital	?	?	?

Table 6-2.   Analog and Digital DSLs

HDSL AND T1

Before there was DSL in any form, there was T1 (T-carrier, level 1 multiplexing). T1 started as a local loop digital system to aggregate voice channels and reduce analog noise, but quickly came to be used as a digital trunking system intended to carry many voice calls between switching offices. Because many DSLs are based on at least some components of T1 circuits, a brief review of the basic T1 components is in order.

Because it was designed as a trunking, or "carrier," system, T1 expects some multiplexing to take place. In 1984, after years of internal use in the PSTN, T1 was made available for customer installation. This multiplexed access arrangement is shown in Figure 6-2.

When used to carry analog voice conversations, a channel bank was installed as customer premises equipment (CPE) at the customer site, which was typically owned and controlled by the customer. Up to 24 analog voice inputs (and outputs) could be attached to the channel bank, which digitized each one into 64 Kbps full-duplex. There was then a *Channel Service Unit* (CSU) and *Digital Service Unit* (DSU) to multiplex the conversations and encode the bits for transmission. The CSU/DSU was attached to a *network interface unit* (NIU), which essentially formed the demarcation point between the service provider's network and the CPE.

At the other end of the link, a similar CSU/DSU was attached to the central office channel bank, which, in turn, fed the digitized voice channels into a variety of switching or alternate arrangements.

In between each end of the link, the T1 was carried on two pairs of copper wire. One pair was for sending, and the other was for receiving. Transmission took place at 1.544 Mbps (1.536 Mbps for user traffic) in both directions. Naturally, the T1 might extend for miles from the customer premises to the central office. In that case, special devices known as line repeaters, which "cleaned up" and repeated the digital signal, were placed at intervals along the line. The most common spacing was 6,000 feet (about 1.8 km), which made sense because analog loops typically had H-88 loading coils at 6,000 foot intervals anyway. Therefore, conversion from two analog wire pairs to a single T1 was easy. An outside plant crew removed the loading coils and installed line repeaters. Capacity on the two pairs now went from only two analog conversations to 24 digital conversations. Bridged taps also had to be removed, if present.

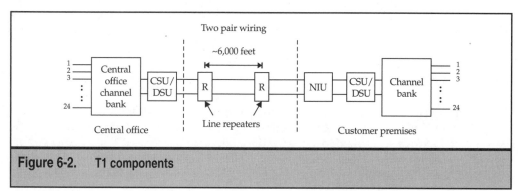

Figure 6-2.   T1 components

T1 was wildly successful for digital access to the PSTN. In fact, the popularity of T1 led directly to the adaptation of HDSL as a way for service providers to deploy T1 circuits more quickly and cost effectively for both internal use and to deliver to customers.

HDSL AS "REPEATERLESS T1/E1"

The T1 multiplexed trunking system formed the basis of the Integrated Services Digital Network. As the ISDN *Primary Rate Interface* (PRI), a T1 provided a business customer with 1.536 Mbps of access to an ISDN switch. In most cases, however, this 1.536 Mbps was provided as exactly twenty-four 64 Kbps channels. As the ISDN Basic Rate Interface, conceptually, a "piece" of a T1 was provided to a residential customer (or small office/home office [SOHO] user) with 144 Kbps of access to an ISDN switch. However, this 144 Kbps was in most cases provided as two 64 Kbps channels and one 16 Kbps signaling/data channel. Usually, the two 64 Kbps channels could be bonded to yield 128 Kbps.

So far, so good; but as analog modem speeds grew closer and closer to 64 Kbps on their own, ISDN speeds—once blazingly fast—seemed relatively slow for many services. Even PRI was most often channelized into 64 Kbps chunks.

Outside of North America and Japan, the PRI was supplied on an E1 and not a T1. The E1 is similar to the T1 in that there are 64 Kbps channels present. However, where the United States T1 supplied 24 channels, the E1 supplied 32 channels, 30 of which could be used for customer information. The E1 line rate was 2.048 Mbps.

Additionally, a T1/E1 for PRI required two pairs of wires and active line repeaters (which "cleaned up" and repeated the digital signal) every 6,000 feet (about 1.8 km). This made widespread use of T1 for residential services too expensive because each circuit installation had to be carefully engineered. However, if a T1 could be made to use one pair of wires or to function without repeaters for longer distances, perhaps T1 (and E1) would have new life as a residential DSL technology. HDSL was designed specifically to meet these conditions.

Bellcore (now Telcordia) proposed HDSL in the mid-1980s to address ISDN channel rate limits and the physical restrictions of T1s for PRI (and BRI, for that matter). HDSL was called a "non-repeatered T1/E1 replacement model" at the time. What Bellcore adapted was a scheme that still used the same 2B1Q (two binary, one quaternary) line code from the ISDN BRI DSL, but was now defined to attain up to 784 Kbps on one pair up to 12,000 feet (about 3.7 km) without repeaters. (As a word of caution, some older HDSL products use other line codes, such as CAP or even DMT. Details on the operation of and differences between these line coding techniques are discussed in Chapter 8.) The 784 Kbps represents twelve 64 Kbps "channels," along with the normal BRI 16 Kbps signaling/data channel.

When the original HDSL was duplexed onto two separate pairs, HDSL could attain 1.544 Mbps in each direction, the same as a T1, but without the 6,000-foot spaced repeaters. Even some bridge taps were tolerated, as long as they were not too long. The 12,000 feet (about 3.7 km) also applied to loops beyond the *remote terminal* (RT) in a *carrier serving area* (CSA) on 24-gauge wire, or 9,000 feet (about 2.7 km) on 26-gauge wire. However, the HDSL access line still needed two pairs of copper wire, which limited the applicability of HDSL to home or SOHO situations, where a second pair of wires was harder to come by. HDSL2 will address this limitation.

The reason that 12,000 feet (about 3.7 km) was so important for HDSL was that at 12,000 feet the CSA rules placed a DLC. So each T1 run to this DLC required at least one pair of repeaters. HDSL removed this requirement, which made the link more reliable and simpler to install.

Initial versions of HDSL at 2.0 Mbps (E1 speeds) just reused the available 784 Kbps chipsets and therefore needed a third pair to deliver the E1 line rate of 2.048 Mbps. Because the two pairs in HDSL ran at 784 Kbps, the third pair was needed to move beyond this speed. Newer versions of HDSL products can actually run at 1.168 Mbps on each pair, making it possible to support E1 speeds. Therefore, HDSL needed no special engineering of the loop with repeaters for distances up to 12,000 feet (about 3.7 km), and the speed is still equal to T1. Confusingly, a technology called "half-duplex HDSL," but really a form of SDSL, attains a speed of 384 Kbps on distances up to 18,000 feet (about 5.5 km).

So HDSL was intended as a way to furnish T1 speeds in a duplex fashion to bandwidth-hungry customers. The HDSL architecture is shown in Figure 6-3.

In the figure, a typical HDSL system is used to provide DS-1 private line services at 1.544 Mbps to two customers. Before HDSL, this would have required the use of two pairs of wires and repeaters every 6,000 feet (about 1.8 km) to provide the classical T1 architecture. With HDSL, the service provider purchases two units, which currently must be from the same vendor, because interoperability has not been a goal or concern before the latest version of HDSL, known as HDSL2. The two units are the *HDSL Termination Unit* (HTU) in the service provider's central office (the HTU-C) and the HTU-R, which is the remote unit placed as close to the customer's premises as possible. The HTU-C is typically a rack-mounted series of units placed close to the central office's wire center or main distribution frame. The lines from the HTU-C are cross-connected to provide channelized or unchannelized T1 service as usual.

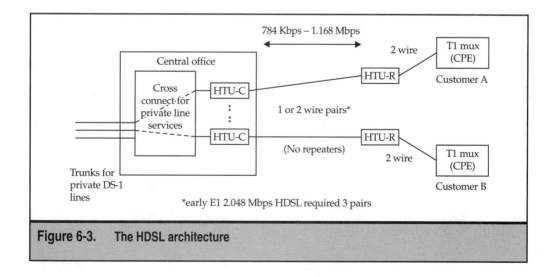

Figure 6-3.   **The HDSL architecture**

The HTU-R unit connects to the HTU-C over a single pair of wires in most cases. Beyond a certain distance, which varies from HTU vendor to vendor, two pairs of wires are still needed. Even so, the advantage of HDSL is that no repeaters are needed. With some older HTU equipment, a third pair of wires was required to provide E1 service at 2.048 Mbps outside of the United States. Newer HTU equipment provides E1 speeds over two pairs, or even one pair over very limited distances in some cases.

The HTU-R is still very much service-provider equipment and not customer-premises equipment. From the HTU-R to the customer premises, two pairs of wires are still needed. The customer-premises equipment is still a common and relatively inexpensive T1 multiplexer (technically, the endpoint of the HDSL link is the CSU). The two-pair interface to the customer is still retained mainly for backward compatibility. The current HTU-Rs are not standardized enough yet to allow customers to purchase just any vendor's HTU-R and expect it to work with the service provider's HTU-C unless the two HTUs are from the same vendor. This will change with HDSL2, but for now this is the case.

The main advantage of HDSL is that it allows the service provider to provision T1 service more quickly and cost effectively. Almost any local loop wire pair will do. No repeaters or special engineering is needed on the line, and one pair of wires should always be available. This simplicity allows service providers to lower the monthly costs of what is essentially still T1 service, but internally provisioned on HDSL.

Note that HDSL does not run to the PSTN switch. HDSL is just a point-to-point private line solution. The other end of an HDSL link is another access line used for private line service. This access line could be a "real" T1 or, in many cases, another HDSL link. It is still a leased line service, not a switched service. The HDSL HTU-Cs and HTU-R are not modems—they are properly DSU arrangements; the same as for T1 lines, which HDSL basically mimics. No analog signals exist on a digital line using 2B1Q unless an extraordinary effort is made to do so. Therefore, a local loop employed for HDSL cannot be used with analog telephones at the same time, except in rare HTU arrangements.

No provision at all in HDSL is made for backward compatibility with existing analog telephone handsets. Presumably, a special digital coding unit would be attached to these telephones so that they could utilize one of the 64 Kbps digital channels. Either that or the analog telephones would be removed and regular office-type digital telephones used in their place. Often, a small user *private branch exchange* (PBX) or *key telephone system* (KTS) would be installed.

Sometimes, especially recently, the single-pair basic 784 Kbps "version" of HDSL has been used at this lower speed (usually called SDSL by various vendors), which extends up to 22,000 feet (about 6.7 km) or even 24,000 feet (about 7.3 km). However, trials have shown that the most reliable 22,000-foot-distance is more like 272 Kbps in most cases. All this may change, however, with the introduction and standardization of HDSL2 (SHDSL). More details on HDSL and HDSL2 are presented in the following chapter.

2B1Q AND ISDN

It might be a good idea to take a closer look at the relationship between 2B1Q line encoding and ISDN. Line codes will become very important in later discussions of ADSL and VDSL, so this is a good place to start. The 2B1Q (two binary, one quaternary) line encoding was intended for the use of the ISDN DSL. 2B1Q is a four-level line code (so known as quaternary) that represents two binary bits (2B) as one quaternary symbol (1Q). As such, 2B1Q is a member of a whole family of digital line codes known collectively as *pulse amplitude modulation* (PAM) codes. The four levels used in 2B1Q make it 4-PAM. So 2B1Q is just a special name for 4-PAM used with ISDN. Other members of the family include 2-PAM, 16-PAM, 256-PAM, and so on.

The 2B1Q line coding was seen as a major enhancement over the original T1 line coding, which was something called *bipolar alternate mark inversion* (bipolar AMI), because 2B1Q encoded two bits instead of just one with every signaling state (baud). This also means that a modern ISDN link running the BRI at 160 Kbps (two 64 Kbps B channels plus a 16 Kbps D channel plus another 16 Kbps overhead) operates at 80 thousand symbols per second (80 kbaud) instead of 160 kbaud. As a point of clarification, it is only on the two-wire local loop (called the U interface in ISDN) that ISDN operates with a 16 Kbps signaling D channel. In all other places, even on the user premises, the D channel is 64 Kbps and the 2B+D aggregate rate is a full 192 Kbps.

2B1Q line encoding was intended to deliver ISDN BRI speeds (144 Kbps, plus the line overhead) through local loops up to 18,000 feet (about 5.5 km). This was done on only one pair of wires, though, and basically gave 144 Kbps full-duplex in each direction using the same frequency range.

So what's the point about 2B1Q and DSL? As it turned out, 2B1Q is not sophisticated enough to achieve multimegabit speeds at long distances, and although 2B1Q required less bandwidth than bipolar AMI, 2B1Q still used the frequency range that analog voice would normally use on a purely analog local loop. Figure 6-4 illustrates this. Note that carrierless amplitude/phase modulation-based (CAP-based) HDSL uses much less of the available spectrum on twisted-pair loops than either bipolar AMI or 2B1Q. Also, CAP, as any other passband modulation method, at least holds out a chance to preserve the 300 to 3,300 Hz passband for analog voice service on the same wires (a later chapter covers CAP in more detail). Not only that, the 2B1Q technology was becoming dated by the mid-1990s. Perhaps for newer DSL technologies, 2B1Q is no longer the best way to go.

Now may be a good time to discuss various line coding techniques for DSL technologies. The familiar 2B1Q code for ISDN is a member of a family of line codes known as PAM codes. To make PAM codes more suitable and efficient for DSL uses, however, it is common to try to "optimize" the PAM code with a technique known as spectral shaping. All this means is that the PAM code should not "go all the way down to 0 Hz frequency," or add a direct current component to the access line. It is important to realize that a PAM code, whether optimized or not, is not a CAP or *quadrature amplitude modulation* (QAM), but something entirely different. PAM is still very much a "baseband" digital line code,

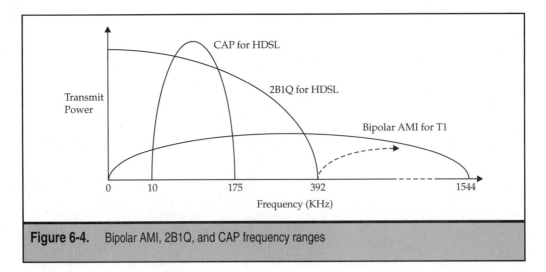

Figure 6-4. Bipolar AMI, 2B1Q, and CAP frequency ranges

although a formal definition of "baseband" in this context is sometimes difficult to put in nontechnical terms.

Every time a new "flavor" of DSL technology is proposed, a great deal of debate revolves around the use of one type or another of PAM, CAP, or QAM as a line code. The important point is that these debates concern three main factors:

▼ Whether the code naturally uses all of the available bandwidth

■ How efficient the code is in terms of speeds and distances

▲ Whether the code is susceptible to outside interference from other line code schemes used in nearby wire pairs

To make things even more complicated, because the ISDN BRI used one pair of wires, a special hybrid arrangement was necessary because the rest of ISDN essentially decreed two-pair wire operation. This operation necessitated the use of special echo canceller devices within the DSL device. (Generally, whenever full-duplex operation is needed with the same frequency range over the same wires, echo cancellation is required.) Note that echo cancellation is required for full-duplex, long distance voice conversations, as well as full-duplex, shared frequency digital links—this applies equally to analog voice, ISDN, or DSL, which is a key point. Whenever the same frequency range is to be used for signals in both directions at the same time, some form of echo cancellation must be used, whether the signal is analog or digital.

Moreover, all modems since the early 1980s have used their own echo cancellation techniques to achieve full duplex operation on the same passband voice frequency range (300 to 3,300 Hz) across the single pair analog local loop.

In any case, it was not the case that HDSL products used 2B1Q line coding exclusively. Other codes, such as *carrierless amplitude/phase modulation* (CAP) or *Discrete Multitone* (DMT), were used as well, and this limited HDSL interoperability severely, of course. HDSL2 (SHDSL) uses two forms of PAM line code, one for initial handshake sequences (to establish interoperability parameters) and the other for bit transfer.

SDSL

SDSL is usually defined as Symmetric DSL, but because HDSL (and several other DSL variations) are also symmetric, this definition loses something and is less than helpful. SDSL started out be taking half of a two-pair T1 HDSL system, but using the same chipsets as HDSL. So SDSL ran at 784 Kbps on a single pair of wires.

Lately, SDSL has begun to be defined as "single pair" HDSL, which is better because SDSL shares much in common with HDSL, and yet, functions on a single pair of wires. (The original HDSL requires two pairs in most cases, and originally needed three pairs in some other cases, such as when supporting E1 speeds of 2.048 Mbps.)

Of course, using multiple pairs of wires for residential service is not an ideal situation. Digitizing the analog local loop would be better if a DSL scheme used the existing single pair of wires—hence, SDSL.

Once the SDSL idea caught on, vendors started to get very creative. Some vendors made versions that operated faster (1.5 and 2.0 Mbps), but with limited transmission distance. Other vendors made versions that operated at lower speeds (384 Kbps), intended to support longer distances. This lower speed SDSL version is sometimes known as MDSL, but this is not a generic DSL at all.

In spite of the popularity of HDSL and the much anticipated HDSL2 standard, SDSL might still have a place in the DSL world. Figure 6-5 shows an SDSL system used in a pairgain system.

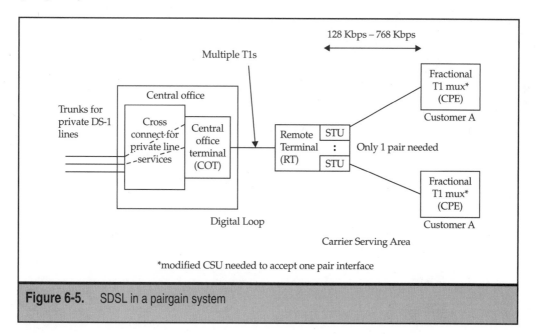

Figure 6-5. SDSL in a pairgain system

In the figure, a central office terminal and remote terminal are connected by multiple T1 links to form the common carrier serving area architecture. Two pairs of wires normally would need to be run from the RT to the customer to support the T1 speed of 1.544 Mbps. The most common RTs, however, support only four T1s themselves, so the architecture is limited if the whole idea is to deliver full T1 rates to each customer. But in a SOHO or residential neighborhood, perhaps just a fraction of a T1 is okay, as long as the price is a fraction also. Maybe the customer could make do with 256 Kbps (four 64 Kbps channels) or even 128 Kbps (two 64 Kbps channels)—the T1s servicing the RT would not be exhausted so readily. This is what SDSL and its variations are for.

Note that there is really no such thing as an "STU," at least not as a standard device. Whatever it is called, the STU is housed at the RT. Only one pair of wires is needed to deliver services to the customer site, which needs a modified fractional T1 multiplexer (technically, a modified CSU) to accept a one-pair interface. Also, "half-duplex" translates to 128 Kbps in some cases, and up to 768 Kbps in others, over only one pair of wires (hence the term "single pair"). The 128 Kbps represents the bandwidth of two 64 Kbps channels. The speed of 768 Kbps represents the bandwidth of twelve 64 Kbps channels.

SDSL allows a service provider to provision DSL service based on three basic parameters: the cost, reach, and speed of the service. Based on performance needs, distance from the local exchange, and budget considerations, customers can choose from a number of the SDSL options. Service providers typically price different services on a staggered scale.

The most common currently supported speeds and distances of SDSL are shown in Table 6-3. This table holds a key as to why SDSL has been so often treated poorly by standards organizations and some telephone companies. If a signal is stronger in terms of power, the signal will carry farther and can be used to carry more information. In old adage of "go slower to go farther" it is assumed that the transmitter power remains the same. But to go farther at the same speed, just turn up the power. Many SDSL implementations did just that: they boosted the transmitter power at each end to achieve very high speeds at very long distances. The problem was that the increased power caused all sorts of interference in the form of crosstalk noise in any wires that happened to be nearby,

SDSL Data Rate	Maximum Distance
128 Kbps	22,000 feet (6.71 km)
256 Kbps	21,500 feet (6.56 km)
384 Kbps	14,500 feet (4.42 km)
768 Kbps	13,000 feet (3.97 km)
1,024 Mbps	11,500 feet (3.51 km)

Table 6-3.   SDSL Speeds and Distances

even other SDSLs! In a common 50-pair bundle of access lines, one powerful SDSL link could cause problems for all of the other services in the same bundle, from other DSLs to analog voice. So SDSL got a kind of bad reputation among the telephony service providers.

As it turns out, SDSL interference was sometimes exaggerated and occurred only at high SDSL speeds (more than 1.5 Mbps). But such SDSL links were rare. But even the possibility of SDSL interference was apparently enough to keep SDSL out of the international standardization process.

SDSL should continue as a common variation of HDSL running at speeds below 1.5 or 2.0 Mbps, even as HDSL2 (SHDSL) appears.

ENTER ASYMMETRIC DSL (ADSL)

Asymmetric DSL addresses some of the limitations that HDSL, HDSL2, and their variations imposed on the newly digitized local loops.

First and foremost, HDSL, SDSL, and the others rarely made allowance for analog voice. (However, nothing prevents vendors today from supporting analog voice, along with techniques like SDSL, especially with line codes other than digital line codes such as PAM, and in fact, some do.) Most people had, and have, analog telephones in their homes, yet HDSL, HDSL2, and its variations carry only digital signals, so pure HDSL or SDSL required users either to purchase special conversion units (known as terminal adapters, or TAs) or to purchase digital telephones. However, neither alternative appealed to users or to the telephone companies. Maybe some way could be found to allow the continued use of analog phones on the newly digitized loop with less expense.

Secondly, a concerted effort was made on the part of the telephone companies in 1992 to deliver digitized video (and the accompanying audio) to the home. Many technologies were explored, and the attraction of ADSL at the time was the promise of delivering these services over the same loop used for analog voice, so ADSL could form a basis for video-on-demand services and so-called "video dial tone" systems that were heavily promoted at the time. Such video service required large amounts of bandwidth downstream (that is, inbound to the home), yet not much bandwidth was needed upstream (that is, outbound from the home). After all, brief commands to start, stop, fast forward, or freeze a video stream were small data packets in the first place.

It also turned out that many home-based activities followed this asymmetric model. Internet Web server access, normal client/office server actions, and even home shopping are inherently asymmetric in the same fashion. Any DSL technique that supports larger bandwidth in one direction and smaller bandwidth in the other is, by definition, an Asymmetric DSL—ADSL is just that.

For a while, it seemed that ADSL would become an "umbrella" term for a variety of asymmetric DSL techniques, including RADSL and VDSL. However, the terminology seems to be moving back toward a more uniform and specific approach to individual DSLs. Oddly, the ADSL Forum Web site (www.adsl.com) is now identified as the DSL Forum on its main Web page and has information about many DSL technologies, not just ADSL.

More detailed information on ADSL is presented at length in Chapters 8–13. For now, it is enough to position ADSL within the DSL technology family. Figure 6-6 shows the general ADSL architecture.

In the figure, the two distinguishing features of ADSL compared to other DSLs are apparent. The splitter is a device that comes between the local exchange and the customer premises; its function is twofold. First, the splitter allows existing analog voice telephone and other equipment, such as fax machines, to continue to operate as before on the customer's premises. Second, the splitter allows the long holding time data traffic to be rerouted around the PSTN voice switch (where it is carried on circuits) onto an IP router or ATM switch network (where this data traffic is carried in packets). This alleviates pressure on the PSTN and cuts down on user costs because not all customer equipment needs to be changed or interfaced with special adapters (as with HDSL and others). The routers or ATM switches carry user traffic to servers presumably located on the Internet or a corporate Intranet, although many other variations are allowed and envisioned.

Next, ADSL is asymmetrical. The downstream speed is much greater—sometimes ten times greater—than the upstream speed. A maximum speed downstream of 8.192 Mbps is defined for ADSL. However, this speed may be quite difficult to achieve in practice, not due to ADSL limitations, but due more the limitations in throughput given the current Internet architecture and backbones. This would be a little like having a 100-mile-per-hour on-ramp to a 55-mile-per-hour freeway. So for the foreseeable future, most equipment vendors and service providers would settle for 1 Mbps to 4 Mbps as a maximum.

Also note that the ADSL link is not switched in and of itself. In other words, the ADSL link forms another type of leased private line from a user's PC or LAN to one other place in the world, but while the ADSL link is not circuit switched, the content of the ADSL link is packet or ATM cell switched. That is, if the service-provider end of the ADSL link ends at an IP router or ATM switch connected to the Internet, the traffic on the ADSL link may still be able to find its way almost anywhere that the switched telephone network does today.

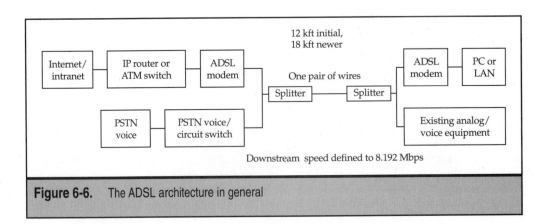

Figure 6-6.   The ADSL architecture in general

RATE-ADAPTIVE DSL (RADSL)

Rate-Adaptive DSL addresses a possible limitation of some early ADSL devices, especially those based on carrierless amplitude modulation/phase modulation (CAP). Basically, once the first generation of ADSL equipment was in place on a formerly voice bandwidth analog local loop and connections were made, the newly digitally enabled line operated at a fixed speed upstream and downstream for the duration of the link's lifetime. The downstream speed especially might vary from location to location and wire pair to wire pair, typically in increments of 32 or 64 Kbps. One of these limited ADSL lines on a block could achieve 640 Kbps, for example, but a neighbor's might be limited to 608 Kbps or even 576 Kbps. The speeds had to be set by technicians working by hand at each end of the line, usually by trial and error. This made early ADSL installation a time-consuming, complex, and expensive endeavor.

The problem is that line conditions on local loops vary all the time. Line conditions may improve or deteriorate, usually depending on rain conditions, or even as the solar radiation on, and overall temperature of, the wire rises and falls seasonally or from day to night.

RADSL is theoretically able to adapt to these changing conditions on the fly, even during active sessions, although there is currently no provision to do so in current RADSL products. For example, users might receive traffic that starts at 576 Kbps in the morning, goes to 640 Kbps in the afternoon, and then drops to 608 Kbps in the evening. But at least RADSL-enabled devices can find their own maximum speed level when they are turned on without a technician.

The whole concept is similar to the idea of "self-equalizing" modems. Modems of 20 years ago needed line conditioning to maximize performance, which was relatively easy to do for leased private lines, but next to impossible to guarantee on switched dial-up connections. A key parameter in line conditioning was equalization, which balanced attenuation over the whole frequency spectrum used. When newer modems could perform self-equalization, it was possible for modems to "drop back" to lower speed operation if line conditions were poor (which is why even new 33.6 Kbps modems still connect at 28.8 or 14.4 Kbps). After the 14.4 Kbps connection was made, however, that is where it stayed, even if the connection lasted hours and line conditions improved. RADSL speeds might someday be able to vary on the fly, moment by moment.

All of the other properties of RADSL essentially mimic ADSL in terms of maximum speeds and distances. RADSL is a natural progression of ADSL, and all ADSL equipment is essentially RADSL-capable today. For equipment based on Discrete Multitone (DMT), RADSL is an inherent capability that appears as a result of the way the technique functions. Although RADSL operation is not impossible in CAP-based ADSL, such operation is difficult to achieve and adds a lot of circuitry and procedural overhead to the CAP devices. The biggest difference is that with CAP, the signal spectrum changes. Nevertheless, several CAP-based ADSL vendors have introduced RADSL equipment.

The basic architecture of a RADSL link is exactly the same as the basic ADSL architecture. When first installed, the RADSL devices find their optimal speeds based on current line conditions. However, if conditions improve (a wet line dries in the sun), RADSL will go no faster. And if line conditions deteriorate (a storm comes through), RADSL will fail if there are too many errors at the line speed. But when the devices see each other again after a failure, the RADSL speed-setting procedure is repeated. Some early RADSL users were not above turning their devices off and then on again on sunny days to force this retraining of the devices and see if their line speeds improved.

G.LITE: ADSL/RADSL WITHOUT THE SPLITTER

Early trials with ADSL and RADSL uncovered a rather serious situation regarding the customer's premises. ADSL/RADSL required the installation and maintenance of a remote device—the splitter introduced earlier in the chapter. The premise's splitter's main function was to allow the continued use of existing analog telephony and faxing devices in the home or SOHO location. However, besides introducing complexity, the presence of the splitter also raised issues about the premise's wiring and configuration, issues that are detailed in a later chapter. For now, it is enough to point out that there was a real concern about the continued care and feeding of the remote splitter device.

The first concern involved the need for the service provider to make an appointment with the customer for splitter and possibly wiring installation. This may not sound like much of a concern, but in today's highly mobile world and everybody-works-who-can environment, finding someone at home during business hours can be a chore. A customer often had to take a day off from work, and missed installation appointments lead to complaints about missed work time on the part of the customer, and so forth. A further concern was the need to dispatch a truck and technician, the "truck roll," which added considerable cost to service initiation and slowed deployment. In many cases, the remote splitter was provided as part of the service, adding to the service provider's capital costs. The associated wiring issues only added to the service delay, expense, and complexity. Clearly, if a way could be found to install and configure ADSL or RADSL speeds and distances, and at the same time support existing analog devices without the need for the remote splitter, this would be a very attractive alternative to pure ADSL/RADSL.

Near the end of 1997, several vendors proposed a variation to full ADSL/RADSL that came to be called G.lite. In fact, the only significant difference between ADSL/RADSL and G.lite besides the absence of the premises splitter and wiring concerns is a restricted operating speed range (most importantly, 1 Mbps downstream as opposed to about 8 Mbps with ADSL). By the end of 1997, Nortel, Microsoft, Compaq, and Intel had all made G.lite support announcements.

The local exchange (central office) side of the link is unchanged. That is, a splitter is still needed there to separate high-speed data packets from voice conversations if analog voice support is included (some forms of G.lite use a second line and support only data applications). The local exchange splitters are housed in the same type of equipment as in the ADSL/RADSL architecture. The big change is on the premises. The overall architecture of G.lite in shown in Figure 6-7. The remote splitter is eliminated.

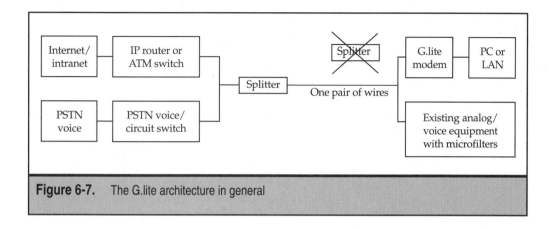

Figure 6-7. The G.lite architecture in general

G.lite has a number of characteristics that make it attractive both to customers and service providers. The major ones follow:

▼ **Adequate access speeds for most users**   Faster Internet access might have the effect of slowing down other portions of the network. This "moving bottleneck" could be a serious problem with many users expecting data at 6 to 8 Mbps downstream. Can servers and the Internet backbone keep up? G.lite adds only 1 Mbps at a time to the load, which would still be more than enough for what users currently expect on the Internet and Web.

■ **Easy installation at the premises**   Perhaps this should be #1. With G.lite, there is no need to arrange an appointment for installation, worry about wiring, supply a remote device, or be concerned about future tech support. G.lite service providers are concerned as much about the premises as telephone companies are concerned about user's modems, which is to say minimally.

■ **Simultaneous voice and Internet and Web access**   In spite of the absence of the remote splitter, it is important to point out that G.lite still allows simultaneous voice telephone calls and Internet and Web access over the same local loop. There is no requirement for "alternating" use, as with present modem versus telephone arrangements on a single local loop.

■ **Low cost to customers**   Without the need for added splitter electronics, G.lite modems should be less expensive than ADSL or RADSL remote modem devices, whatever the packaging. Costs should be no more than current prices for high-end modem packages running at 56 Kbps. It should never be overlooked that customers often mentally add the price of required equipment to the cost of a service (they are technically separate) when deciding whether a service is too expensive. G.lite minimizes this impact.

■ **Low provisioning costs**   Even with the use of G.lite, a service provider must still install equipment in the local exchange (central office). However, the cost of the equipment can be offset by economies of scale, and the fact that the installers and management are local. The advantage of G.lite is that there is no need for a "truck roll" to the premises or concerns about managing remote devices scattered all over a service area.

▲ **Standards initiative**   The DSL vendors have had a high degree of success attracting service providers and equipment vendors to the G.lite architecture. The ITU now has a standard for G.lite or splitterless DSL.

In view of the interest in G.lite, we should take a final summary look at this promising DSL technology. G.lite modems will not replace 56 Kbps or other modems, but they will offer a higher-speed evolutionary path with fewer service provider concerns than any other DSL solution. However, G.lite is still DSL and, as such, is not just a seamless up-grade from 56 Kbps modem operation. G.lite is still dependent on line conditions for maximum speeds (1 Mbps downstream), but the more modest maximum should allow the maximum to be reached in more varied circumstances than ADSL/RADSL. More importantly, G.lite is not intended as a *replacement* for ADSL/RADSL at all. G.lite's intention is to *complement* service providers' ADSL/RADSL deployments into places where higher speeds are not particularly mandated or feasible, and where there are concerns about remote splitter and wiring installation. ADSL/RADSL are still expected to be popular services in and of themselves.

G.lite still requires the same equipment arrangements as ADSL/RADSL in the local exchange or central office. Eventually, G.lite could be put directly into the voice switch port itself, but this option is available to ADSL/RADSL as well. Because G.lite is a variation of ADSL/RADSL, there should be minimal issues associated with tariffs or contracts for G.lite service. G.lite will not see ADSL/RADSL as much of a "rival" as other technologies altogether, such as cable modems or even 56 Kbps modem products.

VERY HIGH DATA RATE DSL: NEWER AND BETTER?

VDSL is an effort to address at least three issues with regard to broadband services delivered over the local loop. First, the telephone companies are employing more and more fiber feeders, as pairgain and digital loop carrier (DLC) systems. It is not unusual for a neighborhood, especially new housing developments or condominium complexes, to be served by fiber optic feeders, with copper local loops handling only the last few thousand feet to a home. Second, the services that the telephone companies and others want to provide seem to require more and more bandwidth, almost year to year. Third, as DSL systems evolve to include more mixtures of voice/video/data traffic all in one, some accommodation for *asynchronous transfer mode* (ATM) cell transport may be desirable because ATM easily combines voice/video/data services on the same physical network. Although it is true that ADSL is also aimed at ATM networks, but accommodates IP as well, VDSL as currently planned will virtually demand ATM networks.

Figure 6-8 shows how the basic VDSL concept addresses these three issues. First, VDSL includes an optical network unit to convert and concentrate VDSL signals onto a fiber feeder system, which might be part of the "next generation" of DLC systems (called

NGDLC). These NGDLC systems are distinguished by a great deal of distributed intelligence, links to Sonet fiber-optic ring configurations instead of RT to COT point-to-point digital links, and easy access to broadband services at the central office.

Second, the bandwidths downstream are far above those defined for ADSL. Speeds of around 13 Mbps will be achievable for 4,500 feet (about 1.4 km), and 1,000 feet (about 0.3 km) of copper supports a whopping 50 Mbps or so. Upstream, VDSL offers a minimum of 1.5 Mbps and could even be deployed in symmetric configurations.

Third, VDSL is intended to carry ATM cells; that is, VDSL is intended to form a Physical Layer for a full service ATM network. In this case, the home VDSL hub is actually an ATM switch, and the services are on a variety of ATM-attached servers. Other modes of operation are allowed, however. The combination of ATM cells transported on VDSL links is intended to be used to support *switched digital video* (SDV) systems. Of course, a wide range of additional services could be offered as well, including very high speed Internet and Web access.

VDSL is sometimes portrayed as an evolutionary step up from ADSL. For now, VDSL viability seems tied to fiber availability and ATM popularity.

Note that VDSL still offers backward compatibility with existing analog phones via a splitter, as does ADSL. Also, VDSL fully expects there to be multiple devices on the customer premises, in many variations. These are all *terminal equipment* (TE) to VDSL and can include PCs, LANs, television sets, and even refrigerators or air conditioners.

So VDSL is more of a "full service network" strategy for service providers. ADSL is more of a "data overlay network" strategy. This is not to criticize either VDSL as a pie-in-sky dream or ADSL as a half-way measure. This just reflects what each is really intended for.

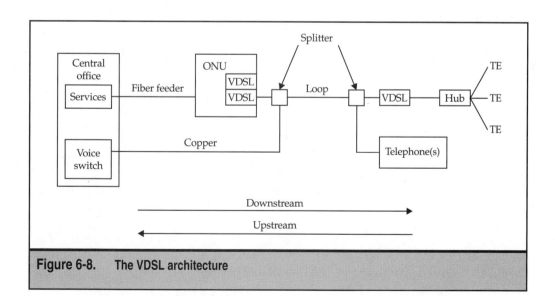

Figure 6-8.   The VDSL architecture

IDSL AND THE INTEGRATED SERVICES DIGITAL NETWORK (ISDN)

What does the future hold for the Integrated Services Digital Network (ISDN) in a DSL world? ISDN was intended to digitize the analog local loop many years ago and formed the very first DSL. Is there still a place for ISDN?

It is worth remembering that the whole movement toward -DSL began with the bit-rate limits that channel structures and Basic Rate Interfaces (BRI) imposed on digital loops. The need for higher bandwidths, especially for video services, led first to HDSL and then to ADSL and VDSL.

Now, ISDN channels could be grouped, or bonded, together to provide higher bit rates, and so 768 Kbps is definitely possible on the Primary Rate Interface (PRI) (but not the BRI). It was the search for a better (more cost-effective) way to provide PRI bandwidths on analog local loops that led to HDSL in the first place.

The whole problem with ISDN is not BRI or PRI speed limits. It is the fact that ISDN still runs through the voice switch. Here is where trunking and switching tie-ups become serious. After all, these are not just voice services anymore; and when we consider that the upgrade to every central office switch costs about $500,000 to convert to ISDN, ISDN looks more and more like a technology people could live without.

The end result of many years of ISDN conversions and service offerings have been spotty availability (sometimes literally by street address number), price worries (will it go up? down? stay the same?), and continued trunk- and switch-blocking woes.

The history of ISDN deployment in the United States has not been a happy one. However, its future is not necessarily bleak. ISDN consists of two things: a *digital network* (DN) and *integrated services* (IS). In most residential or SOHO applications, the DN is supposed to be a digital subscriber line running at 144 Kbps (BRI) using 2B1Q (two binary, one quaternary) coding. In addition, the ISs are supposed to be switch-based. The real attraction of ISDN was to be in the integrated voice, video, and data services.

However, in some places, ISDN DSLs for BRI service were often just used as 144 Kbps "bit pipes." That is, the two B-channels were permanently bonded and used like a 128 Kbps "leased line" running to the switching office. The 16 Kbps signaling channel was still present, but usually messages went into the "bit bucket." At the central office, the line ended up on a cross-connect that led to other DSLs at other sites. Therefore, there were no real ISDN services involved or available.

Why bother with the ISDN DSL then? Well, it turns out that this was a common scheme in many areas of the world outside of the United States for carriers to "bootstrap" themselves into the digital leased line marketplace without running new cable. (Leased lines were scarce and expensive outside of the U.S., mostly for economic and policy reasons.) This plan allowed the use of standard ISDN DSL gear to digitize the loop, and that was all that was needed to supply a Digital Data Service (DDS). The plan required no ISDN switch software upgrade and was very popular in South America. How could these ISDN DSLs be converted to -DSL service architectures?

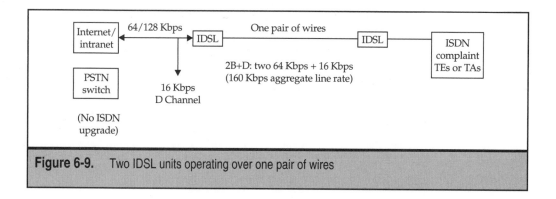

Figure 6-9. Two IDSL units operating over one pair of wires

Perhaps DSL standards could be redefined to use ISDN DSL, and use the newly planned DSL services (such as video and Internet access) as the "IS" part, optionally of course. This idea makes a lot of sense and is the plan behind what is sometimes known as IDSL (ISDN DSL over DSL). Such a process would require a modest standards rewrite, but this rewriting goes on all the time anyway.

Figure 6-9 shows two IDSL units operating at 160 Kbps over one pair of wires. The line is organized into normal BRI channels as 2B+D service at 144 Kbps aggregate. At the premise's end, the user can keep or purchase any ISDN-compliant TAs or TE equipment for excellent backward compatibility. At the local exchange, all D-channel signaling messages are ignored because IDSL is still essentially a private line service. This should not be a limitation because most residential users currently employ ISDN to access the Internet or a corporate intranet. Either one is reachable with ISDL.

Another advantage is that the new DSL services available through IDSL would no longer require the conversion of a central office switch to ISDN (at about $500,000 per switch).

DSL ADVANTAGES

Now is the time to examine the advantages that DSL has with respect to the other access technologies explored in the last chapter. The attraction that DSL technologies have for service providers is summarized here.

First of all, DSL goes in only when a customer requests service. A service provider need not spend millions of dollars and then wait for customers to sign up. Initial costs are expected, of course, but they are generally much lower than other competing technologies.

Any DSL requires no change to central office switch software. In most cases, a splitter carries normal analog voice into the switch, but all other services are handled through separate servers and routers.

Also, DSL can be used for residential users, SOHO users, and large organizations alike. The DSL technology may be different (HDSL, for example), but the service should be essentially the same, with the possible exception of streaming video services.

Another nice thing about DSL is that some versions, especially ADSL/RADSL and VDSL, can interface with a number of different premises arrangements. Individual set-top boxes and PCs are supported, as well as entire home LANs, such as Ethernet. Even newer electrical wiring LAN schemes are allowed at the home end of an ADSL/RADSL or VDSL line.

DSL will even provide an infrastructure for asynchronous transfer mode (ATM) cell transport (especially Very high bit rate DSL, but Asymmetric DSL also). This is important because ATM, in turn, forms the basis for the international standard set of broadband services known as *Broadband ISDN* (B-ISDN). It is hard to think of many other technologies accommodating ATM as well as DSL is able to, especially ADSL/RADSL and VDSL.

Finally, DSL is not a future technology—it is available here and now!

COST COMPARISON OF VARIOUS BROADBAND AND OTHER TECHNOLOGIES

Before examining the more important members of the DSL family in more detail—especially ADSL—we will take a brief look at comparative prices. Some preliminary pricing information is given in Table 6-4. This is a general attempt to allow prospective customers to get an idea of how competitive the pricing will be for the newer broadband access services.

Table 6-4 was compiled from prices appearing in industry sources, such as *Network World*, *Communication Week*, and so on. These are only generalizations; particulars vary widely. Some details on speeds have been sacrificed for brevity. Prices may drop rapidly for new technologies; these are from mid-2001.

The table starts by looking at the price of a 64 Kbps DS-0 link, like that commonly used to link routers or other network devices in a purely private line network. The cost is modest, but so is the speed. The other costs involved are minimal, but the price rises with distance.

Above the DS-0 speed in many parts of the world, it is possible to use ISDN links configured according to the Basic Rate Interface (BRI). This provides two B channels running at 64 Kbps each and a 16 Kbps D channel, which can be used for signaling and other purposes. Usually, the B channels can be bonded together to form a single channel running at 128 Kbps, but above and beyond the monthly recurring charge of about $185, the user must purchase an ISDN terminal adapter (TA) for about $100 to $300. The less expensive units are just PC boards, and the more expensive TAs are standalone units for voice and data use. Also, the service provider must upgrade the switch software and other hardware, which can cost up to $500,000.

A DS-1 runs at an impressive 1.5 Mbps but can cost anywhere from $1,000 to $2,000 per month, depending on distance between end points. The equipment needed at each site will also cost the user about $500. These links are also commonly used to link routers or other network devices in a private line network. Note that this entry does not reflect the impact of HDSL on the pricing of the DS-1. That is, this entry is basically a "traditional" T1 running on two pairs of wires using bipolar AMI line coding equipment with repeaters.

Technology	Speed(s)	Monthly Cost	Other Costs	Comments
DS-0	64 Kbps	$150 average	Minimal	Router connections
BRI (2B+D)	128 Kbps	$185 average	$100–300 average for TA	$500,000 switch upgrade to ISDN
DS-1 without HDSL/HDSL2	1.5 Mbps	$1,000–2,000	$500 average	Router connections
PRI (23B+D)	1.5 Mbps	$1,300 average	$350 average for TA	$500,000 switch upgrade to ISDN
HDSL (HDSL2)	1.5 Mbps	$175 ($60–120 from RBOCs)	$1,200 for equipping line	New DS-1?
ADSL	500 Kbps+	$40+	$250 average ATU	Cost going down
G.lite	500 Kbps+	$40 average	$250 average	No remote splitter
IDSL	128 Kbps	$200 average	$100–300	Preserves ISDN for TA CPE
Cable modems	1–10 Mbps	$25–40	$200–500	Cost going down
LMDS	50 Mbps	$50	$1,000 CPE target	2-way? Video also?

Table 6-4.   Broadband Access Cost Comparison

Running at the same speed as a DS-1, the ISDN Primary Rate Interface (PRI) consists of twenty-three 64 Kbps channels and one 64 Kbps D channel for signaling and other purposes. The PRI is just a better BRI in some sense, although it is usually harder to bond the B channels together for higher speeds. The $1,300 monthly cost is high, but the $350 for the ISDN terminal adapter (TA) is not too bad. However, many TAs are usually needed. Of course, the same ISDN switch upgrade is still necessary.

HDSL runs at the same speed as the DS-1 and ISDN PRI. The cost is typically much less than a "pure" DS-1, however—usually about $175 a month. Some former Bell system companies even offer HDSL at $60 to $120. It does cost up to $1,200 to equip a line for HDSL, however, mainly due to tighter physical requirements. HDSL may turn out to be the "new" DS-1 because it looks exactly the same to users. Of course, the temptation is to continue charging DS-1 prices for HDSL provisioned links. This situation is discussed in more detail in the next chapter.

ADSL can run anywhere from 128 Kbps to 6 Mbps and even up to 8 Mbps downstream to a home, but slower upstream. The table lists the typical downstream speed. The $40 (or more) monthly cost is based on the fact that service providers expect users to pay more for more bandwidth. The relatively expensive $250 cost for the *ADSL termination unit* (ATU) or DSL modem that each user must purchase is sometimes bundled with the service, but not often. Some ADSL service providers have astonished the industry by offering ADSL service for as low as $20 a month, but this service includes many restrictions, such as lack of "always-on" access.

G.lite shares many of the properties of ADSL, but runs at a lower top speed (1 Mbps). In practice, the speeds are usually the same as full ADSL. These DSL modems require no splitter, but costs have been about the same as ADSL. In fact, many customers have G.lite and are not even aware of the fact. Prices are $40 per month and up (again depending on speed) for the G.lite service, and about $250 for the G.lite DSL modem itself.

IDSL is essentially "ISDN without the switch." It provides the same 128 Kbps bonded B channel speed as the ISDN BRI, but does not require any expensive local exchange switch upgrade or changes to the local loop. The monthly charge is targeted at $200, and metered service might add a lot to the total cost of IDSL. Because IDSL appears to the user as a type of ISDN link, the same ISDN TA equipment can be used, which costs about $100 to $300, again depending on the exact type of TA. However, some service providers require a special TA.

Cable modems are a popular alternative to DSL techniques, even among the telephone companies for some strange reason. Unfortunately, in most implementations, cable modems are a sort of half-way solution. That is, cable modems often still need a normal modem telephone connection to function in the upstream direction to the service provider. The 1 to 10 Mbps is usually only on the cable TV system in the downstream direction. However, the faster speed is typically seen as just another premium cable TV service and is priced from only $25 to $40 per month. The $200 to 500 for the cable modem is a drawback, but prices will fall rapidly, as well. Finally, the *local multipoint distribution system* (LMDS) often described as "cellular cable TV" is the latest entry in the broadband access sweepstakes. LMDS offers an astounding 50 Mbps, but typically only downstream to the home. The targeted monthly price of $50 is unsurpassed, but again, it is only a target. Even the customer premises equipment price is targeted in the $1,000 range. LMDS may yet turn out to be two-way and be as popular as cable TV for general video services. However, the auctioning of bandwidth to service providers has been very expensive, and the long-term future of LMDS as a broadband access technique is not a given.

CHAPTER 7

HDSL and HDSL2

The original digital technology that gave the world T1 (1.5 Mbps) and E1 (2.0 Mbps) was used as a trunk technology for many years. On a T1, trunks could carry 24 voice channels on two pairs of copper wire. On an E1, trunks could carry 30 voice channels on the same two pairs. Throughout the 1970s, both T1 and E1 were used almost exclusively within the network, and customers could only get 64 Kbps DS0 digital channels at the user interface.

However, both T1 and E1 have been offered for "last mile" (12 kft, 2.3 miles, 3.6 km) access line applications since the mid-1980s. The divestiture process spurred offerings in the United States, and user pressures around the world for higher speed access spurred deployment elsewhere. Usually, the T1 or E1 offered users 24 or 30 channels at 64 Kbps each, but unchannelized versions ran at a full 1.5 Mbps (1.544 Mbps to purists) or 2.0 Mbps (2.048 Mbps).

However, until recently, fully 80 percent of existing T1 and E1 circuits were still provisioned on two pairs of copper wire. This is in spite of alternate technologies such as wireless, coaxial cable, or, lately, fiber optics based on SONET or the *Synchronous Digital Hierarchy* (SDH).

The persistence of two-pair copper T1 and E1 is mostly due to few premises fiber interfaces, but lots of copper interfaces. Computers and PBXs expect some form of copper wire to interface with their serial or parallel ports, and it has been expensive to use anything else, even today. Even when the T1 or E1 is provisioned on a SONET or SDH or some other form of fiber ring such as FDDI, the last 1 km or half-mile or so is still almost always two copper wire pairs. It makes no real difference whether the T1 or E1 is intended for a customer's private networks or used for an access interface to a public network service like ISDN; the link is still a T1 or an E1.

THE TROUBLE WITH T1/E1

In spite of its popularity with service providers and customers, both T1 and E1 have some real drawbacks that are mostly the result of the age of the technology upon which both are founded. Although improvements have been made to both over the years, in many respects T1 and E1 remain very much 1980s technologies with features that date back to the 1960s in some cases.

Many of these limitations revolve around the repeaters. In most T1 and E1 installations, repeaters are used every 1 km or 4 to 6 kft to regenerate the signal. This had to be done at the time because the electronics affordable then were not as sophisticated as they are today when it comes to recovering weak signals.

It was mainly the repeaters that made the provisioning of a T1 or E1 so labor-intensive to design, install, and condition the line. Loading coils had to go, the repeaters went in, and bridged taps were taken out as well. Mixed wire gauges needed to be avoided if at all possible for maximum performance, and so on.

Without careful attention to the line characteristics and electrical parameters, the T1 or E1 would not work, so the entire process of provisioning a T1 or E1 could take weeks (or months in some cases) to complete. It was not uncommon for projects to be delayed for weeks at a time due to the unavailability of the links between sites.

Repeaters were a major headache to the service providers because they were numerous (most lines needed at least one pair of repeaters: one in each direction on each wire pair), unsophisticated (they had to be affordable), hard to troubleshoot (none ran network management software), and difficult to maintain (many were in conduits or buried embankments).

It took a while, but the bright idea of the late 1980s was, "Let's try to get rid of the repeaters!" Advances in electronics made this not only feasible, but actually more efficient.

HDSL IS BORN

Naturally, nobody meant for T1 and E1 to have any drawbacks at all. Both were considered state-of-the-art technology. The trouble is, the time for the core T1 and E1 technology was the early 1960s! In fact, only the familiar EIA-232 serial interface is as old as T1 and E1 when it comes to common standards in the telecommunications field.

But the end electronics have come a long way since the 1960s. Consider the changes in the desktop PC since 1990 alone. End telecommunications electronics today can take advantage of increased processing power, the low cost and availability of a lot of memory, and overall advances in *Digital Signal Processing* (DSP) chipsets, which make it easy to massage bits almost any way one wishes to.

The telecommunications philosophy beginning in the 1980s was more along the lines of "don't adapt line conditions to end electronics; adapt end electronics to line conditions." After all, analog modems have used this concept for years, especially since around 1982. Modern modems do their own equalization across the frequency range (called *self-equalization*), instead of relying on a technician to do it for the whole line (difficult on a dial-up!). Echo cancellation to minimize self-crosstalk (near-end crosstalk or NEXT) circuits were added, and so on.

Now apply this intelligent end device approach to T1 and E1. The first result was High bit-rate DSL (HDSL). HDSL needs no repeaters (usually) or special line conditioning to reach the digital loop carrier (DLC) or remote terminal (RT) location at 12,000 feet (about 3.7 km). Even some bridged taps are okay on HDSL links in most cases, as long as there are no more than two and their lengths are limited.

The nice thing about HDSL is that in addition to lowering provisioning costs, HDSL makes copper "look like fiber" in terms of performance, which means that the reliability of the link and bit error rates are much better than copper with T1 or E1. This reliability is a real plus when there is more fiber than ever within the network, but little directly to the customer premises.

HDSL FOR T1

When HDSL is used to provision T1s for customers, what does the customer see? A T1, or, technically, a DS1 running at 1.5 Mbps, of course. It may be HDSL inside, but the label still says "T1." Figure 7-1, based on the original Bellcore (now Telcordia) HDSL documentation, shows why this is so.

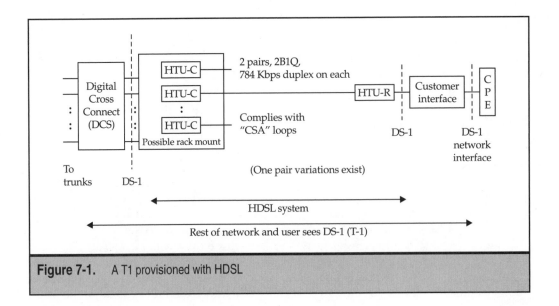

Figure 7-1. A T1 provisioned with HDSL

Note that T1 leased private lines do not go through the service provider's circuit switch. Private lines are routed through a *digital cross connect* (DCS) or *digital access cross connect* (DACS) out to a trunking network. In its simplest form, the DS1 input to the DCS at the local exchange does not come directly from a T1 link, but rather from a special card called an *HDSL Termination Unit – Central Office* (HTU-C). The HTU-Cs may be individual units or gathered into a rack or unit. The advantage is that several HTU-Cs (numbers vary from vendor to vendor) can be supported with common power supplies, battery backups, management methods, and so on.

The opposite end of the HDSL link is the *HDSL Termination Unit – Remote* (HTU-R). The HDSL really takes place between the HTU-C and the HTU-R. In its simplest and original form, the interface between the HTU-C and HTU-R is two pairs of twisted copper wire. The runs may have some mixed gauges or bridged taps, but must comply with international standards for carrier serving area (CSA) local loops.

Each of the two pairs runs at 784 Kbps full-duplex upstream and downstream. The 2B1Q line code from ISDN, a type of pulse amplitude modulation (PAM) line code, is used. Note that this differs from traditional T1, where each pair carried bits in only one direction using a line code called bipolar AMI. In the original T1, this unidirectional operation was done mainly for simplicity of design of transmitters, receivers, and repeaters.

Note that vendor-specific one-pair variations on HDSL existed, some of which are known by different names. Some of these eventually gave birth to SDSL. But the point to be made here is that the network and customer still see T1s. All user equipment and service provider operations function just as before.

HDSL FOR E1

When used to provision an E1, HDSL seems to customers to be exactly that—an E1 running at 2.0 Mbps. The same as with T1; it may be HDSL inside, but the link it still an "E1."

Figure 7-2, based on the original HDSL European Telecommunications Standards Institute (ETSI) documentation, shows why this is the case. The figure shows an E1 entering the HDSL equipment at the service provider's office on the left of the figure, and an E1 emerging to connect the user's CPE on the premises. In between, the HDSL carries the E1. It should be noted that three pair E1 versions are virtually all phased out today. The limitation originally came from the fact that existing chipsets all ran at the T1 halfrate of 784 Kbps. So, three chipsets were needed to get line speeds up above 2 Mbps.

HDSL components are gathered into either *line termination units* (LTUs) at the service provider or *network termination units* (NTUs) at the customer premises. Each termination unit consists of four major components:

▼ The HDSL transceiver itself

■ Some common circuitry used in all versions of HDSL: one-, two-, or three-pair systems

■ A mapping module to map the E1 frame bits into the HDSL frame structure and back

▲ An interface module to accept a standard E1 connector

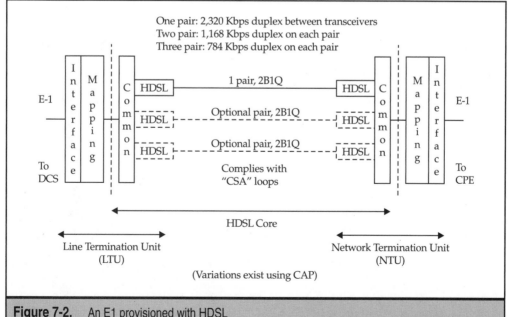

Figure 7-2. An E1 provisioned with HDSL

The components, from common circuitry to HDSL transceiver at each end, form the HDSL core of the whole system.

Keep in mind that E1 leased private lines, as with T1 leased private lines, do not go through the service provider's circuit switch. Private lines are routed through a DCS (or DACS in an access arrangement) out on to a trunking network. This is not shown in the figure, nor was it the common practice until recently to gather the LTUs into common racks or equipment.

In its simplest form, the HDSL link between the LTU and NTU is one pair of twisted copper wire. The runs may have some mixed gauges or bridged taps, but they must comply with international standards for CSA local loops. This pair uses the 2B1Q line code from ISDN, exactly the same as HDSL supported the T1 line speed, and runs at a total of 2.320 Mbps between HDSL transceivers. The "extra" bits above the E1 2.048 Mbps speed are used for overhead and compatibility with SDH signal formats.

It is more common to see multiple pairs, especially two pairs, on HDSL E1 systems. With two pairs in place, each of the two pairs runs at 1.168 Mbps full-duplex upstream and downstream. Note that the total bit rate is the slightly higher (2×1.168 Mbps = 2.336 Mbps) due to increased overhead. Also note that this differs from traditional E1, where each pair carries bits in only one direction, as with T1. And also as with T1, this was done mainly for simplicity of design of transmitters, receivers, and repeaters. When three pairs are used, each runs at 784 Kbps full-duplex upstream and downstream. There is more overhead (3×784 Kbps = 2.352 Mbps), but each pair runs at a lower bit rate and so might reach farther distances. All of the initial HDSL links running at E1 speeds used three pairs because it was felt within the industry that using existing chipsets running at 784 Kbps would offset the cost of needing a third pair of wires. As it turned out, faster chipsets were easy to fabricate once there was a sustained demand for them, so three-pair E1 speed HDSL systems more or less disappeared in favor of two-wire-pair systems based on customized chipsets.

Note that variations of E1 HDSL exist using CAP, an analog line code, as a line coding technique, but these remain more or less curiosities. The point is that whatever the coding between compatible LTUs and NTUs, all user equipment and service provider operations function just as before.

THE HDSL FRAME FOR T1

HDSL, just like almost every other transport link used in high-speed telecommunications (including ADSL), is a *framed transport*—that is, the link sends a series of frames, one after the other, with no pause in between. If there is no "live" used information to send, special idle bit patterns are sent inside the frames. Keep in mind that these transmission frames are distinct from the data link level frames that are used in LANs and some other wide-area protocols.

Figure 7-3 shows the content of an HDSL frame when used for T1 transport on two pairs of wire, based on Telcordia documentation. An HDSL frame is sent once every 6 milliseconds, or about 167 frames per second. A frame consists of a special synchronization symbol in 2B1Q that is 14 bits long. Technically, it is not proper to speak of bits in 2B1Q

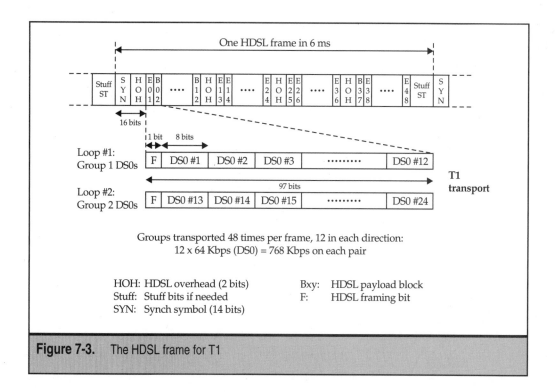

Figure 7-3. The HDSL frame for T1

but rather *quats*, which are 2 bits long. For the sake of simplicity, however, this "bit" terminology will be used throughout this discussion. There is some *HDSL overhead* (HOH), 2 to 10 bits in length, spread throughout the frame.

The rest of the HDSL frame between the synchronization symbols is divided into 48 HDSL *payload blocks*, which are divided by the HOH bits into 4 units, each containing 12 HDSL payload blocks. These blocks are numbered B01 through B48, as shown in the figure. The figure also shows the content of any payload block. One pair (Loop #1) carries what are called Group 1 DS-0s (64 Kbps channels), and the other pair (Loop #2) carries the Group 2 DS-0s. The Group 1 DS-0s are numbered 1 through 12, and the Group 2 DS-0s are numbered 13 through 24. Together, they yield the T1 transport capacity of 24 DS-0s. A framing bit (not the same as the T1 framing bit) precedes the payload block, making each payload block 97 bits long ($8 \times 12 + 1 = 97$).

The groups are transported 48 times per HDSL frame, one per payload block, or 48 times in 6 milliseconds. Each pair carries bidirectional traffic, and because 12×64 Kbps = 768 Kbps, the line rate of 784 Kbps on each pair represents the 12 DS-0 channels plus 16 Kbps of HDSL overhead.

Recall that vendor-specific one-pair variations on the basic two-pair HDSL exist, but are not part of the original Bellcore HDSL standard. The international standard for one pair T1/E1 speeds is ITU Recommendation G.991.2, commonly called G.shdl. In the United States and among many vendors of DSL equipment, the one-pair version is called HDSL2.

THE HDSL FRAME FOR E1

One of the things that makes HDSL suitable for both T1 and E1 is the fact that both use exactly the same HDSL frame format. This format is a brilliant thing to do, but there are some differences between an HDSL frame carrying a T1, and an HDSL frame carrying an E1.

When used for an E1 transport, just as with a T1, the HDSL link is a framed transport sending a series of frames, one after the other, with no pause in-between. Special idle bit patterns are sent if there is no "live" used information to send.

Figure 7-4 shows the content of an HDSL frame when used for E1 transport on three pairs of wire, based on ETSI documentation. An HDSL frame is sent once every 6 milliseconds, or about 167 frames per second. A frame consists of a special synchronization symbol in 2B1Q that is 14 bits long (although, as with T1, it is not technically proper to speak of bits in 2B1Q, but rather quats, which are 2 bits long). There also is some HDSL overhead (HOH), two to ten bits in length, spread throughout the frame.

The rest of the HDSL frame is divided into 48 HDSL payload blocks, which are divided by the HOH bits into four units, each containing 12 HDSL payload blocks. These blocks are numbered B01 through B48, and each is 97 bits long. So far, this frame structure is exactly the same as the HDSL frame structure used for T1.

The content of any payload block also is shown in the figure. The major differences between T1 and E1 frame structures and terminology are most obvious here. The F bit of the

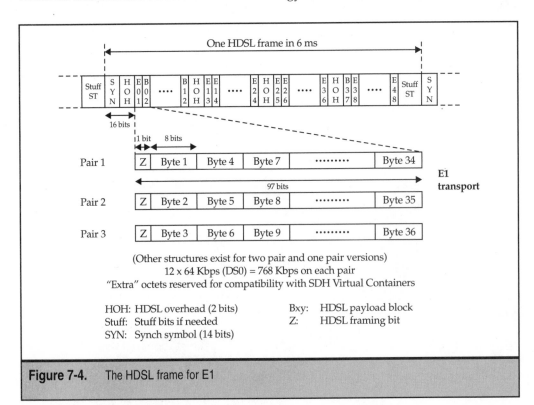

Figure 7-4. The HDSL frame for E1

T1 transport is now the Z bit in the ETSI E1 version. The "bytes" (oddly enough, the ETSI documentation calls them bytes rather than octets, which is more common in international specifications) are loaded into the HDSL payload blocks on each of the three pairs in a simple pattern. Bytes 1, 4, 7...34 are in the first pair payload block sequence, bytes 2, 5, 8...35 are in the second pair payload block sequence, and bytes 3, 6, 9...36 are in the third pair payload block sequence. Now there are only 32 bytes in the E1 frame structure. The "extra" bytes are used for compatibility with the SDH E1 Virtual Container structure.

The groups are transported 48 times per HDSL frame, one per payload block, or 48 times in 6 milliseconds. Each pair carries bidirectional traffic, and because 12×64 Kbps = 768 Kbps, the line rate of 784 Kbps on each pair represents the 12 DS-0 channels plus HDSL overhead.

Two-pair and one-pair variations exist and are part of the ETSI HDSL standard, but these are not shown in the figure. Basically, the payload block sizes and the line rates increase to keep the frame time at 6 milliseconds. For the sake of completeness, these frame structure variations are shown in Figures 7-5 and 7-6. In each case, 16 Kbps must be added to the derived transfer rate for frame overhead, and so the two-pair system runs at 1,152 Kbps + 16 Kbps = 1,168 Kbps on each pair. The one-pair system runs at 2304 Kbps + 16 Kbps = 2,320 Kbps = 2.304 Mbps.

Now, of course, it is widely anticipated that HDSL will gradually be phased out in favor of single-pair versions such as HDSL2, G.shdl, or G.991.2 (all three really mean the same thing).

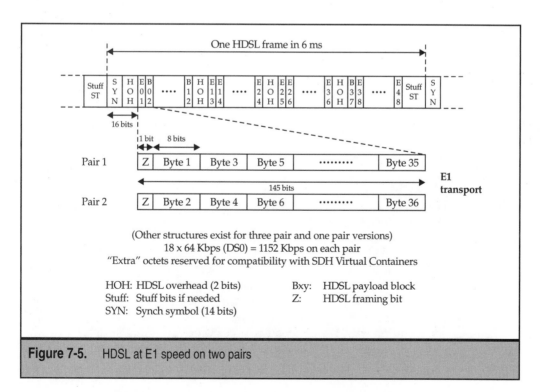

Figure 7-5.   HDSL at E1 speed on two pairs

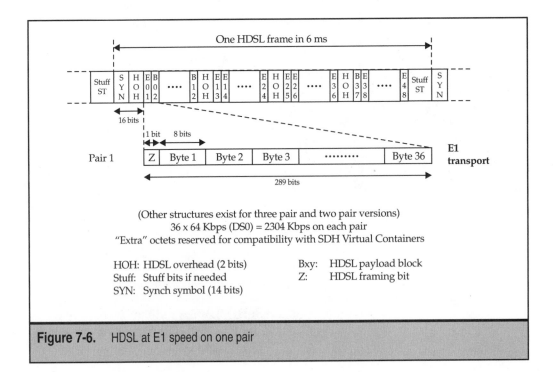

Figure 7-6. HDSL at E1 speed on one pair

BENEFITS OF HDSL

Using HDSL to deliver either T1 or E1 services benefits service providers and customers. Most of these benefits are on the service provider side of the ledger, but customers have indirect benefits, as well.

First and foremost, only the local exchange card (the HTU-C for T1 or the LTU for E1) and customer card (the HTU-R for T1 or the NTU for E1) are needed for the simplest form of HDSL service.

The loops do not need repeaters every few thousand feet or fraction of a kilometer. Instead, HDSL allows 24 AWG (0.5 mm) copper pairs to provide services for up to 12 kft (2.3 miles or 3.6 km). On 26 AWG (0.4 mm) copper, services extend up to 9 kft (1.7 miles or 2.7 km). Usually, two bridged taps are allowed if the length of each is less than 5,000 feet (1.525 km).

HDSL can be extended to reach about 26 kft (5.0 miles or 7.93 km) by using heavier wire gauges (22 AWG or 0.63 mm) or HDSL repeaters, which are more often called doublers in HDSL, because they "double" the length achievable. It sounds odd that HDSL repeaters even exist, but these newer versions are much more efficient and powerful than their early T1/E1 cousins.

Remote HDSL equipment, HDSL repeaters or customer premises cards, can be powered from the local exchange itself. This is a common practice known as *wet HDSL*, from the way technicians tested for the low voltage current on the pairs (wet your fingers by licking them and grab the wire…).

Another important benefit is that HDSL can be monitored with the same *operations system support* (OSS) equipment and software as before. After all, outside of the HDSL core system, the link is still T1 or E1.

Finally, HDSL can be used almost anywhere. Some 80 to 90 percent of copper cable plant is appropriate for HDSL, whether based on CAP or PAM line codes.

USES OF HDSL

It may seem obvious just what an HDSL link can be used for. Basically, anywhere a T1 or E1 makes sense, HDSL makes even more sense. However, this might be a good time to list the main uses of HDSL to provision T1 and E1 services from the customer perspective. Because much of this book deals with the uses of ADSL and its related technologies, keeping symmetrical applications in mind is a good idea. The main uses of HDSL are here:

▼ Internet access to servers, not just from clients

■ Private campus networks with installed copper cable plant

■ Extend central PBX to other office park locations

■ LAN extensions and connections to fiber rings

■ Video conferencing and distance learning applications

■ Wireless system base station connections

▲ *Primary rate access* (PRA) for ISDN

HDSL is widely used to provide companies and individuals with fast Internet access to their *servers*, not just from clients. Asymmetrical DSL variations limit upstream traffic to a fraction of the downstream rate. The placement of servers in homes or small offices becomes difficult because, in this case, the downstream traffic from remote clients will be a fraction of the upstream traffic from the local server. Yet the bandwidth is just the opposite. In fact, one of the biggest selling points for HDSL versus ADSL is found in the simple question, "Do you have a server on-site?"

The symmetrical speed issue is shown in Figure 7-7. Most organizations with servers such as Web sites linked to the global public Internet have T1 or E1 links that run at 1.5 Mbps or 2.0 Mbps. Many organizations might have two such links for redundancy, and there might be a router capable of load balancing on the premises, but usually a stream of packets from any client to any server will always use one link or the other for routing stability, sequencing, and other reasons. So each user sees only a 1.5 or 2.0 link anyway.

The fact that most Internet servers have only T1 or E1 access speeds has two major implications. First, for an asymmetrical DSL, downstream speeds above 1.5 or 2.0 Mbps are mostly wasted. The link cannot deliver traffic faster than a server can deliver it to the Internet. Now, there are times when multiple clients or multiple client applications are

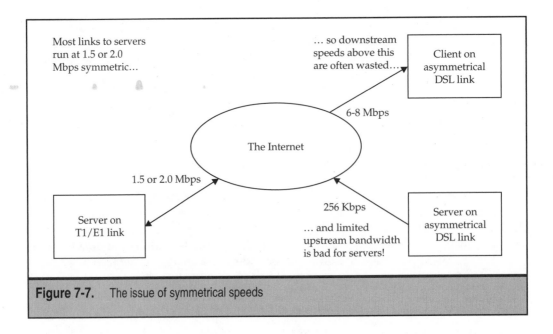

Figure 7-7.   The issue of symmetrical speeds

running, and then these higher speeds can be used. But this examines the simplest case. Also, when asymmetrical DSLs are used for servers, restricting upstream bandwidth below 1.5 or 2.0 Mbps is a bad idea. So symmetrical DSLs, like HDSL, still make a lot of sense for any location, whether home or office, that has a server on-site.

HDSL is also used for private campus networks with installed copper cable plant. Many companies, colleges, and universities have many copper pairs between buildings. Until HDSL came along, it was expensive and difficult to squeeze more than 64 Kbps out of each pair.

Organizations with a central PBX can extend 24 or 30 voice channels to other office park locations easily and simply with HDSL. The same applies to LAN extensions and connections to fiber rings, although, in this case, the HDSL (really the T1 or E1) is probably unchannelized into its raw bit rate of 1.5 Mbps or 2.0 Mbps.

Video conferencing and distance learning applications are natural for HDSL. Again, the link in this case would probably be unchannelized. The position of the client and server are unimportant, due to the symmetrical nature of the HDSL link.

Moving back to channelized applications, wireless systems have lots of copper wire in use. Every base station transceiver tower must be connected to a central operations center and to switches and trunks, as well. HDSL makes a lot of sense when used for wireless system base station connections.

Finally, HDSL can be used to provide more cost effective PRA for ISDN services. The cost of provisioning the high-speed digital line for ISDN has long been a reason cited for higher-than-anticipated ISDN pricing structures. HDSL makes this process of PRA DSL provisioning simpler and cost effective. Of course, one of the key barriers to the more widespread use of HDSL for T1 service to SOHO and residential users is the fear among service providers that this move would dramatically erode today's high tariff rates for traditional T1 service. Indeed, this has already happened in a number of areas.

HDSL FOR ISDN

Mention has just been made of the use of HDSL to supply ISDN Primary Rate Access. The link in this case would operate at the Primary Rate Interface (PRI) speed in the country it was deployed in. In the United States, the PRI is 1.5 Mbps, and in most other places it is 2.0 Mbps.

Figure 7-8 shows how HDSL might be used to provision ISDN PRA to a customer premises from an ISDN local exchange (LE). The figure shows HDSL on two pairs of wire, each running at 1,168 Kbps, giving the E1 transport rate of 2.336 Mbps when HDSL overhead is included. In cases where the premises exceed a certain distance, HDSL repeaters can be added to each pair. Note the bidirectional functioning of each pair.

In this configuration, the LE side of the link is the *Local Exchange Primary Access – HDSL* (LEPA-HDSL) equipment. This equipment can be gathered into racks, of course, but the term DSLAM is not really appropriate here. In any case, an E1 input and output port can be found on the ISDN switch.

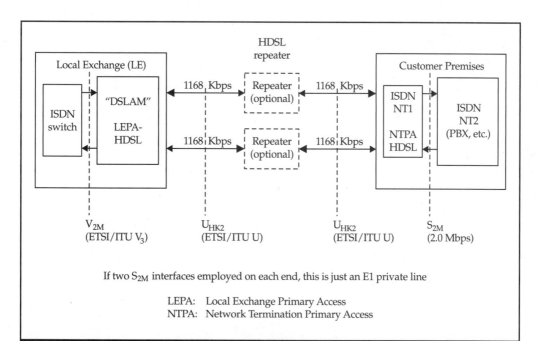

Figure 7-8.   HDSL for ISDN PRA

On the customer premises, the *Network Termination Primary Access – HDSL* (NTPA-HDSL) equipment basically forms the ISDN Network Termination Type 1 (NT1) unit (technically, a functional grouping). This may in turn be attached to an ISDN Network Termination Type 2 (NT2) unit, such as a PBX or other ISDN-compliant piece of equipment. Note the unidirectional operation of these interfaces.

The interfaces themselves are also shown in Figure 7-8. At the LE, the LEPA-HDSL to switch interfaces conforms to the ETSI/ITU V_{2M} interface specification at 2.0 Mbps (E1). At the customer premises, the NT1 to NT2 interfaces conform to the ETSI/ITU S_{2M} interface specification, also at the E1 rate.

The HDSL links themselves conform to the ETSI/ITU U_{HK2} interface specification. HK2 essentially means "HDSL on two pairs" in this case. The point is that ETSI and the ITU have made HDSL an official part of the ISDN PRA interface.

Note that if two S_{2M} interfaces are employed on each end of the link, or the V_{2M} interface leads to a digital cross-connect and not an ISDN switch, this is now just an E1 private line. Of course, this private line would still conform to all ETSI and ITU ISDN standards.

THE DOWNWARD SPIRAL

In North America, the great majority of all new T1s and E1s are now provisioned with HDSL. The customer is still buying a T1 or E1, but delivered with HDSL. This is not to say, however, that the customer does not benefit from HDSL at all. Figure 7-9 shows the impact that HDSL has had on customers. The main benefit to and effect on customers is reduced T1 and E1 pricing.

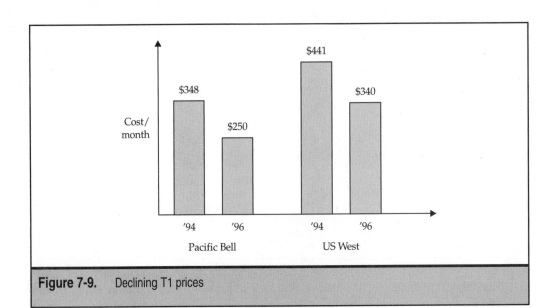

Figure 7-9. Declining T1 prices

This figure might look as if it is in need of updating. However, the timeframe involved shows the initial impact of HDSL on T1 pricing. Prices have continued to drop somewhat, but nothing like the 1995 introduction of HDSL on a large scale for T1 provisioning.

The figure illustrates the typical pricing of only two United States service providers—and two very large ones at the time—but the effect has been the same all around the world. In 1994, Pacific Bell (now part of SBC) and US West (now Qwest) priced T1s at about $348 and $441 per month, respectively. That year saw little HDSL. In 1996, when some 60 percent of T1s provisioned in the United States were on HDSL, the prices had fallen to $250 and $340, respectively.

This is a 28-percent drop in the case of Pacific Bell ($98/$348) and a 23-percent drop in the case of US West ($101/$441). In the future, prices should continue to drop (but not as fast) as HDSL and the even more efficient HDSL2 becomes even more widespread.

HDSL LIMITATIONS

In spite of the enormous appeal of HDSL and its wide-ranging benefits to service providers and customers, it still labors under limitations that made HDSL less than ideal for all situations. These limitations led directly to the emergence of HDSL2, of course.

One of the biggest concerns was that only the barest essentials of HDSL were covered in the specifications, which has meant that there are a number of proprietary implementations of HDSL with wide variations in features that do not allow vendor interoperability at all. This concerns service providers, who would like more consistency between product offerings, and who always like more vendor independence.

Another aspect of HDSL is that the more technical benefits are almost invisible to customers. There might be some better performance in terms of bit rates and some reduced costs, but to the customer it is still a T1 or E1. Customers expect high-performance in any case, and always expect prices to fall. In terms of HDSL CPE, perhaps customers need even more direct benefits.

Whether configured for T1 or E1, HDSL is still supposed to use the 2B1Q (4-PAM) line code. The problem is that 2B1Q itself has real limitations of bandwidth efficiency and distances.

In addition, requiring the use of multiple pairs at least halved, and in some cases with older three-pair HDSL, reduced by two-thirds, the availability of T1 and E1 service in a given area. If HDSL could be standardized on only one pair, as some early variations have done on a proprietary basis, this would maximize the use of the existing cable plant.

The fact is that HDSL equipment has dropped in price so much that the cost of the two pairs of copper wires are significant cost factors.

Finally, in spite of its efficiencies, HDSL can still be slow to deploy. After two or three pairs are located, the rest of the process can take only hours. Some regions have a shortage of copper pairs. However, one still needs to find two or three pairs running between the proper locations. One pair, in contrast, always runs between the same end points.

BEYOND HDSL: HDSL2

All of the limitations of current HDSL products are addressed in the "next generation" of HDSL, commonly known as HDSL2. The "2" refers to "HDSL, the second generation" and the name has stuck.

The HDSL2 specification is now completed, and the first products based on ITU Recommendation G.shdl or G.991.2 are starting to appear. This availability is the result of a lot of hard work by a number of organizations, with the primary work done by the ANSI T1E1.4 committee. However, ANSI's work on HDSL2 was endorsed by both ETSI and the ITU.

There were three primary goals for HDSL2:

▼ 12 kft (2.3 miles or 3.6 km) coverage (which ANSI calls a "full CSA").

■ It must work even with other services on adjacent cable pairs (ANSI calls this *spectral compatibility*).

▲ Vendor interoperability (HTU/LTU/NTUs may be mixed and matched among vendors).

In addition, HDSL2 was the first major standard to be developed with the idea of *power management*. This was part of the spectral compatibility goal, and this new feature held up HDSL2 for a while until all of the details of this new addition to the technology were worked out.

Developing technologies, especially new technologies, is a study of applied engineering. Engineering in turn applies science to the real world. So much of HDSL2 concerns taking electrical formulas and making compromises in order to embody them in equipment that functions reliably in the everyday world under less-than-ideal conditions. This is easier said than done. HDSL2 has already faced its share of compromising situations.

For instance, higher speeds mean smaller bit intervals, which means more errors if all else stays the same. But T1 and E1 specifications limit bit errors, and regulators or customers may demand or be entitled to rebates for elevated error rates. If there is in "plain" HDSL a need to do full T1 or E1 speeds on only one wire pair, one way to make this possible is to add some *forward error correction* (FEC) to the HDSL frame. These extra bits can both detect and correct some errors in the HDSL frame, perhaps just enough.

However, adding a FEC function to the HDSL2 device adds to the latency end to end.

The HDSL2 standard was much slower to complete than anticipated. There were many issues and challenges that the HDSL2 developers were faced with. The basics of HDSL2 were never in doubt: only one pair needed for up to 12 kft (3.7 km) maximum span between H2TU-C and H2TU-R on CSA-complaint loops for either 1.5 Mbps or 2.0 Mbps service. The link still appears to be a T1 or E1 for management purposes, and both CPE and switching office equipment receive two pairs, mainly for backward compatibility.

But there were also some more controversial points regarding HDSL2. There is and will be no analog voice support in HDSL2, only some form of VoDSL. The best way to explain this lack of analog voice support is that the PAM line code used in HDSL2 leaves no place to put the 4 KHz voice channel "under" the digital signal (PAM has a strong signal component all the way down to 0 Hz). Some vendors and service providers are concerned about the impact on HDSL2 for residential services because of this lack of analog voice support.

There will be doublers (also called "line extenders," but just repeaters in disguise) in HDSL2, just as in HDSL. Many candidate loops for HDSL2 exceed 12 kft (3.7 km). So, HDSL2 will allow the use of HDSL2 line extenders to reach out to 24 kft (about 7.5 km). Some vendors are trying to stretch things even more with an "extreme" form of HDSL2 that will allow links up to 36 kft, or about 11 km.

An issue in HDSL2 involves the power to these line extenders. The service provider powers the line extender. But this can only be done with one line extender. If there is a second line extender unit at 24 kft and not CPE, this line extender must be powered from the customer premises. This is not attractive for a number of reasons, so "extreme" HDSL2 might not be common at all.

One of the biggest issues with HDSL2 is dealing with interference (noise). In multipair bundles, most of the interference will come from HDSL's 2B1Q and ADSL's DMT or CAP line codes. HDSL2 was designed to expect ADSL in one form or another to be running nearby. But the original HDSL 2B1Q is a considerable source of interference to HDSL2, so service upgrades must be carefully planned.

The HDSL2 in Detail

The basic architecture of an HDSL2 link is shown in Figure 7-10. The H2TU-C and H2TU-R are still serviced by the two wire pairs of a normal T1 or E1, naturally. But in between, there is only one pair of wires up to 12 kft (about 3.7 km) long that comply with CSA guidelines.

The line code used in HDSL2 is pulse amplitude modulation (PAM), but this requires some words of explanation. As already mentioned, a major goal of HDSL2 was to make equipment from different vendors interoperable. So HDSL2 runs an initial "handshake" protocol between H2TU-C and H2TU-R to determine basic features and supported options. The handshake protocol uses a simple 2-level version of PAM called 2-PAM. Once this negotiation is completed, the units transfer data using a version of PAM that includes 16 levels (4 bits per line code change, or baud) and trellis coding for error detection and correction. This is called 16-TCPAM. Three of the bits are used for data transfer, and the fourth bit is used as a forward error correction (FEC) that is trellis-coded.

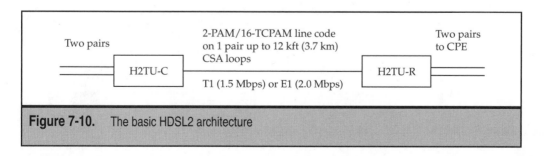

Figure 7-10.   The basic HDSL2 architecture

When it comes to applications for HDSL2, running 1.5 Mbps or 2.0 Mbps on one pair of wires opens up a whole new world of applications for higher speed data connectivity. Figure 7-11 shows only some of the applications of HDSL2, but these are the ones that are expected to be the most common. Keep in mind that without HDSL2, the process of provisioning T1 or E1 links can be labor-intensive (to find, qualify, and engineer the two pairs) and time-consuming (multiple truck rolls, coordinated testing, and so on).

It should be pointed out that even with HDSL2, the user CPE and the service end of the link still see a normal, 2-pair T1 or E1 interface. But between H2TU-C and H2TU-R, and even through the main distribution frame (MDF), there is only one pair involved.

As shown in the figure, HDSL2 can make the process of linking customer frame relay access devices (FRADs) to the service provider's frame relay switch more cost-effective and timely. A lot of frame relay user-network interfaces (UNIs) still operate at 56 or 64 Kbps, although 128 Kbps has become very popular. But frame relay at 56/64 Kbps runs the risk of being left out of the broadband sweepstakes (and even 128 Kbps is only an improvement, not a cure) as users put aggregated traffic from many low-speed voice circuits, video, and other high-bandwidth data streams across the frame relay UNI. HDSL2 could make higher speed frame relay UNIs attractive and extend the life of the frame relay technology.

Also, as cellular services grow, people sometimes forget that each and every cell site needs to be linked to a central *mobile telephone switching office* (MTSO). Usually, these links are multiple landlines at T1 or E1 speeds, the better to cut down on additional wireless equipment and noise at the cell sites. HDSL2 requires only 1 pair for 24 or 30 voice calls to and from the cell site, and even more calls can be carried with voice compression.

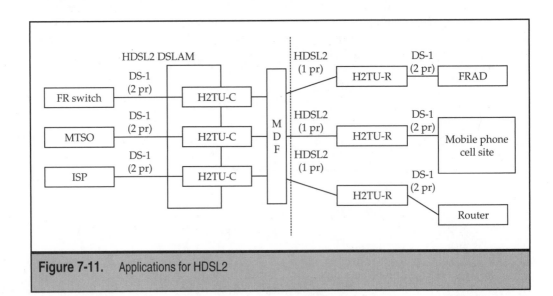

Figure 7-11. Applications for HDSL2

Finally, and probably most important for this DSL discussion, HDSL2 can link a router on the customer premises to an ISP using only one pair. Many DSLs can do the same thing, but HDSL2 is the fastest *standard* symmetrical way to do this. When there are servers on the customer premises, only symmetrical solutions really make a lot of sense.

These are only a few of the applications for HDSL2, of course. Anywhere a T1 or E1 can be used, HDSL2 could be used instead.

One of the best potential uses of HDSL2 is to offer a method to deliver high-speed residential service when SONET/SDH rings are used as a DLC technology. Mention has been made in Chapter 3 of SLC-96 and related DLC technologies using multiple pairs of T1s (or E1s, for that matter). Many, if not all, other forms of DSL struggle when it comes to DLC arrangements that service CSA loops instead of straight wire home runs from switching office to subscriber. Until DSL came along, DLCs cared only about the voice passband (analog) and/or 64 Kbps digitized voice (and some cared only about 32 Kbps or even less). But HDSL2 offers a much cleaner evolutionary path from limited DLC arrangements to more permissive environments. If HDSL2 is ever to become a widespread residential service offering, some accommodation for DLCs must be made.

A traditional DLC service can gather up to 96 DS-0s at 64 Kbps (if the user lines are analog, the DLC will digitize them at 64 Kbps) from users. These 96 channels are mutliplexed onto 4 T1s (E1 DLCs will have slightly different numbers, of course), which are provisioned on 10 pairs of wire (8 pairs for the T1s, and 1 pair for OAM&P). These "pairgain" systems (so named because they "gain back wire pairs," up to 86 of them, for other uses) have been around for some time, and are commonly exemplified by the "slick-96" ("subscriber line carrier," SLC-96) equipment used in the old AT&T/Bell System. So with a DLC in use, 10 pairs leaving the switching office can potentially be sold as 96 DS-0s from the service provider perspective.

However, when HDSL2 meets the next generation DLC (NGDLC) architecture based on SONET/SDH rings, much more bandwidth is available to deliver to the customer. The H2TU-Cs could be housed in a DLSAM that is part of the SONET/SDH DLC. In this case each NGDLC can support up to 96 DS-1s at 1.5 Mbps, not DS-0s running at 64 Kbps as before (E1 supports 2.0 Mbps links). Now, if the SONET ring runs at the OC-3 speed of 155 Mbps (STM-1 in SDH), only 84 DS-1s can be supported on the ring. But if the ring runs at the OC-12 speed of 622 Mbps (STM-4) or the OC-48 speed of 2.4 Gbps (STM-16), then the SONET/SDH ring can support 336 DS-1s (3 or 4 NGDLCs) or 1,344 DS-1s (14 NGDLCs) respectively (E1 support numbers will differ slightly).

So a single fiber can be sold as hundreds or even thousands of symmetrical, high-speed circuits directly to homes in residential areas.

Final Thoughts on HDSL2

Keep in mind that HDSL2 is not SDSL or ADSL. There are important similarities, however. HDSL2 provides 1.5 Mbps or 2.0 Mbps transmission on a single pair, and many current SDSLs can provide 1.5 Mbps or 2.0 Mbps transmission on a single pair as well. And some versions of ADSL run at 1.5 Mbps or 2.0 Mbps on a single pair also. But there are important differences among these DSL technologies, so it might be a good idea to go over the differences in some detail.

First compare HDSL2 to SDSL. SDSL equipment has been vendor-specific and although there is some more or less accidental compatibility, SDSLs are not interoperable. ETSI does have some specifications regarding SDSL, but this documentation is not enough to declare SDSL a standard by any means. In contrast, HDSL2 is both an ANSI and ITU-T project.

Also, HDSL2 has full CSA coverage out to 12 kft (3.7 km) while trying to run itself along with other DSLs in the same cable bundle. In many if not all cases, SDSL can only reach as far by using a lot of power that disrupts other services in the same 50-pair cable binder (even 200- or 300-pair cables are organized into four or six 50-pair "binders"). HDSL2 is very respectful and accommodating of other DSLs running nearby, especially ADSL.

But at the same time, HDSL2 is very different from ADSL. ADSL can have higher speeds than 1.5 Mbps or 2.0 Mbps (at the price of more stringent operating conditions). Most importantly, ADSL is asymmetrical (there is a very low-speed symmetrical option that few seriously entertain) with only high speeds in the downstream direction to the user. Upstream is a fraction of what HDSL2 can do.

It is also true that symmetrical services fit better with most current networking equipment (asymmetrical links are relatively new) and for most services (especially those where servers at the wrong end of the asymmetry, like in the home). And trying to effectively multiplex 100 asymmetric links with 1.5 Mbps in one direction and 128 Kbps in the other can be a challenge because most backbone links are distinctly symmetrical.

Moreover, HDSL2 has no speed variability or guesswork as ADSL does. HDSL2 is set to operate at 1.5 Mbps or 2.0 Mbps, and that is that. It either runs at the standard speed, or the link fails. ADSL, especially RADSL versions, can run at such a multitude of speeds that it is often hard to figure out if ADSL/RADSL is running at the right speed or not, let alone what the pricing structure should be (one neighbor or another with very good—or very bad—wires could actually end up paying more than the other for less speed).

The main point here is not that HDSL2 is in some sense better than SDSL or ADSL. The point is that service providers, equipment vendors, and even customers should know what they are getting into with one or the other form of service.

HDSL2 Power Control

The really distinctive feature of HDSL2 is not symmetry, but *power control*. Because power control is such an important new feature of HDSL2, and a feature that might become common in many if not all DSLs, this aspect of HDSL2 deserves a short section of its own.

All DSLs are concerned with the amount of signal power they generate, naturally, in order to maximize their speed and reach. The signal strength must be kept at a high enough margin above the noise (the SNR) in order to assure proper operation. The problem is that the more power that is pumped onto a link, the more likely it becomes that the link will interfere with (disturb) adjacent pairs in the same wire bundle (usually called a binder).

Most DSLs can do nothing about interfering with DSLs running on nearby pairs in the same 50-pair binder. However, HDSL2 cares. HDSL2 can even throttle back power if the signal between H2TU-C and H2TU-R is more than adequate, letting other DSLs in the same 50-pair cable binder deal with lower potential interference power, even other HDSL2 pairs.

This approach makes a lot of sense, but it is very complex to perform in practice. The power level of any DSL (or any other network technology) is set by standard or vendor for "worst case" CSA or line conditions. The CSA rules allow for a variety of conditions, and some of these conditions that are fine for voice allow DSLs to perform only marginally. By way of analogy, consider if everyone were required to shout loudly at all times because some people wear earmuffs when it's cold out (this would be the "worst case" for listening). But when it were summer, or if whispers were all that were needed, it would be impossible to converse in a large group with all the shouting going on. The result would be crosstalk.

Because of this crosstalk risk, one of the key goals of HDSL2 was to achieve *spectral compatibility* with as many DSLs as possible. In other words, HDSL2 is supposed to be a "good neighbor" to other DSLs running in the same 50-pair binder. In particular, it turned out that the downstream signal range of HDSL fully overlapped that of upstream ADSL, causing many problems with crosstalk. And because the upstream ADSL signals come from the ATU-R, low power is desirable due to consumer cost and operational considerations. Studies with HDSL showed that when HDSL and ADSL were provisioned in the same binder, ADSL might fall to as low as 96 to 192 Kbps through no fault of its own. This is far below the speed needed to guarantee ADSL customer acceptance.

It would be best to isolate different services by binder, but this is just not possible. So the idea of HDSL2 power control was a good one. ANSI studies have shown that about 60 percent of all CSA loops can use reduced HDSL2 power and still work properly with adequate margins. In some cases, power could be dropped from 3 to 6 dB—not an inconsiderable amount. This reduced power level for HDSL2 could make a significant difference in the amount of interference or disturbance added to any ADSL running nearby.

And HDSL2 does expect ADSL to be running nearby. So HDSL2 adjusts power accordingly. The H2TUs adjust their sending power by actively monitoring both the *bit error rate* (BER) and signal-to-noise ratio (SNR) on the pair of wires running between them. If the BER is good enough when the units first see each other, the signal power level can be slowly dropped. If the BER rises past some critical value (typically 10^{-7}), then the signal power level can be increased again until the errors are within the performance boundaries.

Specifically, HDSL2 signal power adjusts 0.01 dB per minute (3 dB in 5 hours) until the optimum configured combination of BER and SNR is achieved. The adjustments can be made to increase as well as decrease power. Power control could be added to many DSLs as time goes on.

By way of summary, Table 7-1 compares the major characteristics of HDSL with the newer HDSL2.

Note that when it comes to the most important physical loop characteristics, such as lengths and bridged taps, HDSL and HDSL2 are identical in terms of applicability.

Characteristic	HDSL	HDSL2
Overall line speed	T1 (1.544 Mbps) or E1 (2.0 Mbps)	T1 (1.544 Mbps) or E1 (2.0 Mbps)
Pairs needed	2	1
Line code	2B1Q	16-TCPAM, 2-PAM for handshake
Line rate	784 Kbps or 1024 Kbps on each pair	1.552 or 2.0 Mbps on one pair
Bits/baud	2 bits/baud ("quat")	3 bits/baud plus 1 for FEC
Bandwidth	196 KHz on each pair	Varies
Load coils?	No	No
26-AWG loop length max	9 kft (about 2.7 km)	9 kft (about 2.7 km)
24-AWG loop length max	12 kft (about 3.7 km)	12 kft (about 3.7 km)
Maximum bridged tap length	2 kft (about 0.6 km)	2 kft (about 0.6 km)
Maximum total bridged tap length	2.5 kft (about 0.75 km)	2.5 kft (about 0.75 km)

Table 7-1.   HDSL and HDSL2 Characteristics

CHAPTER 8

The Asymmetric Digital Subscriber Line (ADSL) Architecture

In spite of the confusion over the relationships between other DSLs and ADSL, one thing is clear: ADSL is the most standardized of all, in terms of available documentation, residential service availability, and open specifications. When it comes to residential high-speed Internet access, many service providers will begin with ADSL and may even end with ADSL.

Figure 8-1 shows the original ADSL system architecture, as documented by the ADSL Forum (now just the DSL Forum). Although the figure may seem confusing at first, the overall arrangement of ADSL components is straightforward. The need to standardize makes many of the details necessary. Like most architecture diagrams, the architecture establishes a number of standard interfaces between major components. In between the interfaces, various *functional groupings* are defined that may be gathered by equipment vendors into products that embody the required functions, along with any options or enhancements that the manufacturer feels are necessary. The internal functioning of these components is left to individual vendors.

The acronyms in the figure mean the following:

ATU-C	ADSL termination unit, CO side
ATU-R	ADSL termination unit, remote side
B	Auxiliary data input (such as a set-top box)
DSLAM	DSL access multiplexer
POTS-C	Interface between PSTN and splitter, CO side
POTS-R	Interface between PSTN and splitter, Remote side
T-SM	T-interface for service module
T	May be internal to SM or ATU-R
U-C	U interface, CO side
$U-C_2$	U interface, CO side from splitter to ATU-C
U-R	U interface, remote side
$U-R_2$	U interface, remote side from splitter to ATU-R
V_A	V interface, access node side from ATU-C to access node
V_C	V interface, CO side from access node to network service

One of the reasons that this figure is so important is that the basic ADSL architecture has changed somewhat over the past few years. Before showing the current architectural versions (there are now two) for ADSL, it is important to see how ADSL was first envisioned in order to understand how things have changed.

Several features of this original ADSL vision are quite important. First of all, note that provision is made to support analog voice service (designated *plain old telephone service*, or POTS). A special splitter device carries the 4 KHz analog channel from switch to premises "under" the digital bandwidth on the ADSL link.

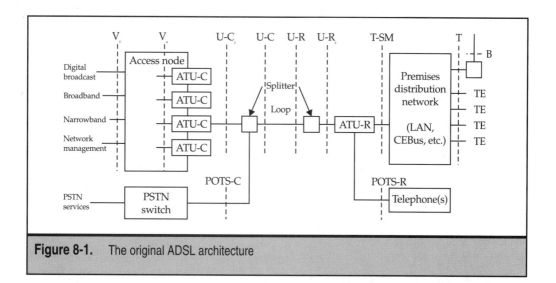

Figure 8-1.   The original ADSL architecture

Next, consider that many services are envisioned for ADSL systems, including digital broadcast and broadband (that is, video and Internet access) services, as well as network management. One of the great issues with this initial ADSL vision was the inclusion of access to specialized services that did not yet exist. All of these services were to be accessed outside of the normal *central office* (CO) or *local exchange* (LE) switch, neatly solving the trunking and switch congestion problems. Many ADSL links are serviced by a single ADSL access node in the central office (or local exchange). This access node is sometimes called a DSLAM, or *DSL access multiplexer,* but this is somewhat misleading. Although a DSLAM can certainly supply service access to ADSL lines, the full architecture of a DSLAM is much more encompassing than the simple arrangement shown here. More details on the DSLAM architecture are discussed later in this book.

The interfaces listed in the figure might not be self-explanatory. The B interface is straightforward and just indicates a possible auxiliary input, such as a satellite feed into a set-top box. The T-SM interface between the ATU-R and service module (everything else besides the ATU-R itself) might be the same as the T interface in some cases, especially if the service module is integrated into the ATU-R. If the T-SM interface does exist, it can be more than one per ATU-R and more than one type per ATU-R. For example, an ATU-R might have both 10Base-T Ethernet and V.35 connectors. Likewise, the T interface between premises distribution network and terminal equipment might also be absent if the terminal equipment is integrated with the ATU-R in some fashion.

The various U interfaces might not exist if the splitter is made an integral part of the ATU devices, or if the splitter entirely disappears. The presence or absence of the remote splitter became a real issue in ADSL and led directly to the development of G.lite. It is even possible to entirely do away with the remote splitter or some other form of low-pass filter that allows only the 4 KHz voice band to pass through, but this makes support of existing analog telephones on the same line impossible. However, if the DSL service provider is just an ISP and not a traditional voice telephone company, this "voiceless" DSL arrangement might be just fine. Also, the V interfaces might be logical rather than physical, which is especially true of the V_A interface if the DSLAM or ADSL access node performs some concentrating or switching tasks. If the V_C interface to the service networks is physical, as it likely will be, this interface is allowed to take a variety of forms appropriate for TCP/IP, ATM, or other types of service networks.

Finally, the implementation of ADSL on the customer premises can take a variety of forms. Under the architecture, these schemes form the Premises Distribution Network. This could be as simple as individual wire pairs running to devices, such as TV set-top boxes or PCs, or as elaborate as a full local area network (LAN), such as Ethernet in the home. A potential drawback would be the need to run new premises wiring to attached ADSL devices. Newer standards using home electrical wiring or existing telephone wires for the same purpose are supported and very attractive for this reason.

CURRENT ADSL ARCHITECTURES

As mentioned, the detailed ADSL architecture developed by the DSL Forum is now mainly of historical interest. It is still worth discussing the original model, however, to see how things have changed with experience. However, a visit to the Forum's Web site now refers those interested in ADSL architectures to the new ITU-T standards for ADSL (G.992.1) and G.lite (G.992.2). This section covers the current architectures for ADSL and G.lite.

The current basic ADSL architecture is shown in Figure 8-2. Although based on the figures in ITU-T Recommendation G.992.1, this figure adds many explanatory details while preserving the overall intent of the original. The local loop itself is shown in bold, as are the currently defined standard interfaces. The interface acronyms mean the same as in the Figure 8-1. There are only two new interface acronyms. The T-R interface is established for equipment that separates the DSL modem (ATU-R) function from the switching (or routing) function on the customer premises (few vendor's ADSL equipment performs only DSL modem functions, so this interface is usually internal). The T/S interface terminology is borrowed from ISDN (and B-ISDN) and essentially separates the DSL modem from the customer's *premises distribution network* (PDN). It all depends on whether the PDN links directly to a user device such as a PC (then this is a T interface) or to a LAN hub (S interface). Now, this architecture covers only the simplest case of direct PC or LAN hub connection. Chapter 11 explores the details of ADSL premises arrangements and points out that it is much more common today to connect a device to the DSL modem that will route IP packets and provide some security against hackers and viruses. In fact, some if not all of these functions are often built into the DSL modem itself.

The *network termination* (NT) acronym is also borrowed from ISDN and is used to describe the device that forms the customer end point of the service, in this case ADSL service.

Note the presence of the central and remote splitter as before. The splitter is a combination high-pass filter (frequencies above the voice band can pass) and low-pass filter

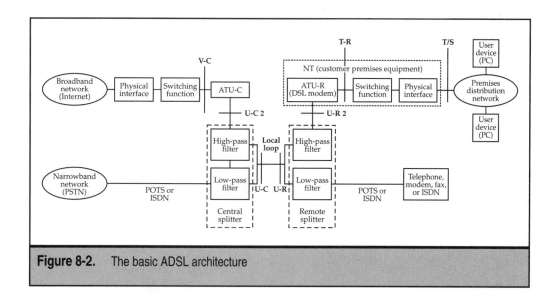

Figure 8-2.   The basic ADSL architecture

(frequencies in the voice band can pass) that allows voice and data to share the local loop in their own frequency ranges and yet allows one or the other range to be split off when needed. Both analog (POTS) or digital (ISDN) voice services can share the DSL loop, but the filters have different *passbands* for each type of voice service. The remote splitter allows the customer to continue to use a telephone, modem (or ISDN card), fax machine, and so on without any change to the wiring or equipment at all. The central splitter passes the voice off to the voice switch, or course, which might be an ISDN switch or not. Voice services then pass on to the PSTN as before.

Naturally, it is the data part of the figure that is of the most direct interest for ADSL. The NT block is the customer premises equipment (CPE) for ADSL. The splitter handles the data–voice split, so there is no need for a filter in the DSL modem (ATU-R) portion of the device. There can be, and usually is, some other switching and/or routing functions in the DSL modem along with a physical interface. Most often, the physical interface is an RJ-45 10Base-T Ethernet port, and the switching function includes at least some manipulation of IP packets. These IP features might include *network address translation* (NAT) or even some simple security filters. Even at its most rudimentary, it is not unusual for the DSL modem to at least touch the IP packets and perform some type of encapsulation on them. For example, the DSL modem might take IP packets inside of the Ethernet frames on the PDN and place the packets inside a stream of ATM cells. These different types of IP packet *encapuslations* are discussed later on in this book in Chapter 10.

Note that there can be more than one user device (almost always a simple PC) connected to the DSL modem device. At its simplest, the PDN can consist of a single Ethernet twisted-pair cable with the send and receive pairs reversed. This *crossover* cable can be plugged directly into the *network interface card* (NIC) or LAN port on a PC, but only one PC can be linked directly to the DSL modem in this fashion. Because more and more homes have multiple PCs, it is becoming more common to link the DSL modem to a 10Base-T LAN hub. This requires a straight-through or *pin-through* LAN cable, not a crossover LAN cable, but the hub often has four or five ports for linking PCs, printers, and other LAN-enabled devices to each other and to the DSL link (all devices linked to the hub require pin-through cables). Chapter 11 provides more details on premises ADSL arrangements.

In the central location, the figure shows a single ATU-C behind the central splitter. However, it is much more common to see ATU-Cs gathered into DSLAMs, and this aspect of the architecture was one of the reasons for keeping the original diagram of the ADSL architecture. ATU-Cs are still gathered into racks.

DSLAM racks might also include some switching and/or routing functions as well. Because most ADSL links transfer a stream of ATM cells between ATU-R and ATU-C (and use of ATM is required in G.lite), it is most common for the DSLAM to include ATM switching functions. If the DSLAM also includes routing functions, the ATM cells might even stop here. So the physical interface shown at the upper left of the figure could be part of the DSLAM itself, and typically is. If ATM cells are sent out of the DSLAM, this physical interface would be an ATM interface. If the DSLAM is also a form of router, the interface requirement is more relaxed.

Finally, the end result is to link the user device(s) to the broadband network, which is just the global public Internet. No matter that much of the Internet does not always fit the definition of broadband, even the "streaming video" definition. The Internet is all there really is for now.

This architecture is not the only one allowed today. There is also the variation shown in Figure 8-3. This is G.lite, or ITU-T Recommendation G.992.2. Much of the figure repeats elements from the full ADSL figure and need not be repeated here. The emphasis is on the differences between what is often called "full ADSL" and G.lite.

It is not an exaggeration to say that the reason G.lite exists is because of the issues surrounding the remote ADSL splitter. A full discussion of these issues is best left to the premises chapter of this book, Chapter 11. For now, it is enough to note that full ADSL suffered from splitter issues such as user installation, ownership, power, wiring, and so on. G.lite addresses these remote splitter issues by eliminating the remote splitter entirely.

The splitter is gone in G.lite, but not the high-pass and low-pass filters, of course. Without these filters, voice operation on the loop would be impossible. The splitter is essentially an integrated high-pass and low-pass filter inside the same cover. G.lite does away with the splitter and distributes to high-pass and low-pass filters on the premises. The only difference between the NT in full ADSL and the NT in G.lite is the need to add the high-pass filter function to the NT.

The low-pass filter now becomes packaged as something commonly called a *microfilter*, although most are as large, if not larger, than a cigarette lighter. One microfilter is needed for each and every telephone or ISDN device on the premises. The microfilter usually plugs in between the device and wall socket, although there are special microfilters for wall-mounted telephones and related devices that attach directly to walls. Microfilters

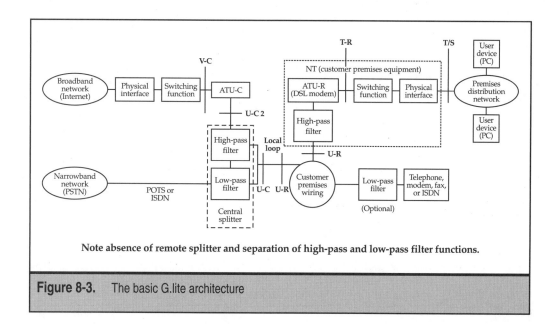

Note absence of remote splitter and separation of high-pass and low-pass filter functions.

Figure 8-3. The basic G.lite architecture

are designed to be customer installable. The use of the microfilter is optional in G.lite. It is still possible to deploy a "data-only" G.lite link.

Because the splitter function is now distributed, it is possible to connect the local loop to the customer premises wiring as before, without the need for a specially installed remote splitter. So the figure shows this use of the customer premises wiring to connect all devices from DSL modem to telephones and beyond. But it is important to realize that the PDN shown in the upper right of the figure is *not* necessarily the same as the customer premises wiring at the end of the local loop. It could be the same wire, but only if the PDN does not use the 1.1 MHz and below frequencies used by the voice and ADSL service. As mentioned previously, it is common that the PDN be some form of special LAN cabling run through the customer's premises and so entirely separate from the wiring used by ADSL and voice services. This might sound confusing, but this issue will be dealt with in detail in the chapter on ADSL premises arrangements.

AN ADSL NETWORK

Technically speaking, there is no such thing as an ADSL *network*. ADSL by itself cannot provide all that is needed from Web site to PC. It is more accurate to say "a network with ADSL links" but this section speaks loosely about the true situation.

ADSL technology is more than just a faster way to download Web pages onto a home PC. ADSL is part of a whole networking architecture that has the potential to supply residential and small business users all types of new broadband services. In this context, "broadband" services means services that need network links fast enough to support Internet streaming video.

Figure 8-4 shows just what a network based on ADSL links would look like. In the simplest version of this architecture, customers would essentially need only a new ADSL modem. This device would have some routing capabilities and one 10Base-T Ethernet port for attaching a LAN hub or single PC. All of the ordinary RJ-11 jacks that would support the existing analog telephone(s) in the home office/small office (SOHO) would require microfilters as shown in the figure, although these are needed only when other devices use the voice passband on the ADSL link. In many cases, additional premises wiring may need to be run for the ADSL modem, but this is beyond the scope of the ADSL network because inside wire belongs to the owner of the premises, both in the United States and in many other countries.

In the central office (CO), or local exchange, the analog voice service is passed to the CO voice switch with a central splitter arrangement. The ADSL local loops now terminate in an ADSL access node instead of leading directly to the CO switch as when analog

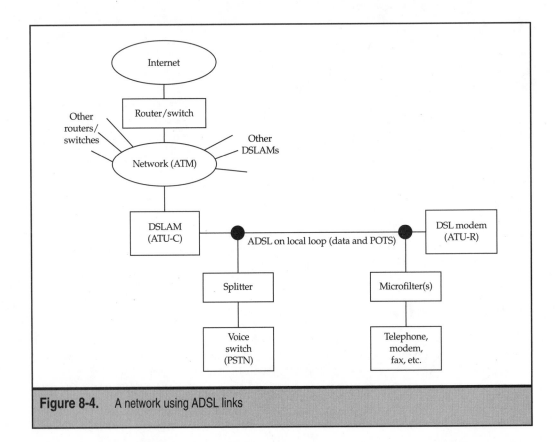

Figure 8-4. A network using ADSL links

modem PC access is used. The access node (which is a type of DSLAM) multiplexes the ATU-Cs from many ADSL links together. On the "back end" of the DSLAM, links to TCP/IP routers or ATM switches can be maintained.

There is a network (usually ATM) that is used to tie together many DSLAMs from many central offices. Many central offices might have multiple DSLAMs if a lot of DSL customers are serviced from that site. Other IP routers or ATM switches can be accessed by this network as well, although only one Internet access router or switch is required. The use of multiple switches and/or routers can provide for redundancy and also allow customers of the ADSL service provider to sign up for more than one ISP.

This point about multiple ISPs and ADSL is an important one that should not be forgotten. Cable modem users almost always are locked in to use the ISP that the cable service provider has an exclusive contract with. But many telephony regulators demand open access for services that local telephone companies provide. So a local ADSL service provider might allow their ADSL customers to sign up for more than one ISP (although only the one subscribed to is reachable by that customer, of course) to satisfy regulatory issues. In the United States at present, in cases where the ADSL service provider is essentially the same corporate entity as the network service provider (ISP), careful bookkeeping is needed to keep the unregulated ISP line of business providing the Internet access separate from the regulated local access line of business providing the ADSL service.

The reason that ATM is so attractive as a way to link DSLAMs with ISPs is that the use of ATM requires only one physical interface on each DSLAM and router or switch to link them all together using ATM *virtual circuits*. Linking all of this equipment with a physical mesh or point-to-point leased lines can be expensive and inefficient. Even a small ISP with a limited budget can install a single ATM interface on a router and so become reachable (at least in theory) by all customers on any DSLAM linked to the same ATM network.

Now, the same DSLAM to ISP router/switch connectivity could be achieved not by putting ATM on the ISP router, but by making a router out of the DSLAM. However, because the use of ATM is common on the ADSL link in full ADSL, and required in the case of G.lite, the DSLAM is usually an ATM switch. This is because the ATM function has to be there anyway, and there is no real need for the DSLAM to examine the IP packets themselves. The reality is a little more complex, but these issues are best discussed with DSLAM architectures in Chapter 12.

These ISP switches and routers enable users to access the services of their choice. Note that these services may also be housed in the CO, either provided by the local exchange carrier (LEC) or a competitor under a collocation agreement. In many cases, the servers may be located across the street within a short cable run from the CO. But in any case, the point is that the ATM network is not always necessary, nor is much of a network even required to link DSLAM to Internet.

Typically, services provided by the ADSL service providers would include Internet access through an ISP (of course), access to a work-at-home server (or the corporate intranet) through some form of virtual private network (VPN), video-on-demand (rarely, but ADSL providers are always willing to try video), and even servers from advertisers of information services such as on-line brokerages (also uncommon right now). Note that access to these services may be either through TCP/IP or ATM—ADSL allows for both, even though ATM-based services are quite rare.

ADSL AND LINE CODE STANDARDS

As with any other technology, ADSL is in need of standards. All technologies go through a phase of exploration and experimentation—early airplanes and automobiles, for example, took on many bizarre shapes and sizes. Before consumers will accept any new technology and pay hard-earned cash for it, however, the technology must become standardized enough to satisfy everyone. People want technology-based products that are consistent in appearance and performance, independent from a particular vendor, and will work with other devices in the same category.

In the United States, an ADSL standard for physical layer operation was first described in *American National Standards Institute* (ANSI) standard T1.413-1995. In other words, this document describes exactly how ADSL equipment communicates over a previously analog local loop. The document does not, nor is it intended to, describe the entire ADSL network architecture and services, or the internal functioning of the ADSL access node, and so forth. It specifies such ADSL fundamentals as *line coding* (how the bits are sent) and the *frame structure* (how the bits are organized) on the line.

ADSL products have been manufactured that use *carrierless amplitude/phase modulation* (CAP), *quadrature amplitude modulation* (QAM), and *discrete multitone* (DMT) technology as line coding techniques. Other line codes have been tried in a lab, but these are the most common. Whatever line coding technique is used, whenever the same two wires in a pair are used for full-duplex operation, either the frequency range must be split into upstream and downstream bandwidths (simple frequency division multiplexing, or FDM) or echo cancellation must be used. (Echo cancellation eliminates the possibility of a signal in one direction being interpreted as a "talker" in the opposite direction, and so "echoed" back to the origin.) In ADSL, both FDM and echo cancellation techniques can be combined and typically are, which means that due to the asymmetric nature of ADSL bandwidths, the frequency ranges may overlap, but do not coincide. So both FDM and echo cancellation are used in a sense at the same time.

In any case, T1.413 specifies that ANSI standard, compliant ADSL must use DMT coding with either FDM or echo cancellation to achieve full-duplex operation. That said, it should be noted that FDM is the simpler method to implement. Echo cancellation is always vulnerable to the effects of near-end crosstalk, where a receiver picks up signals that are being transmitted on an adjacent system. The other system may be another pair of wires, or even the transmitter of the same system, because obviously the closest transmitter to a given receiver is its own transmitter in the opposite direction. FDM avoids near-end crosstalk by allowing the receiver to totally ignore the frequency range that the near-end transmitters are sending on. Of course, FDM cuts down on the total amount of bandwidth available in either direction. So echo cancellation makes more efficient use of the bandwidth, but at the price of complexity and sensitivity. Also, echo cancellation enables the lowest possible frequencies to be used, which maximizes performance.

As far as line coding is concerned, ADSL could have used any one of a number of widely used methods. The familiar 2B1Q from ISDN DSLs and HDSL was a possibility, as were other choices, such as CAP or QAM. DMT was chosen for full ADSL and G.lite for a number of reasons, not the least of which was the inherent rate adaptive nature of DMT devices, which means that DMT devices can easily adjust to changing line conditions, such as moisture or interference. Also affecting the choice were DMT's resistance to noise (mostly AM radio) and the presence of digital signals on adjacent wire pairs (crosstalk). However, CAP products have been successful in many ADSL deployments. Today, the use of CAP (which is essentially a form of QAM) has all but disappeared from ADSL products. However, CAP/QAM is still worth discussing. This is because CAP/QAM still exists in older ADSL equipment and has been proposed as the line code for many other forms of DSL still under development. But it is best to consider echo cancellation in a little more detail first.

ECHO CANCELLATION AND DSLS

Some form of echo control is required whenever the same frequency range is used for sending signals in both directions at the same time on the same physical path.

Echoes commonly arise from impedance mismatches on the signal path. In other words, some of the signal is reflected back to the sender at these points. When the same frequency range is used in both directions, this signal reflection could easily be mistaken for a signal originating at the remote end of the circuit. Echo cancellers electronically subtract the signal sent from the signal received, allowing any signal actually sent from the remote end to be distinguished more easily.

One way to accomplish echo control is to separate the frequency range into upstream and downstream bandwidths (FDM). Now, no echo control is needed in the end devices.

The top part of Figure 8-5 shows what happens when no echo control is applied to ADSL. A 4 KHz baseband signal range for the analog voice passband is shown, along with a typical ADSL 175 KHz bandwidth for upstream traffic (from the home) and about a 900 KHz bandwidth for downstream traffic (to the home). This is asymmetrical, and this straight FDM approach does away with the need for echo control circuitry in the ADSL end devices.

However, pure FDM is not the most efficient use of the available bandwidth. The lower part of the figure shows a more effective approach where the upstream and downstream bandwidths actually overlap. Now, even though there is only a partial overlap in frequency range, echo control circuitry is needed in ADSL devices that use this approach.

Note that CAP-based ADSL devices typically use the FDM approach (no echo cancellation), whereas DMT-based ADSL devices typically use the echo cancellation approach, although there are exceptions. This echo cancellation approach is called an echo-FDM "combination" due to the asymmetrical nature of the devices. Basically, there are "FDM ADSL" and "echo cancelled ADSL" systems and equipment.

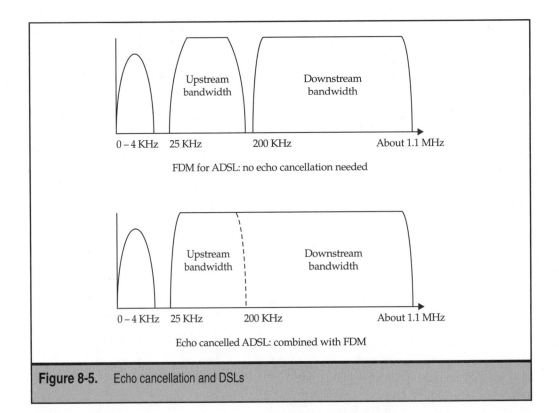

Figure 8-5.   Echo cancellation and DSLs

CAP AND DMT

CAP and DMT are the line codes that have been most frequently used in ADSL products (ATU-R and ATU-C). Line codes determine how the 0s and 1s of the digital signal are sent and received. One is CAP and the other is DMT.

At one point in time, no other controversy existed in ADSL (or DSL in general) that is at once so vital to the technology and yet so confusing to non-specialists. Sometimes the debate about the relative merits of each assumed almost religious fervor, but both CAP and DMT line codes work just fine in ADSL devices. The debate revolved around other issues, such as performance, cost-effectiveness, and signal processing delay (that is, how long it takes a bit to make its way through a DMT or CAP device). In other words, which is somehow *better* from a price/performance standpoint?

CAP is closely related to QAM, and many treat the two as virtually indistinguishable. DMT is more complex than CAP/QAM and fairly new as a line code. Some see CAP as a type of "last gasp of over-engineered QAM" and others see DMT as needlessly complex and a drain on resources.

In spite of the presence of DMT in the T1.413 standard, there is little in the standard that is tied to DMT exclusively. Both CAP and DMT can be used in ADSL, sometimes even in the same device.

HOW CAP WORKS

Several major vendors of ADSL equipment have used CAP as the line code. The line code simply determines how the 0s and 1s are sent from the ATU-R to and from the ATU-C, but this is not to downplay the line code's importance. A perfect line code should function well under less-than-perfect line conditions, including the presence of noise, crosstalk, and impairments, such as bridged taps and mixed gauges.

CAP is a close relative of a coding method known as quadrature amplitude modulation (QAM). In fact, they are almost completely mathematically identical, and engineers frequently refer to them in the same breath. The difference is in the implementation. The best description of CAP is probably "carrier-suppressed QAM." Because a carrier relays no information, it is common in a number of coding methods not to bother sending the carrier at all and reconstructing it electronically at the destination. This makes the technique "carrierless," or, more properly, "carrier suppressed."

Carrier suppression requires more circuitry in the end device than QAM does, but circuitry is now much cheaper than even a few years ago (a Hayes 9600 bps modem cost $799 in 1991, for example). So think of CAP as an "improved" QAM.

QAM established "constellations" based on two values of a received signal: amplitude and phase difference. Any point fixed by a given phase difference and amplitude represents a defined sequence of bits (for example, 0001 or 0101, and so on). CAP is essentially a type of QAM in which the QAM "constellation" is free to rotate (because there is no carrier to "fix" it at an absolute value). An element in the CAP circuitry called a *rotation function* can still determine the points of (and so the bits values in) the QAM constellation. So to get CAP from QAM, add a rotation function to the receiver and suppress the carrier at the sender.

CAP uses the entire local loop bandwidth up to about 1.5 MHz (except for the 4 KHz baseband analog voice "channel") to send bits all at once. In other words, there are no subcarriers or subchannels to worry about. Full-duplex operation is achieved by FDM, echo cancellation, or both, but almost all CAP products have used FDM exclusively. CAP, mostly due to its QAM roots, is a mature, stable, and well-understood technology.

CAP/QAM OPERATION

This section presents a few details on the operation of CAP and QAM.

All analog carrier signals are characterized by amplitude, frequency, and phase. Any one, or even all three, may be changed, or *modulated*, from the form generated by the carrier and so used for signaling the 0s and 1s that make up digital information content. As an example, consider phase. Phase may be used in a *differential phase shift keying* (DPSK) scheme to convey digital information. In a differential phase shift keying situation, the phase of the carrier wave is changed by a specific angle to represent each baud (change in line signaling condition) in the transmission. The use of the word *differential* means that the phase shift is measured from the current phase of the carrier, not from some absolute or reference phase.

In a simple example, two phase shifts can be employed. If the transmitter wishes to send a 1, it simply changes the phase of the carrier (from its current phase) by 1800. If the transmitter wishes to send a 0, the phase shift introduced is 00 (the current phase of the carrier is not changed). As a technical aside, the spectrum of the signal contains a strong amplitude at the carrier frequency with sideband signals appearing as a consequence of the phase shifting.

A step up from simple DPSK is *quadrature phase shift keying* (QPSK), which can be understood at two levels. On the surface, it is simply DPSK with 2 bits per signaling condition, called a baud. The phase shifts are measured from the current phase of the wave.

Although this level of understanding may suffice for many, it is misleading when considering more complex situations. In QPSK, the signal that is actually sent is a combination of a *sine wave* and *cosine wave* at some carrier frequency F. Because a sine and cosine wave function are always 90 degrees out of phase, they are said to be in *quadrature*, which means that two quantities are perpendicular, or 90 degrees different from one another, so the phrase is well used in this situation.

In fact, QPSK really has nothing to do with phase shifting. The phase shift is a consequence of the modulation of the amplitudes of two waves in quadrature. So if modulation increases the amplitude of one wave while the amplitude of the other wave decreases, this effectively shifts the phase of the resulting signal. If one understands QPSK at this level, the topic of QAM will be trivial. If the reader is not mathematically sophisticated, then a view of QPSK as shifting phases is sufficient. This presents a more detailed (and more correct) view for those who need to understand the concepts of digital communication more deeply.

QUADRATURE AMPLITUDE MODULATION (QAM)

Just as QPSK can be viewed at two levels, so can QAM. On the surface, a QAM modulator creates 16 different signals by introducing both phase and amplitude modulation. By introducing 12 possible phase shifts and two possible amplitudes, it is possible to obtain the 16 different signal types. These 16 signal conditions can represent 4 bits per line condition, or "4 bits per baud" signaling.

But as with DPSK, there is a deeper level of electronics involved. At a deeper level, QAM simply modulates the amplitudes of two waves in quadrature. Instead of using amplitudes of ±1, a simple QAM example uses four different amplitudes for each of the two waves. If the four amplitudes are labeled ±A1 through A4, then the 16 different signal types are simply obtained by using all possible pairs as amplitudes combined with simple sine and cosine wave functions (for example, A1 sin (F_t) + A1 cos (F_t), A1 sin (F_t) + A2 cos (F_t), ...) to create the necessary signal types. This creates the characteristic *constellation* pattern reproduced in all modem manuals based on QAM (and CAP is basically the same). A sample constellation is illustrated in Figure 8-6. Although it is much more common to represent these constellations as simple points, arrows have been used to emphasize the phase angles and amplitude differences.

Note that although the example in the figure uses 12 different phase angles, and 4 have two amplitudes of the sine and cosine components (a four-level system) to encode 4 bits per baud (change of signal, in this case phase and amplitude together). Other systems may use greater (or fewer) phase and amplitude level combinations. The greater the number of phase and amplitude levels, the greater the number of constellation points and correspondingly greater numbers of bits per signal. So, there is 16-QAM with 16 point constellations, 64-QAM with 64 point constellations, and 256-QAM and even 1024-QAM as well. Like all multibit encoding systems, QAM is limited to the number of levels that can be discerned at the receiver in the face of noise.

Consider how QAM/CAP may be used to deliver digital video information. The output bit stream from a digitizer is broken into 4-bit nibbles (one-half byte is one *nibble*) because the example uses QAM/CAP with 16 points in the constellation, as in the previous figure. The nibbles are fed (pardon the pun) to the encoder that selects the proper amplitudes for the waves in quadrature.

The output from the encoder is next fed to the actual modulator, which creates the correct output signal. Note that the modulator actually combines the appropriate amplitudes of the sine and cosine of the carrier frequency, thus creating the phase and amplitude shifts associated with the correct constellation point. Finally, the signal is filtered to assure that it does not interfere with other channels. Note that the resulting signal contains substantial energy at the carrier frequency; this should not be surprising. Both CAP

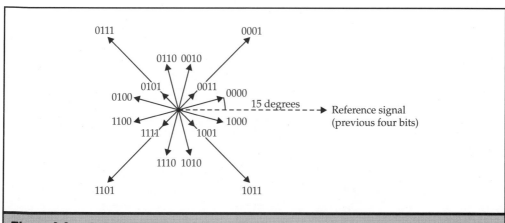

Figure 8-6. CAP (16-QAM or 4x4 QAM) modulation

and QAM are firmly analog line codes, meaning that it is easy to "slide" a 4 KHz voice channel "under" the frequencies used for CAP/QAM.

When used to deliver ADSL, CAP usually takes on the FDM form shown in Figure 8-7. Different CAP implementations have used different frequency ranges to carry upstream and downstream traffic, as shown in the figure as well. Note also that the signal power upstream is slightly greater than the power used downstream. The more bandwidth that is used for the downstream data, the faster the link can run, of course. In the extreme case, the downstream bandwidth is limited to be equal to the upstream bandwidth. But the upstream frequency always starts at 35 KHz and the downstream frequency always starts at 240 KHz. The maximum downstream frequency is about 1.49 MHz (1491.2 KHz) in all cases.

In CAP, there is only one carrier for upstream and downstream transmission. This differentiates CAP from DMT, as is shown shortly. The bits that CAP transfers are organized into transmission frames (different than, for instance, Ethernet frames) and a forward error correction (FEC) code can be added to minimize resends necessitated by noise and crosstalk on the line.

CAP actually defines two types of traffic: Class A and Class B. Class A traffic is organized into ATM cells or IP packets and can be delayed while a FEC is added without the client or the server failing. Class A traffic is not "latency-sensitive" in CAP talk. Because the use of the FEC involves the interleaving of data bits, the FEC is applied only to bits in

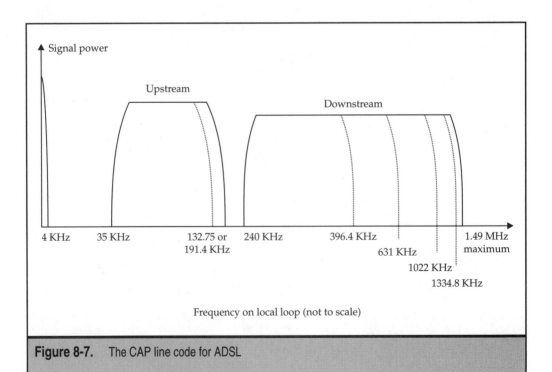

Figure 8-7. The CAP line code for ADSL

the "interleaved data" buffer. Class B traffic, such as a digitally encoded video channel, cannot be held up for FEC considerations, and so such "latency sensitive" traffic must be sent from a "fast data" buffer. DMT also follows this fast and interleaved buffer structure, although only an interleaved buffer with FEC is required in the standards.

DMT FOR ADSL

ADSL devices (the ATU-C and ATU-R) have been built that use QAM, CAP, and DMT technology as line codes. However, the official standard line code for both full ADSL and G.lite is DMT. Although DMT is often said to be "newer" than CAP or QAM, DMT was actually invented years ago by Bell Labs. DMT was never implemented until recently for many reasons, not the least of which was that CAP and QAM were sufficient for all telecommunications purposes common at the time.

The main reason that DMT languished was because DMT essentially puts 256 analog modems on a chipset and runs them all at the same time. Needless to say, this required breakthroughs not only in circuit miniaturization, but also in power delivery and heat dissipation. These breakthroughs have now all been made, of course, although DMT chips do run very hot.

Before getting into the details of DMT, which can be quite complex, it might be best to examine just how the concept of "256 modems on a chip" relates to DMT. Figure 8-8 shows how two analog modems transfer bits over a local loop. Now, there is a lot more than the local loop between the two modems, but that is all that needs be shown to understand DMT.

The modems in the figure use the 0–4 KHz voice bandwidth (actually more like 300–3400 Hz) to achieve a maximum bit rate of 33.6 Kbps in both directions on the wire. Going faster would require more bandwidth given the same levels of power and noise on the line, but this is just not possible for many reasons. For example, any digital loop carriers between the modems will only digitize the 0–4 KHz passband, and there are also voice switch bandwidth limits to deal with.

How can someone use normal V.35 analog modems to get faster Internet access? Figure 8-9 shows one way. This arrangement shows four modems grouped together in a form sometimes called *analog modem bonding* because the result is similar to bonding

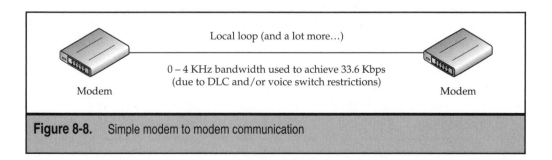

Figure 8-8.   Simple modem to modem communication

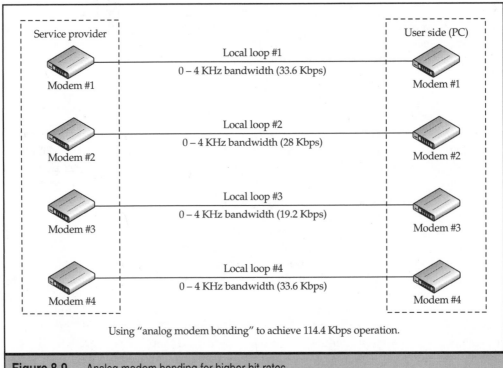

Using "analog modem bonding" to achieve 114.4 Kbps operation.

Figure 8-9. Analog modem bonding for higher bit rates

64 Kbps channels in ISDN. Commercial products using this concept with analog modems have been introduced and marketed in the past few years.

The user and service provider must have compatible units, of course. The ones shown in the figure have four modems, each using a separate telephone line. The user can now dial four separate numbers and have each pair of modems *train* (connect) based on the local line conditions. For example, the figure shows modem pair #1 on local loop #1 training at 33.6 Kbps (very good line conditions), modem pair #2 on local loop #2 training at 28 Kbps (not as good line conditions), modem pair #3 on local loop #3 training at 19.2 Kbps (worse line conditions), and modem pair #4 on local loop #4 training at only 33.6 Kbps (very good line conditions). Each modem pair uses the 0–4 KHz voice passband, of course, due to the limitation outlined earlier in this section.

So what will be the throughput that the user sees upstream and downstream? The total comes to 114.4 Kbps, which is comparable to ISDN BRI. Note that on bad days, the throughput might be lower, but on good days the throughput could be as high as 134.4 Kbps (4 × 33.6 Kbps, but no higher).

What's wrong with this picture? Nothing, except that four separate lines could be expensive, and the multiplexing and demultiplexing of four separate streams of bits that would all represent one serial IP packet sitting in a buffer (and each stream could carry a different number of bits per unit time) would require a lot of complexity and processing power at the chip level.

Nevertheless, this is the idea behind DMT. The only step needed to derive DMT for ADSL from this series of modems is to realize that the limiting factor here is not the complexity of chip buffer design, but rather the need for each modem pair to use the same frequency range of 0–4 KHz. Suppose there were no DLCs to deal with, and the loops were all home runs to the central office location. And further suppose that the loops terminated not only at the voice switch, but at a DSLAM as well. Congratulations. This is DMT.

Figure 8-10 shows how the modems now use a separate frequency range (the subcarrier) to share a single local loop and achieve multimegabit operation. The single loop is shown in bold, and there are now 256 modems, the last of which is using the 1020 to 1024 KHz frequency range. The service provider and user side modems are now the ATU-C and ATU-R.

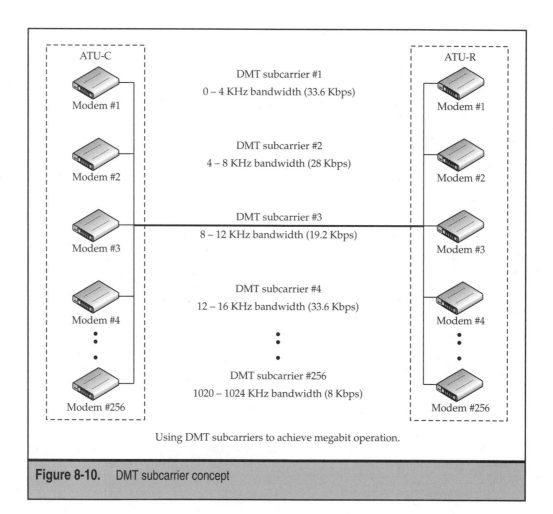

Using DMT subcarriers to achieve megabit operation.

Figure 8-10.   DMT subcarrier concept

All of the other concepts from analog modem bonding now apply to these modems, but on a single local loop. The modems on the chipset still train independently, based now not on individual line conditions, but on noise and crosstalk present in the particular subcarrier frequency range. Some subcarriers might be shut off altogether, and the higher subcarriers can be very inefficient (for instance, in the figure, subcarrier #256 only trains at 8 Kbps). But even if each subcarrier trains at only a modest 24 Kbps, this yields a respectable 6.144 Mbps in each direction.

This figure shows only the concept of DMT, and not the reality, which is detailed shortly. But many of the ideas used in DMT can be understood from this simple conceptual diagram. For example, powering and getting rid of the heat generated by all these chips and modems was and is an issue in DMT. The buffering, feeding of bits onto many paths that are all potentially running at different speeds (or not at all), and gathering the arriving bits back into the original packet was the real complexity and challenge of DMT. Some early DMT boards had five main chips on them. Only one handled all of the modem subcarriers, and the other four were for buffer and memory management.

Note that conceptually, each DMT subcarrier is bidirectional. That is, if subcarrier #3 is trained at 19.2 Kbps, this subcarrier is capable of carrying bits at this speed in both directions. One way to try to simplify the bit management task is to dedicate some subcarriers to strictly downstream operation and dedicate others to strictly upstream operation. This is what is done in real DMT for ADSL. When more subcarriers are dedicated for downstream operation than for upstream, this gives ADSL its distinctive asymmetrical nature. Symmetrical, optional operation limits the number of downstream subcarriers to be equal in number (and speed) to the number of upstream subcarriers, and so on.

Rather than try to squeeze more appreciation for DMT out of a conceptual diagram, the time has come to look at real DMT in detail.

DMT BINS

DMT works by first dividing the entire bandwidth range on the formerly analog passband limited local loop into a large number of equally spaced subcarriers. Technically, they are called subcarriers, but many people still call them subchannels. Another common term for a subcarrier in DMT is *bin*. Above the preserved baseband analog signaling range, this bandwidth usually extends to 1.1 MHz. The entire 1.1 MHz bandwidth is divided into subcarriers, starting at 0 Hz and numbered from 0 to 256. Each subcarrier occupies 4.3125 KHz (the spacing provides a minimal *guardband* between the subcarriers), giving a total bandwidth of 1.104 MHz on the loop. Some of the subcarriers are special, and others are not used at all. For example, channel #64 (bin 64) at 276 KHz is reserved for a pilot signal.

The structure of the DMT bins and some of the more important bins are shown in Figure 8-11.

Bins 0 and 256 are never used for data. Bin 0 will end up with whatever direct current there is on the loop and so cannot be used for data. Bin 256 is at something called the *Nyquist frequency* and so cannot be modulated in any meaningful way for data (modulating this bin would push the frequency past the 1104 KHz limit). Most DMT systems can

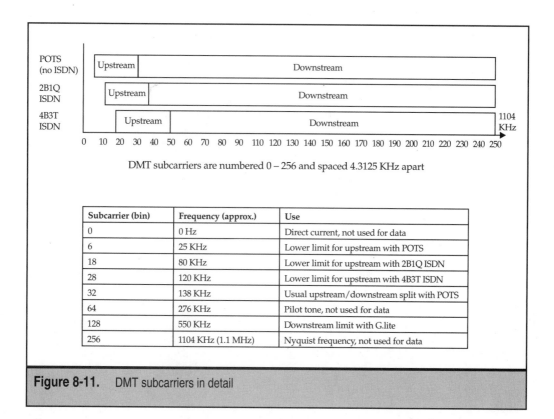

Figure 8-11.   DMT subcarriers in detail

use only 249 or fewer subcarriers for information. The lower bins, 1 through 6 in most cases, are reserved for 4 KHz passband analog voice. Because 6×4.3125 Hz = 25.875 KHz, it is common to see 25 KHz as the starting point for ADSL services. Note that a wide guardband is used between the analog voice and the DMT signals. In addition, the signal loss at the upper bins, such as 250 and above, is so great that it is difficult to use them at all for information transfer on long loops.

There are two other possible starting points for upstream traffic in ADSL DMT. Bin 18 at about 80 KHz is used as the starting point when the voice service on the local loop is ISDN using the 2B1Q line code. Bin 28 at about 120 KHz is used as the upstream starting point if the ISDN voice service uses a line code known as 4B3T (4 binary, 3 tertiary). Note that the presence of ISDN simply shifts the upstream starting point in terms of subcarriers and decreases the number of available downstream bins somewhat. For simplicity in the rest of this chapter, and in most of this book, it is assumed that the legacy voice service on the local loop is analog voice.

There are 32 upstream bins, usually starting at bin 7, and 250 downstream bins, which might overlap with the higher upstream bins. Because each of the bins is 4.3125 KHz wide, of course, and only when echo cancellation is used are there actually 250 downstream bins.

When only FDM is used for echo control, there are typically 32 upstream bins and only 218 or less downstream bins because they no longer overlap. The upstream bins occupy the lower end of the spectrum for two reasons. First, the signal attenuation is less there, and customer transmitters are typically lower-powered than local exchange transmitters, which is a concern. Second, there is more noise at the local exchange with the possibility of crosstalk, so it only makes sense to use the lower portions of the frequency range for the upstream signals.

Bin 64 carries a *pilot tone* that helps with bit reassembly at the receiver and so cannot be used to carry data. Finally, bin 128 is the upper subcarrier limit for G.lite, essentially limiting G.lite to downstream speeds of about 1 Mbps.

When ADSL devices that employ DMT are activated, each of the subcarriers is "tested" by the end devices for attenuation. In actual practice, the testing is a complex kind of handshaking procedure, and the parameter used is *gain* (the reciprocal of the attenuation). The noise present in each of the subcarriers is measured as well.

Usually, each of the numerous subcarriers employs its own line coding technique based on QAM. This may strike some as odd, given the fervor that vendors have shown in the past when seeking to distinguish CAP/QAM and DMT. Nevertheless, there obviously is at least some QAM in DMT. The real attraction of DMT is not so much that it is different than CAP and QAM, but rather that based on DMT's performance monitoring, some subcarriers will carry more bits per baud than others. For example, bin 47 might be running 16-QAM (4 bits per baud) while the adjacent bin 48 has to make due with 8-QAM (3 bits per baud). The total throughput is the sum of all the QAM bits sent on all the active subcarriers (some may be completely "turned off").

Moreover, all of the subcarriers are constantly monitored for performance and errors. The speed of an individual bin or group of bins can vary, of course, giving DMT a "granularity" of 32 Kbps. In other words, a DMT device might function at 768 Kbps or 736 Kbps (that is, 32 Kbps less), depending on operational and environmental conditions. Just by way of comparison, CAP devices usually offer 340 Kbps granularity (768 Kbps or 428 Kbps), but pure QAM can offer granularity as fine as 1 bps, which means that nothing technically limits CAP/QAM to one level of granularity but not another. In fact, some vendors of CAP-based ADSL equipment have claimed 32 Kbps granularity, and even RADSL capabilities. It should be noted that these CAP RADSL products modify their spectrum when the rate changes, and now become a real issue to manage with regard to spectral compatibility.

DISCRETE MULTITONE (DMT) OPERATION

Figure 8-12 shows discrete multitone technology in operation in an ADSL device on a typical local loop. The figure actually has two parts. The upper part shows a kind of ideal situation, such as that found in a straight run of 24-gauge copper wire less than 18,000 feet without a lot of outside noise (good luck finding one of those). The only real attenuation effects come from the distances and frequencies involved. The lower part of the figure shows a typical local loop in the real world.

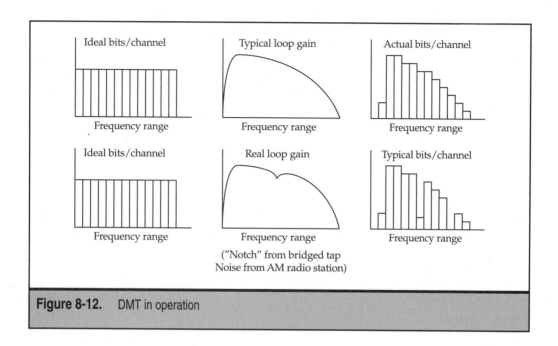

Figure 8-12.   DMT in operation

Consider the ideal design first. Across the frequency range, on the left, there exists a targeted maximum number of bits per second per subcarrier (bin) that the device would like to send and receive. However, the middle figure shows the situation on a typical loop. The gain (the reciprocal of the attenuation) is better or worse depending on frequency. At higher frequencies, distance effects dominate; at lower frequencies, impulse noise and crosstalk dominate. This leaves a broad middle range (about 25 KHz to 1.1 MHz) for signals, with the gain slowly dropping off with increasing frequency.

DMT devices can measure the gain in each subcarrier and adjust the actual number of bits per second per channel to reflect the actual line gain profile. The upper right part of the figure shows this.

Of course, the real world is not so cooperative. The lower series on the figure shows a real loop gain profile in the middle. Two line impairments are added. The first is a distinctive "notch" caused by the effects of a bridged tap. The bridged tap acts like a long delay circuit as the signal travels out and back. The returning signal interferes with the main signal. The placement of the notch in the frequency range depends on the wavelength (length) of the bridged tap itself; however, one frequency or another will be affected by the standing wave. (To interfere in the voice range, the bridged tap would have to be about 60 miles [about 96 km] long.) The second impairment is noise from a nearby AM radio station. These stations broadcast in the same high KHz (or "kilocycles" in AM radio talk) range as that of ADSL devices trying to listen in. Because local loop wires are long antennas, it comes as no surprise that this signal is picked up (analog voice, operating at 4 KHz baseband, is immune to this AM radio interference).

DMT devices can measure the gain in each subcarrier and adjust the actual number of bits per second per channel to reflect the actual line gain profile. The right part of the figure shows this. Note that some channels are even "turned off." Also note that DMT is inherently rate adaptive, as the figure illustrates, as channels increase or decrease their ability to support a given bit rate.

CHAPTER 9

The ADSL Interface and System

The ADSL standard technically defines a "metallic interface" between a network (the PSTN) and customer. However, the ADSL Forum (now just the DSL Forum) has extended this interface into the full architecture introduced previously and even proposed several ways for services to be accessed and delivered over the ADSL interface. This chapter describes the standard ADSL interface in full. This standard is based on ANSI T1.413, the *Network and Customer Installation Interfaces—Asymmetric Digital Subscriber Line (ADSL) Metallic Interface* specification. Many ADSL products are based on the information in the ANSI specification in the United States, from DSL Forum technical recommendations, and from the ITU-T G.992.1 (full ADSL) and G.992.2 (G.lite) and detailed here. The chapter then explores some of the aspects of the overall ADSL architecture and service platforms outlined by the ADSL equipment vendors and service providers.

ADSL UNIDIRECTIONAL DOWNSTREAM TRANSPORT

ADSL interfaces can do more than support a single bit stream to and from the customer premises, although this is certainly an option. ADSL can do a lot more than this. ADSL, like most transports, is a *framed transport*; the bit stream inside the ADSL frames can be divided into a maximum of seven bearer channels (just called bearers in ADSL) at the same time. The bearers fall into two main classes: There can be up to four totally independent downstream bearers that always function unidirectional ("simplex" in the specifications) downstream. That is, these four bearer channels can carry only bits downstream to the customer. The four bearers are designated AS0 through AS3. The *AS* has no real meaning except perhaps to its inventor. In addition to the AS channels, there may be up to three bidirectional ("duplex" in the specifications) bearers that can carry traffic both upstream and downstream. These bearers are designated LS0 through LS2. The *LS* is apparently as meaning-free as *AS*. Note that these bearer channels are logical channels, and that bits from all channels are transmitted at the same time over the ADSL link and do not use dedicated bandwidth.

Any bearer channel can be programmed to carry bits in any multiple of 32 Kbps (the "natural" DMT granularity). Bit rates that are not simple multiples of 32 Kbps (70 Kbps, for example) can be supported, but only by carrying the "extra" bits (in the 70 Kbps example, 6 Kbps) in the shared overhead area of the ADSL frame.

The following discussion of possible combinations of AS and LS subchannels becomes quite complex. So before plunging in, it might be a good idea to consider just what aspects of ADSL this chapter is concerned with.

The ITU-T specifications define in general two ways of sending information inside of ADSL frames. The first way, which most ADSL equipment vendors support, is to use asynchronous transfer mode (ATM) cells to carry the information inside of the ADSL frames. ATM cells are fixed length data units with a 5-octet (byte) header and 48-octet payload or information area. So ATM cells are always exactly 53 octets long. Content such as IP packets are carried in ATM by chopping up the packet into a series of ATM cells. These ATM cells are multiplexed onto *virtual circuits* (VCs) that are represented by a *virtual path identifier* (VPI) and *virtual channel identifier* (VCI) in combination. So an ATM VC might be identified by the VPI:VCI combination 1:48. All cells multiplexed onto a link or physical subchannel carry the same VPI and VCI to make reassembly possible.

The second way defined by the ITU-T for sending information inside of ADSL is *synchronous transfer mode* (STM). STM is sometimes called "bit synchronous" mode because the ADSL equipment is not aware of any structure at all to the bits flowing on the subchannels such as AS0 or LS1. All ADSL knows is that there are bits inside the ADSL frames on the subchannels. There is no multiplexing that can be done on STM traffic in ADSL equipment, because all ADSL knows about STM are bits, not packets or frames. So the main reason that there are more than just AS0 or LS0 in ADSL is to allow STM ADSL equipment to do more than one thing at a time. For example, if an STM subchannel like AS0 is busy downloading a data file, the whole idea of having AS1 is to allow for the simultaneous (multiplexed) downloading of a digital video stream or some other form of information. This is an important point.

So the ITU-T basically divides ADSL operational modes into ATM and everything else (STM). The problem is that most ADSLs are used for high-speed Internet access and the only information format the user (or Internet) cares about is IP packets, not unstructured bit streams (STM) or cells (ATM). So the DSL Forum defined a third type of ADSL operational mode: *packet mode* (also often called *frame mode*). In packet or frame mode, ADSL equipment can operate on (route) the IP packets inside the ADSL frames. The main point here is that with packet or frame mode, as with ATM, only subchannels AS0 and LS0 are required for ADSL.

The structure of the subchannels that are discussed in detail in this chapter with regard to STM for ADSL are shown in Figure 9-1.

Only AS0 for downstream traffic and LS0 for upstream or downstream traffic are required in STM. The STM subchannels run at multiple of 32 Kbps, with the exception of LS0, which is allowed to run at 16 Kbps. Note that these subchannels are defined between the ATU devices and the higher layers of ADSL, where operations such as bit multiplexing, or ATM switching, IP routing, and multiplexing (although there are no ATM or IP functions in STM) are done. In the following sections, various combination speeds for the AS and LS subchannels are explored to give classes of upstream and downstream ADSL speeds. One side, the ATU-C side, handles the interface to the network and would be implemented in the DSLAM, and the other side, the ATU-R side, handles the interface to the customer premises equipment (CPE) and would be implemented in the user's DSL modem.

In addition to the traffic-bearing subchannels, there is an optional subchannel defined for a *network reference timing* (NRT) stream that is used for voice. There is also finally a mandatory management subchannel for what is officially known as *operations, administration, and maintenance* (OAM).

Recall that multiplexing in STM is handled only through the use of additional subchannels beside AS0 and LS0. Also, LS0 is the only way to transfer data upstream to the ATU-C.

In contrast to STM for ADSL, Figure 9-2 shows how ATM is used with ADSL. The figure shows a different subchannel structure than before, mainly because multiplexing can be done with ATM cells and need not rely on different subchannels to allow the mixing of traffic flows.

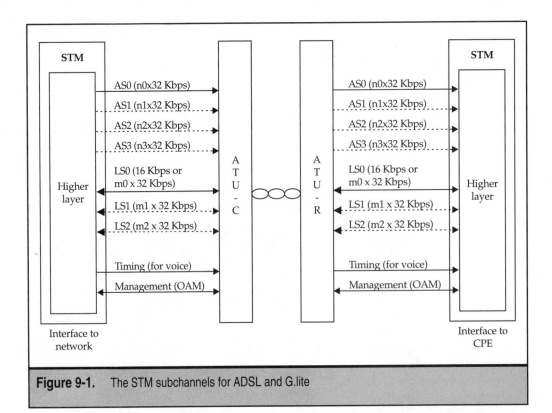

Figure 9-1. The STM subchannels for ADSL and G.lite

Things are simpler when ATM is used with ADSL and G.lite. There can be two ATM ports in the higher layer network interface, although only Port 0 is required. This allowance for two ports might sound odd because ATM provides all the multiplexing needed, but there are other reasons for allowing ATM Port 1 to exist. ATM cells can carry more than just IP packets, but the format of the ATM cell payload is quite different for traffic other than IP packets. So if IP packets are multiplexed onto an ATM cell stream and carried by ATM Port 0, a stream of ATM cells on ATM Port 1 can carry digital voice or video without relying on IP packets to carry this voice or video. Naturally, if the digital voice or video is carried inside of the IP packet, then only ATM Port 0 is needed. ATM Port 1 allows the ADSL link to carry IP packets and something else at the same time. Although not shown in Figure 9-2, the speeds of the ATM subchannels essentially mimics the same 32 Kbps multiple structure as the STM subchannel speeds.

ATM cell data transfer is done at Port 0 by mapping downstream ATM cells to ATU-C and ATU-R subchannel AS0 and mapping upstream ATM cells to ATU-C and ATU-R subchannel LS0, as shown in Figure 9-2. If the optional ATM Port 1 is implemented, then the second port (Port 1) is mapping downstream ATM cells to ATU-C and ATU-R subchannel AS1 and mapping upstream ATM cells to ATU-C and ATU-R subchannel LS1, as also shown in the figure. The ATM cell streams are ATM0 and ATM1 respectively. There are certain ATM management cells that flow in both directions on both the AS and LS subchannels, but the flow for user data transfer is as shown in the figure in all cases.

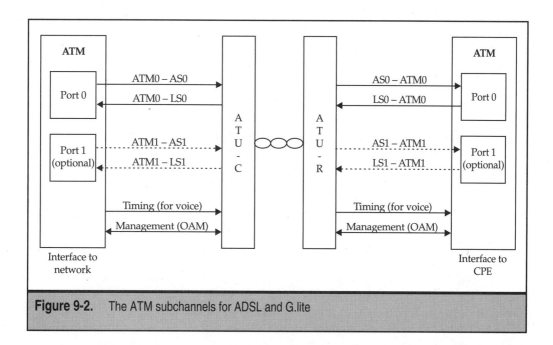

Figure 9-2. The ATM subchannels for ADSL and G.lite

There is an optional timing signal (the *network timing reference*, or NTR) as with STM, and the management OAM subchannel still exists. This management channel is for ADSL and should not be confused with the ATM management cells that flow on the AS and LS subchannels, the cells which manage the ATM layer and not the ADSL layer.

As mentioned previously, the ITU-T basically establishes ATM for ADSL and uses STM for everything that is not ATM cells inside ADSL frames. But if the main type of traffic on an ADSL link is IP packets inside of variable length frames and not fixed length ATM cells, perhaps there should be some accommodation for IP packets besides "other." So the DSL Forum established *packet mode* in addition to ATM and STM mode for ADSL and G.lite links.

Packet mode essentially uses the ATM subchannel structure and concepts for IP packets inside of frames and not the STM subchannel structure. The main idea behind packet mode is that because IP packets can provide their own multiplexing of traffic flows, using ATM to provide this multiplexing is not necessary, and STM is not the best mode because STM has no multiplexing provision at all. Packet mode makes the ADSL and G.lite devices in a sense "IP packet aware" and because ADSL and G.lite are commonly used for fast Internet access and little else, this makes perfect sense. However, IP packets can still be carried inside of ATM cells on ATM subchannels, and this is commonly done on ADSL links.

This overview of ADSL (and G.lite) subchannels will help with understanding the AS and LS speeds and options detailed in this chapter. The point is not to confuse, but rather to allow for the appreciation of the flexibility of ADSL in terms of speeds and device implementation.

So bearer channels to and from the ATUs (AS and LS), with the exception of the allowance for a 16 Kbps LS0 in STM mode, can run at multiples of 32 Kbps. Now, allowing bearers to run at almost any multiple of 32 Kbps might not be the best idea in the world, especially when interoperability is a concern. Accordingly, the ANSI ADSL specification has established four *transport classes* for the downstream simplex bearers. These are based on simple multiples of 1.536 Mbps (the user transfer rate of a T1). The transport classes are 1.536 Mbps, 3.072 Mbps, 4.608 Mbps, and 6.144 Mbps. The duplex bearers can carry a control channel and some ISDN channels (the BRI and 384 Kbps). Please note that ADSL is not limited to any particular transport classes in and of itself. Future specifications may be made for 1.544 Mbps transport (the full T1 line rate) or 2.048 Mbps (the E1 line rate). No maximum speed is defined for any bearer. The upper limit is dependent on the total carrying capacity of the ADSL link.

Consider STM mode first. ADSL products have established various subchannel data rates for the default bearer bit rates. The maximum transport class speed of 6.144 Mbps is not allowed on all AS bearers at the same time. The limitations are shown in Table 9-1.

Support for at least AS0 is mandatory, regardless of mode. The maximum number of subchannels that can be active at any given time and the maximum number of bearer channels that can be transported at the same time in an ADSL system depends on the transport class. Transport class support depends on the achievable line rate of the specific ADSL loop and the configuration of the subchannels, which may be configured to maximize the number of subchannels or the line rate. Switching between configured subchannel speeds and numbers is intended for future study. For now, whatever structure and rate is put in place on an ADSL link remains fixed.

The transport classes themselves are numbered 1 through 4. Support for transport classes 1 and 4 are mandatory, and support for transport classes 2 and 3 are optional. In addition, a series of optional transport classes prefixed by *2M* are intended for use with 2.048 Mbps E-carrier-based systems in common use outside of the United States.

Transport class 1 is mandatory and is intended for the shortest loops, but it offers the highest downstream capacity of any ADSL configuration. This class carries 6.144 Mbps downstream and can be made of any combination of one to four bearer channels running at simple multiples of 1.536 Mbps. Support for at least one subchannel running at

Subchannel	Subchannel Data Rate	Allowed Values of n_x
AS0	$n_0 \times 1.536$ Mbps	$n_0 = 1, 2, 3,$ or 4
AS1	$n_1 \times 1.536$ Mbps	$n_1 = 0, 1, 2,$ or 3
AS2	$n_2 \times 1.536$ Mbps	$n_2 = 0, 1,$ or 2
AS3	$n_3 \times 1.536$ Mbps	$n_3 = 0$ or 1

Note: Rates equivalent to multiple DS1s are also supported.

Table 9-1.   ADSL Subchannel Rate Restrictions for Default Bearer Rates

6.144 Mbps on AS0 is mandatory. Transport class 1 can carry the following optional configurations, all of which add up to 6.144 Mbps:

▼ One 4.608 Mbps bearer channel and one 1.536 Mbps bearer channel
■ Two 3.072 Mbps bearer channels
■ One 3.072 Mbps bearer channel and two 1.536 Mbps bearer channels
▲ Four 1.536 Mbps bearer channels

Transport class 2 is optional and carries 4.608 Mbps downstream. This class may be comprised of any combination of one to three bearer channels running at multiples of 1.536 Mbps. Systems can provide any or all of the bearer rates because none are mandatory. AS3 is never used in transport class 2. Transport class 2 can carry the following configurations, all of which add up to 4.608 Mbps:

▼ One 4.608 Mbps bearer channel
■ One 3.072 Mbps bearer channels and one 1.536 Mbps bearer channel
▲ Three 1.536 Mbps bearer channels

Transport class 3 is also optional and carries 3.072 Mbps downstream. This class may be made up of one or two bearer channels running at simple multiples of 1.536 Mbps. Systems can provide any or all of the bearer rates because none are mandatory. AS2 and AS3 are never used in transport class 3. Transport class 3 can carry the following configurations, all of which add up to 3.072 Mbps:

▼ One 3.072 Mbps bearer channel
▲ Two 1.536 Mbps bearer channels

Transport class 4 is mandatory and runs on the longest loops, but offers the minimum downstream carrying capacity. The bearer channel is just 1.536 Mbps running on AS0.

The ADSL specifications also takes into account networks based on the E-carrier hierarchy of 2.048 Mbps, which is common outside of the United States. In fact, all of the common local loop structures found outside of the United States are addressed in the specification, specifically in Annex H to T1.413. Only AS0, AS1, and AS2 are supported using this 2M structure, as shown in Table 9-2.

Subchannel	Subchannel Data Rate	Allowed Values of n_x
AS0	$n_0 \times 2.048$ Mbps (optional)	$n_0 = 0, 1, 2,$ or 3
AS1	$n_1 \times 2.048$ Mbps (optional)	$n_1 = 0, 1,$ or 2
AS2	$n_2 \times 2.048$ Mbps (optional)	$n_2 = 0$ or 1

Table 9-2. ADSL Subchannel Rate Restrictions for 2.048 Mbps (optional)

As with the 1.536 Mbps structure, support for AS0 is the minimum requirement. The maximum number of subchannels that can be active at any given time and the maximum number of bearer channels that can be transported at the same time in an ADSL system depends on the transport class. Furthermore, transport class support depends on the achievable line rate of the specific ADSL loop and the configuration of the subchannels, which may be configured to the maximize number of subchannels or the line rate. Switching between configured subchannel speeds and numbers is intended for future study. For now, whatever structure and rate is put in place on an ADSL link remains fixed.

For 2M structures, the transport classes themselves are numbered 2M-1 through 2M-3. Support for all 2M transport classes is optional. The configurations of the 2M transport classes closely follow the 1.536 Mbps transport classes. That is, transport class 2M-1 still runs at 6.144 Mbps downstream in total.

Class 2M-1 can be made up of any combination of one to three bearer channels running at simple multiples of 2.048 Mbps. All transport class 2M-1 configurations are optional and can carry the following configurations, all of which add up to 6.144 Mbps:

▼ One 6.144 Mbps bearer channel

■ One 4.096 Mbps bearer channel and one 2.048 Mbps channel

▲ Three 2.048 Mbps bearer channels

Transport class 2M-2 is optional and carries 4.096 Mbps downstream. 2M-2 may be made up of one or two bearer channels running at simple multiples of 2.048 Mbps. Systems can provide any or all of the bearer rates because none are mandatory. AS2 is never used in transport class 2M-2. Transport class 3 can carry the following configurations, all of which add up to 4.096 Mbps:

▼ One 4.096 Mbps bearer channel

▲ Two 2.048 Mbps bearer channels

Transport class 2M-3 is optional and runs on the longest loops, but offers the minimum downstream carrying capacity. The bearer channel is just 2.048 Mbps running on AS0.

Now consider ATM mode. One other point must be made about ADSL transport classes. Support for ATM mode is optional for carrying ATM cells downstream. As already mentioned, ATM cells are fixed-length, 53-octet (8 bits, or a byte) data units. ATM cells each have a 5-octet header and a 48-octet payload. Information is carried in the 48-octet payload according to *ATM Adaptation Layer 1* (AAL1) rules. The AAL defines how information is formatted within the ATM cell payload area. In AAL1, which offers a *constant bit rate* (CBR) and stable delay through the network with connections between endpoints, one payload octet is used for additional overhead, and the other 47 octets carry user data. AAL1 is the easiest and simplest way to make ATM cell streams look and act like traditional circuits. However, the use of AAL5, which is used when ATM cells carry IP packets or other forms of variable length data units or frames, is also allowed (and mandated) by the DSL Forum. When ADSL is used for downstream ATM cell transport, only AS0 is used, so there is only one configuration option—AS0—which can run at one of four different rates. These rates are defined as ATM transport classes 1 through 4 and run at 1.760 Mbps, 3.488 Mbps, 5.216 Mbps, and 6.944 Mbps, respectively. The odd-sounding speeds are a result of the desire to preserve compatibility with existing AAL1 and circuit definitions already entrenched in ATM documentation, and they take into account the effects of ATM overhead on user data rates.

ADSL BIDIRECTIONAL (DUPLEX) TRANSPORT

So much for the unidirectional (simplex) downstream channels. Upstream simplex channels are being studied, but up to three bidirectional (duplex) bearer channels can be transported at the same time on an ADSL interface. One of these is always the mandatory control channel, designated the C channel. The C channel can carry signaling messages for selection of services and call setup. All user-to-network signaling for the simplex downstream channels is carried here, and the C channel can also carry signaling for the duplex channels, if present, as well.

The C channel is always active and runs at 16 Kbps in transport classes 4 and 2M-3. In transport classes 4 and 2M-3, the C channel messages are always carried in a special overhead section of the ADSL frame. All other transport classes use a 64-Kbps C channel, and the messages are transported in duplex bearer channel LS0.

In addition to the C channel, an ADSL system can carry two optional bidirectional bearer channels: an LS1 running at 160 Kbps and an LS2 running at either 384 Kbps or 576 Kbps. The exact structure of the bidirectional channels varies by transport class as defined for the simplex channels, so the easiest way to relate them is by using a table. Table 9-3 relates the duplex channel structures to the simplex bearer channel transport classes.

Transport Class	Optional Duplex Bearers That May Be Transported (See Note 1)	Active ADSL Subchannels
1 or 2M-1 (minimum range)	Configuration 1: 160 Kbps + 384 Kbps	LS1, LS2
	Configuration 2: 576 Kbps only	LS2 only
2, 3, or 2M-2 (mid range)	Configuration 1: 160 Kbps only	LS1 only
	Configuration 2: 384 Kbps only (see Note 2)	LS2 only
4 or 2M-3 (maximum range)	160 Kbps only	LS1 only

Table 9-3.   Maximum Optional Duplex Bearer Channels Supported by Transport Class

NOTE: 1. When the 160 Kbps optional duplex bearer channel is used to transport ISDN BRA, all signaling associated with the ISDN BRA (160 Kbps) is carried on the D channel of the 2B+D signal embedded in the 160 Kbps. Signaling for the 576 Kbps, 384 Kbps, and non-ISDN 160 Kbps duplex bearers may be included in the C channel, which is shared with the signaling for the downstream simplex bearer channels.
2. Whether transport classes 2, 3, or 2M-2 should support the 576 Kbps optional duplex bearer is left for further study.

The two notes mean the following:

1. If the channel carries ISDN, the D channel can still be used for signaling; otherwise the C channel is used.

2. Nobody knows yet whether mid-range ADSL loops can run at 576 Kbps upstream.

As might be expected, the bidirectional channels have an option to transport ATM cells. In addition to AAL1, the channel carries ATM cells formatted according to AAL5, which is just another way of mapping user information into a series of ATM cell payloads. The attraction of AAL5 is that it has the minimal overhead of any AAL. However, AAL5 is intended for *variable bit rate* (VBR) applications and will not guarantee a stable delay through the network (therefore, the application itself must provide the stable delay if one is required). The data rates for the LS channel when used for ATM cell transport is 448 Kbps or 672 Kbps. Again, the odd rates are due to a desire for compatibility with existing ATM documentation and takes into account the effects of ATM overhead on user data rates.

COMBINING THE OPTIONS

The establishment of standard bearer channel structures and speeds at least prevents a free-for-all by ADSL equipment designers and vendors. However, the presence of a set of transport classes for simplex downstream operation and multiple bidirectional bearer channel options does not completely achieve this goal. There still exists a bewildering array of options in both directions that technically comply with the specifications. What is needed is a way to combine AS channel structures with LS channel structures in a way that is both meaningful and standardized. Fortunately, the ANSI ADSL specification does this as well. Each of the simplex and duplex channels can be configured independently, as shown in Table 9-4. The table forms a useful summary for the major points mentioned previously. The exact configuration is specified by certain parameters carried in the ADSL frame for each of the bearer channels. Naturally, there is a corresponding table for bearer channels based on the 2.048 Mbps structures. This is shown in Table 9-5.

Transport Class	1	2	3	4
Downstream Simplex Bearers				
Maximum capacity (in Mbps)	6.144	4.608	3.072	1.536
Bearer channel options (in Mbps)	1.536 3.072 4.608 6.144	1.536 3.072 4.608	1.536 3.072	1.536
Maximum active subchannels	Four (AS0, AS1, AS2, AS3)	Three (AS0, AS1, AS2)	Two (AS0, AS1)	One (AS0 only)
Duplex Bearers				
Maximum capacity (in Kbps)	640	608	608	176
Bearer channel options (in Kbps)	576 384 160 C (64)	* 384 160 C (64)	* 384 160 C (64)	160 C (64)
Maximum active subchannels	Three (LS0, LS1, LS2)	Two (LS0, LS1) or (LS0, LS2)	Two (LS0, LS1) or (LS0, LS2)	Two (LS0, LS1)

*Whether transport classes 2 or 3 should support the 576 Kbps optional duplex bearer is under further study.

Table 9-4. Bearer Channel Options by Transport Class for Bearer Rates Based on Downstream Multiples of 1.536 Mbps

ADSL OVERHEAD

It should come as no surprise that ADSL includes overhead in the channel bit rate figures, as well as the capacity to carry user information. In addition to the upstream and downstream bearer channels, ADSL includes overhead for a variety of functions. One crucial function is synchronizing the bearer channels, which means that the devices at each end of the ADSL link know which channels are configured (the ASs and LSs), at what rate they run, and where their bits are located in the stream of ADSL frames.

Other overhead functions in ADSL include an *embedded operations channel* (eoc), an *operations control channel* (occ) used for remote reconfiguration and rate adaptation, error

Transport Class	2M-1	2M-2	2M-3
Downstream Simplex Bearers			
Maximum capacity (in Mbps)	6.144	4.096	2.048
Bearer channel options (in Mbps)	2.048 4.096 6.144	2.048 4.096	2.048
Maximum active subchannels	Three (AS0, AS1, AS2)	Two (AS0, AS1)	One (AS0 only)
Duplex Bearers			
Maximum capacity (in Kbps)	640	608	176
Bearer channel options (in Kbps)	576 384 160 C (64)	* 384 160 C (64)	160 C (64)
Maximum active subchannels	Three (LS0, LS1, LS2)	Two (LS0, LS1 or (LS0, LS2)	Two (LS0, LS1) or (LS0, LS2)

*Whether transport class 2M-2 should support the 576 Kbps optional duplex bearer is under further study.

Table 9-5. Bearer Channel Options by Transport Class—Optional Bearer Rates Based on Downstream Multiples of 2.048 Mbps

detection by means of a *cyclical redundancy check* (crc), more bits set aside for operations, administration, and maintenance (OAM), and bits used for forward error correction (FEC, carried in "fe" bits) so that some errors can be corrected without a need for retransmitting information. Oddly, many of the ADSL overhead acronyms are not capitalized. Presumably, the role of the overhead bits in ADSL is not diminished by this apparent loss of alphabetic status.

All of the overhead bits in ADSL are sent in both the upstream and downstream directions. In most cases, the overhead bits are sent as 32-Kbps bit streams, but there are exceptions. For higher-speed channel structures, there is a maximum bit rate of 128 Kbps and a minimum bit rate of 64 Kbps, with a default of 96 Kbps downstream and a maximum of 64 Kbps and minimum of 32 Kbps, with a default of 64 Kbps (the maximum) in the upstream direction.

In some cases the overhead bits are embedded in the overall bit rate of the ADSL frames and do not consume additional bandwidth. In other cases, the overhead bits add marginally to the overall bit rate in one direction or the other. For example, transport class 1 running at 6.144 Mbps downstream adds a maximum of 192 Kbps and a minimum of 128 Kbps to the overall bit rate. When coupled with the maximum overhead rates on the duplex channels, the transport class 1 line rate rises from 6.144 Mbps to either 6.976 Mbps (maximum) or 6.336 Mbps (minimum), with 6.912 Mbps being the most typical rate using the overhead default speeds. Other transport classes are similarly affected.

When overhead considerations are added to ADSL channel capacities, the rate determined becomes known as the ADSL aggregate bit rate.

A key concept in ADSL overhead is the need for synchronization among the bearer channels. This concept was touched on briefly earlier as concerning the structure of the bearers on the ADSL link, but ADSL synchronization has another important function. As it turns out, not all bits sent on an ADSL link are considered equal, which makes sense because some of the bits may represent delay-sensitive audio and video services, other bits may represent Web page access, and others may represent simple bulk file transfers of e-mail.

This being the case, the ADSL specification established two major bit categories. The terms chosen to represent these are rather unfortunate, but they are firmly established. All bits transported across an ADSL link come from either a "fast" data buffer or an *interleave data buffer*. A better term for the fast data buffer would have been *low-latency data buffer*, and this is the definition provided in the documentation. The interleave data buffer transports bits that can function properly given a bounded *interleaving delay*, whereas the fast data buffer transports bits that cannot function properly with an additional interleaving or buffering delay. The term *interleaving* refers to how the bits in this buffer are protected from errors, rather than how the buffer area is serviced by the sending device, as might be expected. Better terms might have been *delay-sensitive buffer* and *delay-tolerant buffer*, but these phrases are never used in ADSL standard documents.

In other words, some of the bits sent on ADSL channels are never buffered at all beyond whatever buffering is needed to format the frame they are to be transported in. Other bits may have some FEC manipulations done to them and calculations added to them, and they may sit in an interleave buffer until there is room in the ADSL frame to transport them. This interleave buffer may be configured in various sizes, as well. The whole system forms a rather neat, but limited, arrangement of raw priorities independent of priorities assigned to traffic streams by the routers or switches in the networks themselves.

Space is assigned in each ADSL frame to transport bits from the fast data buffer separately from the interleave data buffer bits, so there is never any question as to which bits are being sent and where they are located.

THE ADSL SUPERFRAME

ADSL devices, specifically the ATU-C and the ATU-R, exchange bits by using a line code, usually DMT (the standard) or CAP. Bits are just bits, though. It is what the bits represent that is important. How are IP packets represented as ADSL bits? What about ATM cells? Motion Pictures Expert Group (MPEG) video? Dolby digital audio? How is a sender to tell an ADSL receiver what the bits represent when all types may be transported at the same time to or from a multimedia-capable device? It is all done with ADSL superframe.

All protocols today function in layers, and ADSL is no exception. At the lowest level of any protocol, there are bits, which are represented by the DMT or CAP line codings. The bits are organized into frames and are gathered into what ADSL calls *superframes*. Frames are "first order bit structures," and they are both the last things that bits are before the individual bits are sent, as well as the first thing that bits become when they are received. Note that the ADSL superframe has much more in common with the T1 frame or superframe than the familiar Ethernet LAN frame. In fact, an Ethernet frame can form the content of an ADSL superframe. The overall structure of the ADSL superframe is shown in Figure 9-3.

In ADSL, the superframe is broken into a sequence of 68 ADSL frames. Some frames have special functions. Frames 0 and 1, for instance, carry error control information (that is, a cyclic redundancy check, or CRC) and *indicator bits* (ib) that are used to manage the link. Other indicator bits are carried in frames 34 and 35. A special synchronization frame follows the superframe and carries no user information. One ADSL superframe is sent every 17 milliseconds. Because ADSL links are essentially point-to-point links, no frame addressing or connection identifier is needed at this level of ADSL.

Inside the superframe are the ADSL frames themselves. One ADSL frame is sent every 250 microseconds (1/4000 of a second) and consists of two main parts. The first part is the fast data. *Fast data* is considered to be delay-sensitive, yet noise-tolerant (audio and video, for instance), by the equipment vendor, and ADSL tries to keep the latency associated with this to an absolute minimum. The contents of the fast data buffer of the ADSL device are placed here. A special octet called the *fast byte* precedes this section and carries the CRC and indicator bits where necessary. Fast data is protected by a FEC field in an attempt to correct fast data errors (audio frames can hardly be re-sent).

The second part of the frame contains information from the interleaved data buffer. Interleaved data is packaged to be as impervious to noise as possible, at the cost of increased processing and latency. The interleaving of the data bits makes the data less vulnerable to

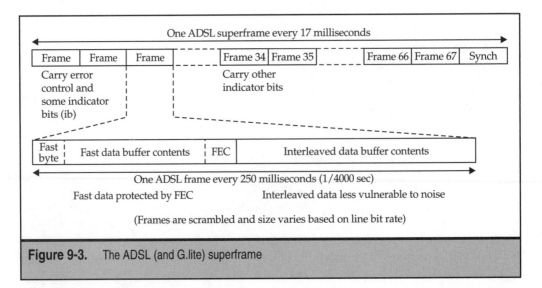

Figure 9-3. The ADSL (and G.lite) superframe

the effects of noise. This part of the frame is mainly intended for high-speed Internet access. All frame contents are scrambled before transmission to minimize the possibility of false superframe synchronization. This is common practice on all framed transports.

Only the interleaved data path and buffer is required in ADSL and G.lite implementations. The fast data path and buffer is optional. One of the functions of the fast data byte in the ADSL frame is to tell the receiver if the fast data path or buffer is active and present or not.

Note that there are no absolute frame sizes for the ADSL superframe. Because ADSL line rates vary and are asymmetrical, the frames sizes themselves may vary. However, the frame size is fixed in the sense that frames must be sent every 250 microseconds (fast and interleaved every 125 microseconds), and one superframe must be sent every 17 milliseconds. Naturally, the maximum line rates for ADSL establish a maximum frame size. The buffer sizes are determined by the speed and structure of the bearer channels when configuration is first done. There is nothing to prevent reconfiguration of buffer sizes during operation of the ADSL link, but no such provision is currently made in ADSL specifications.

Also, there is nothing that determines how or which user bit streams fill the fast and interleaved buffers. This problem is beyond the scope of the ADSL standard. ADSL simply provides the information transfer mechanism.

As was mentioned previously, frames 0 and 1, and 34 and 35, have special roles in the ADSL superframe. These frames carry a cyclical redundancy check for the superframe and also the various indicator bits representing the overhead functions. The other frames, namely frames 2 through 33 and 36 through 67, carry overhead information as well, but overhead representing the embedded operation channel (eoc) and synchronization control (sc). All of this information is carried in the fast data byte position of each ADSL frame in the superframe.

To make things even more confusing, the fast data overhead bits have different structures depending on whether the frame is an even (0, 2, 4,…) or odd (1, 3, 5,…) numbered frame. The structure of all these bits in the fast data byte is shown in Figure 9-4.

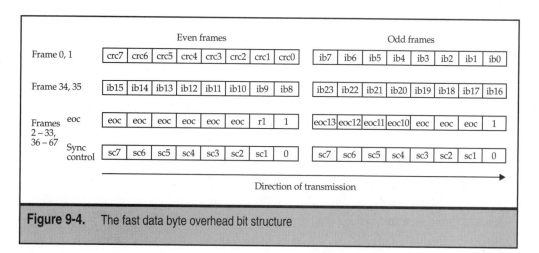

Figure 9-4.   The fast data byte overhead bit structure

Four main overhead functions are detailed in the figure. Bits exist for error detection (crc) and indicator bits (ib) for OAM functions in frames 0 and 1. Bits exist for other OAM functions in the ib bits in frame 34 and 35. The other frames carry the configuration bits (eoc) and synchronization control bits (sc) for determining bearer channel structures and the like.

The "r1" bit is reserved for future use and must be coded to a 1 value. Note that four other bits must also be coded as a 0 value or 1 value. These help to identify the eoc or sync control frames. Also note that this 0 or 1 bit will be the first bit of the fast data byte, as shown in the figure.

The largest portion of the overhead is for the indicator bits. These have a variety of functions that are defined for the downstream direction in Table 9-6.

The indicator bits 0 through 7 and 14 through 23 (the first 8 and the last 10) are reserved for future use. Indicator bits 8 through 13 have explicitly defined functions.

It is easy to understand the los or rdi bits (ib12 and ib13). The los bit is for *loss of signal* and is used by an ADSL ATU to indicate whether the pilot signal in the opposite direction disappears or drops below a certain threshold. The los bit is 1 when there is no los condition to report and 0 when there is. The rdi bit is for *remote defect indication* and is used by an ADSL ATU to indicate whether a *severely errored frame* (sef) is received. A sef is defined as occurring when two consecutive ADSL superframes do not have the expected symbols in the synchronization frame following frame 67. This synchronization frame should not be confused with the synchronization control (sc) bits, which are totally separate. The value of the rdi bit is 1 when there is no sef to report and 0 when there is the same pattern as the los indicator.

The other four indicator bits are ib8 through 11. All four bits represent far-end error conditions, which is just the device (ATU-R or ATU-C) on the opposite end of the ADSL link. Near-end errors exist, as well, but these are beyond the scope of the current discussion. Ib8, called the febe-i, stands for *far-end block error on the interleaved data* in the ADSL

Indicator Bit	Definition
ib0-ib7	Reserved for future use
ib8	febe-i
ib9	fecc-i
ib10	febe-ni
ib11	fecc-ni
ib12	los
ib13	rbi
ib14-23	Reserved for future use

Table 9-6. Indicator Bit Functions in Downstream Direction

superframe. This bit is used to indicate when the *cyclical redundancy check for the interleaved data* (the crc-i) in the received superframe does not match the local calculation result. The bit is 1 when the crc is correct, and 0 otherwise. Ib10, called the febe-ni, performs the same function on the non-interleaved (fast) data, with the same values. Ib9, called the fecc-i, stands for *forward-error correction code on the interleaved data* in the ADSL superframe. This bit is used to indicate when the forward error correction (FEC) code is used to correct errors in the received data. The bit is 1 when there is no error to correct and 0 otherwise. Ib11, called the fecc-ni, performs the same function on the non-interleaved (fast) data, with the same values.

All this might seem somewhat strange considering that the ADSL specification seems to firmly establish the FEC for the primary method of fast data buffer protection and the crc in the superframe as the primary method of interleaved data buffer protection. However, the indicator bits show that both methods are used for fast or interleaved data. Indeed, a crc is calculated as well on the fast data buffer, and a special FEC code is generated, not by each individual interleaved data frame, but rather on a sequence of them. The fast data is just faster through the ATU because the fast data bits are not re-shuffled (interleaved) before crc and FEC generation and then serialized again at the other end of the link.

The structure and meaning of the eoc bits for the embedded operations channel and the sc bits for synch control are quite complex and need not be discussed in detail.

THE ADSL FRAME STRUCTURE

Only one other major topic needs to be discussed regarding the ADSL bit stream between ATU-R and ATU-C—the structure of the bits within the individual frame that make up the ADSL superframe. As it turns out, this structure is easy to describe, but difficult to explain in detail. It is simple because each ADSL frame in a superframe has a fixed structure. For each data buffer, fast or interleaved, the frame simply takes a given number of bytes (octets) for the AS0 bearer channel, followed by AS1, and so on up to AS2. These bytes are followed by LS0 bytes, then LS2, and finally LS3. If there are no bytes for a particular AS or LS, these areas are empty. Finally, there are some added overhead bytes shared by the channels.

The structure is complicated by the fact that ADSL has many line rates that are different in each direction, of course. Only the transport classes add any overall organization to this loose arrangement. Note that any AS or LS bits may be transported in either the fast or interleaved data buffer area within an ADSL frame. Each user data stream is assigned to either the fast data buffer or the interleaved data buffer during an initialization process. However, if the AS0 user stream (unidirectional downstream) is assigned to the fast data buffer area, it cannot be simultaneously assigned to the interleaved buffer. In other words, if an ADSL frame contains bits for AS0 in the fast data buffer area of a frame, the exact same number of bits must exist for AS0 in the interleaved buffer area of the frame.

Configurations for the default number of bytes in the ADSL frame based on transport classes is shown in Table 9-7. These are defaults and can be changed. As well, this information is for ADSL frames transmitted to the ATU-R. Note that if a particular buffer area has a non-zero value, the corresponding value in the other buffer must be zero.

Signal	Interleave Data Buffer				"Fast" Data Buffer			
	Trans Class 1	Trans Class 2	Trans Class 3	Trans Class 4	Trans Class 1	Trans Class 2	Trans Class 3	Trans Class 4
AS0	96	96	48	48	0	0	0	0
AS1	96	48	48	0	0	0	0	0
AS2	0	0	0	0	0	0	0	0
AS3	0	0	0	0	0	0	0	0
LS0	2	2	2	*255	0	0	0	0
LS1	0	0	0	0	5	0	0	5
LS2	0	0	0	0	12	12	12	0

*When coded as an all 1s byte (255), the LS0 bearer channel is a 16 Kbps C channel carried in the synchronization control overhead.

Table 9-7.   Default Buffer Allocations for Transport Classes Based on 1.536 Mbps Multiples

To bring about some closure on the discussion of ADSL frames and superframes, note that in transport class 1, the default configuration allocates 96 bytes to AS0 and AS1 in each ADSL frame. Because there are 8 bits in a byte and 4,000 ADSL frames are sent each second, regardless of total frame size, the bit rate on both the AS0 and AS1 bearer channels will be 3.072 Mbps (96 bytes × 8 bits/byte × 4000/second). And indeed, two downstream bearer channels running at 3.072 Mbps is an established option for transport class 1. In fact, in this case, based on default buffer size, this is effectively the default configuration. Note also that the LS0 channel runs at 64 Kbps in both directions (2 bytes × 8 bits /byte × 4000/second).

The structure of an ADSL frame based on transport class 1 as sent by the ATU-C is shown in Figure 9-5. Note that the fast byte must be present, but all of the data bytes come from the interleaved data buffer.

Of course, there is a companion default buffer allocation for the 2M transport classes based on multiples of 2.048 Mbps services. These are shown in Table 9-8.

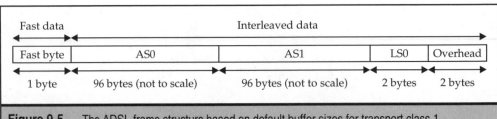

Figure 9-5.   The ADSL frame structure based on default buffer sizes for transport class 1

Signal	Interleave Data Buffer			"Fast" Data Buffer		
	Trans Class 2M-1	Trans Class 2M-2	Trans Class 2M-3	Trans Class 2M-1	Trans Class 2M-2	Trans Class 2M-3
AS0	64	64	64	0	0	0
AS1	64	64	0	0	0	0
AS2	64	0	0	0	0	0
LS0	2	2	*255	0	0	0
LS1	0	0	0	5	0	5
LS2	0	0	0	12	12	0

*When coded as an all 1s byte (255), the LS0 bearer channel is a 16 Kbps C channel carried in the synchronization control overhead.

Table 9-8.   Default Buffer Allocations for Transport Classes Based on 2.048 Mbps Multiples

As before, consider transport class 2M-1 with AS0, AS1, and AS2 all sending 64 bytes in each ADSL frame. This means that there are three downstream bearers running at 2.048 Mbps (64 bytes × 8 bits/byte × 4000/second). And three downstream bearer channels running at 2.048 Mbps is an established option for transport class 2M-1. In fact, in this case, based on default buffer size, this is effectively the default configuration. Note also that the LS0 channel runs at 64 Kbps in both directions (2 bytes × 8 bits/byte × 4000/second) in this configuration.

This technically challenging chapter ends with a look at the ADSL frame structure sent from the ATU-C based on the transport class 2M-1 default buffer allocations (see Figure 9-6). As before, the fast byte must be present.

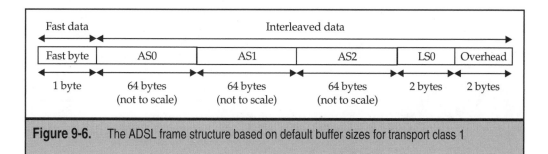

Figure 9-6.   The ADSL frame structure based on default buffer sizes for transport class 1

CHAPTER 10

Inside the ADSL Frames

The previous chapter described the overall architecture of the ADSL physical layer standard. It introduced interfaces and placed emphasis on the flow of information downstream to a user's premises and upstream to the service provider's access node. The chapter introduced the ADSL transport classes and covered the structure and overhead of the ADSL superframe. Finally, it described several ADSL frame structures, based on the transport classes.

The previous chapter investigated one rather obvious feature of the ADSL frames. That is, the flow of bits across the ADSL link as described were just that—bits. An ADSL link is every bit as much a circuit as a 33.6 Kbps modem connection. ADSL just runs faster. When ADSL is configured and operating with an AS0 running at 6.144 Mbps, the ADSL frames carry 6.144 Mbps of bits all the time from ATU-C to ATU-R. These bits may represent useful information, or they may possibly represent a bit pattern that essentially forms an idle channel. The point is that bits are just bits. Circuits in ADSL are the same as circuits in ISDN or the analog content PSTN. Circuits use all of the bandwidth all of the time. Circuit switches always switch bandwidth to bandwidth, point to point, and do nothing else. As described to this point, all that ADSL does is remove the bandwidth used to access the Internet from the PSTN, and to the bandwidth available, and place the bandwidth onto the ADSL access node in the service provider's office. This would be enough if the only goal of ADSL were to relieve PSTN congestion, but ADSL can do more—much more.

The time has come to look inside the ADSL frames and look at some of the possibilities that make ADSL not just an adjunct to the PSTN, but a true alternative to the PSTN. ADSL-specific services might even be added to the basic ADSL fast-Internet access service, especially when the need for fast and secure access to corporate LANs is added for workers at home or in small branch offices in mostly residential areas.

Bits sent across an ADSL link are just an unstructured stream of information. The ADSL link is, loosely speaking, just a "bit pipe" or "bit pump." Bits go in one end and emerge unchanged from the other end of the link, between ATU-C and ATU-R. For ADSL to provide useful services to a customer, however, the bits must mean something; they must represent information according to the service.

Only a small degree of organization is provided for these bits by the ADSL superframe structure. An ADSL superframe is sent every 17 milliseconds (almost 59 per second) and consists of a sequence of 68 ADSL frames. ADSL frames contain both fast bits (audio and video, for example, which is delay-sensitive and needs a stable delay) and interleaved bits (Web data, for example, usually delay-insensitive but error-sensitive). This is as far as the ANSI standard T1.413 goes.

The question is what is inside the ADSL frames? This chapter begins by examining the original vision for ADSL frame content established by the ADSL Forum, and then considers the current state of affairs. Today, what is inside ADSL frames is both simpler (it seems that IP packets are really all that count) and more complex (the ways that the IP packets are *encapsulated* into the ADSL frames) to answer than when the question was first posed when trial ADSL services began.

For this important piece, the ADSL Forum (now just called the DSL Forum, and referred to as such in this book when current activities are discussed) has been the indispensable source of information. The ADSL Forum was founded as an association of service providers and equipment vendors and other interested parties who need to go beyond the low-level technical aspects of ADSL if full-service networks are to be deployed based on ADSL. Forums are not an uncommon practice in the telecommunications industry. The ATM Forum and Frame Relay Forum, among others, have provided ample precedent in this regard and have been instrumental in bringing their technologies to market and into the public consciousness.

Almost all technology forums have some *principal members* who write and vote on *implementation agreements* (often called *technical reports* or *technical recommendations*) regarding the technology under development. These implementation agreements are binding on all members of the forum, including the lesser members in a category, often called *auditing members,* who have access to forum documentation, but who have a reduced role in formulating and approving the implementation agreements. That said, please note that there is no mechanism in a forum for punishing members for not observing the implementation agreements or taking the drastic step of banishing the member from the forum itself. Compliance with implementation agreements is a matter of honor, and most members try scrupulously to comply with forum decisions. However, there is always room for variations in the name of improvement, with or without an implementation agreement in place or anticipated.

The ADSL Forum initially defined four distinct distribution modes for all Digital Subscriber Line (DSL) technologies, ADSL included. The distribution mode determines just what form is taken on by the bits inside the ADSL frames when they are sent. Figure 10-1 shows the main characteristics of the four distribution modes.

At first glance, the figure seems complex and confusing. Part of this is because four distribution modes are combined into one figure, but this is done to make it easier to compare the four. Although there are many similarities between the four configurations, their significant differences must be appreciated to understand what ADSL can do and what other network components and capabilities ADSL needs in order to provide services to customers. I discuss each of the four distribution modes in detail.

The first, and simplest (and also least interesting), distribution mode established by the DSL Forum is known as *bit synchronous mode.* The name is awkward, but it basically means that any bits placed in a buffer (the "fast" data buffer or interleaved buffer) in a device at one end of the link (the ATU-R) will pop out at the buffer in the device at the other end (the ATU-C). In another document, the DSL Forum suggested that the "fast" data buffer operate about 10 times faster that the interleaved buffer in terms of latency. Latencies of about 2 milliseconds for "fast" data and 20 milliseconds for interleaved data are mentioned. The difference is due mainly to the way the bits in the two buffers are handled with regard to error protection.

Implementation of the fast data buffer in ADSL is optional. Only the interleaved buffer is strictly required. The main purpose of the fast data buffer is to carry streaming video (TV shows or movies) from a service provider's video server to the user's ATU-R.

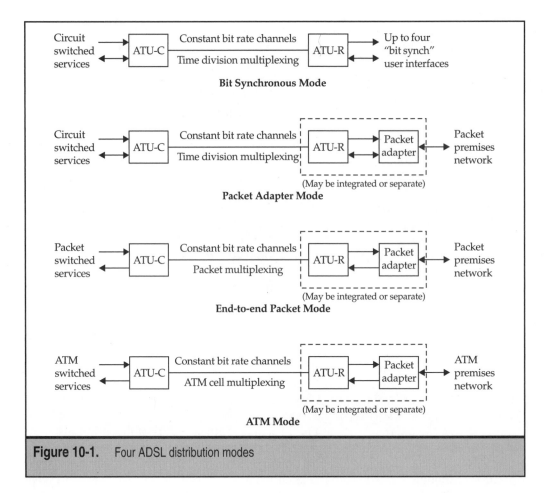

Figure 10-1. Four ADSL distribution modes

This is different than the path for a user viewing video from a Web site on the Internet. All IP packets, which is how this Web video would reach the user, go into the DSL modem's interleaved buffer.

At the top of the figure is the bit sync distribution mode. Up to four "bit sync" symmetric user devices can be hooked up to a given ATU-R at the customer premises when functioning in bit sync distribution mode, which makes sense because there are four downstream bit flows established in ADSL (AS0 through AS3). Naturally, if only AS0 is present on the ADSL link, then only one flow is present at the ATU-R, and presumably there would be only one device. This user device may be a TV set top box or a PC, but all of the bits are delivered to the attached device. The upstream and directional channels must consist of at least the control C channel and may also include the LS structure. In its simplest form, the flow would consist of AS0 downstream at 1.536 Mbps in the United States and 64 Kbps upstream (also used for bidirectional functions).

In the bit sync mode, the ATU-C in the ADSL access node at the service provider's office simply hands off the user's bits arriving on the LS or C channel to some circuit-switched services at the service provider's office. In this mode, the ADSL link is essentially a bit pipe to a fixed end device (just like a leased line). The ADSL link always operates at the line speed, or in what is called a *constant bit rate* (CBR). The ADSL link can be divided into channels, but always with straight *time division multiplexing* (TDM) establishing time slots in the ADSL frames for bits.

It was this simplistic "bit pipe" ADSL link operation that doomed bit sync mode from the start. The bit sync mode was often seen as just a "proof of concept" mode to allow vendors and service providers to show that ADSL could indeed shift traffic off of the PSTN and onto a specialized network. But viewing the path of bits from DSL modem through the DSLAM and on to some service as a simple circuit posed a lot of problems on the backbone network side of the DSLAM. Most of these issues involved the asymmetrical nature of ADSL. For example, the only links available to connect DSLAMs to anything else were symmetrical. If the total downstream bandwidth available to the users through the DSLAM was 45 Mbps (30 users at 1.5 Mbps each, for example), the same 45 Mbps was available upstream through the DSLAM, even if the total amount of bandwidth needed was only perhaps 12 Mbps (30 users at 256 Kbps each upstream). Obviously, the bit streams used between DSLAM and DSL modems had to be made more aware of the structure of the bits inside the ADSL frames and superframes.

The second distribution mode in the figure is *packet adapter mode.* The only change occurs at the customer premises. The main difference from bit asynchronous mode is that the devices at the customer premises are now expected to send and receive packets instead of just a stream of bits. The packets are placed in the ADSL frames by some packet adapter function, which may be a separate device or built-in to the ATU-R. Note that this mode expects to interface with some packet-enabled premises network and not just a simple bit-consuming and generating end-user device.

Why bother? Well, bit sync ADSL requires that each user device be plugged directly into the ATU-R where it can only send or receive bits on one of the ADSL channels, such as LS1. But there may be many devices at a customer's premises that might take advantage of ADSL speeds. Many people have more than one PC. Some even have 10Base-T Ethernet-type LANs in their homes (kits for linking two PCs this way start at a hundred dollars or so). When linked to a SOHO customer, there may be more than just four devices needing to use the ADSL link. Not only traditional LANs may be used as the premises packet distribution network. Newer techniques are allowed as well, such as the use of a building's power distribution wiring to carry information or various forms of wireless networks. In fact, these alternative architectures may actually get a boost from ADSL.

Packet adapter mode allows this to take place. Packets from many sources and destinations at the customer premises can share a single LS1 channel on the ADSL link. Of course, the ATU-R still maps these packets to fixed channels on the ADSL link, but if the other end of the link beyond the ATU-C and access node is an Internet router, the arriving and departing packets can be dealt with effectively.

Also, note that in packet adapter distribution mode the packets are still sent as a stream of bits on TDM channels to wherever the end device is beyond the ATU-C because the endpoints are still reached only by circuits. That is, each packet flow needs its own TDM channel on the ADSL link, and each ADSL channel is still a CBR transport.

The third distribution mode in the figure is the *end-to-end packet mode.* Actually, this is not really a distinct distribution mode at all from packet adapter mode. In fact, both packet adapter and end-to-end packet modes must be used to produce an overall packet service. However, the two types of packet distribution modes are typically presented separately, as they are here. The major difference between this end-to-end packet mode and the packet adapter mode is that packets are now multiplexed on the ADSL channels. In other words, the packets to and from a variety of user devices are not mapped to a sequence of ADSL frames representing ASs or LSs, but are all sent on an "unchannelized" ADSL link running upstream and downstream at a given rate. Note that the big change still occurs at the customer premises, but the user choices are now more constrained; user packets must be the same as those used by the service provider's portion of the link. User devices send packets to and from the packet adapter device. At the service end of the link at the ATU-C, the packets are not all just shuttled to the endpoint represented by the LS channel. In packet mode, the packets are sent to their proper "servers" based on packet address. For fast Internet access, of course, the packets are IP packets.

In this mode, IP packets can be multiplexed and switched (that is, routed) onto the Internet at the service side of the ADSL link. Note that using ADSL to send and receive packets can also be accomplished with bit sync or packet adapter mode. The difference is that the packets traveling on the ADSL system in these two modes are invisible to the ADSL system. In packet distribution mode, the packet switching is part of the ADSL network. The bits in packet mode must be organized as packets, not possibly organized as packets as in bit sync and packet adapter modes. In the case of packet mode ADSL, the ADSL link becomes much like a link between an IS (intermediate system) router and a small office router in an Internet access arrangement. Of course, other types of packets also may be used in this mode. The transport of video packets is just as possible as IP packets, as long as both the client and server understand the type of packet being transported.

The last distribution mode in the figure is *asynchronous transfer mode* (ATM), or more properly end-to-end ATM mode. ATM multiplexes and sends ATM cells from an ATM adapter (at the ATU-R) instead of IP (or some other) packets. At the ADSL service provider side, the ATU-C passes the cells to an ATM network. Remember that the contents of these ATM cells may still be IP packets, and the DSL Forum recently decided to adopt the IP point-to-point protocol (PPP) over ATM for this mode. But the ADSL network deals with the ATM cells, which must form the contents of the ADSL frames. Of all the distribution modes, ATM mode has evoked the greatest amount of interest from ADSL vendors and service providers.

As a final word about the figure, remember that the ATU-C devices shown in the figure are typically part of a DLSAM assembly.

ADSL POSSIBILITIES

ADSL modems (essentially the ATU-C and ATU-R) send bits back and forth inside ADSL frames and superframes. The four ADSL distribution modes establish the contents of the frames as unstructured bits to the network, packets as the source of the bits at the ATU-R (Remote), packets across the ADSL link, or ATM cells across the ADSL link, respectively. In practice, there are really three forms that the bits inside the ADSL frames on an ADSL link can take on: unstructured bits, IP packets, and ATM cells. The use of ATM cells for ADSL transport does not rule out the use of IP packets in the end devices of the network, of course. The nature of ATM cells makes it possible to carry the IP packets inside the ATM cells, as long as devices other than the end-user PC and Internet server (such as the user's ADSL modem and the ISP's router) generate and understand the content of the ATM cells they send back and forth. So information can change the form visible to the network as the information flows from client to server and back. One part of the network might route IP packets, while another section switches ATM cells that still carry the IP packets inside.

The current ADSL end-to-end architecture establishes six major components for a network that includes ADSL links. This architecture is spelled out in the DSL Forum Technical Report 25 (TR-025), and this forms the basis for Figure 10-2. The upper part of the figure shows the reference architecture, and the lower portion of the figure shows an example implementation of the architecture based on the common practices of DSL service providers. Although presented here in an ADSL context, this reference architecture could easily be used for an type of DSL service offering.

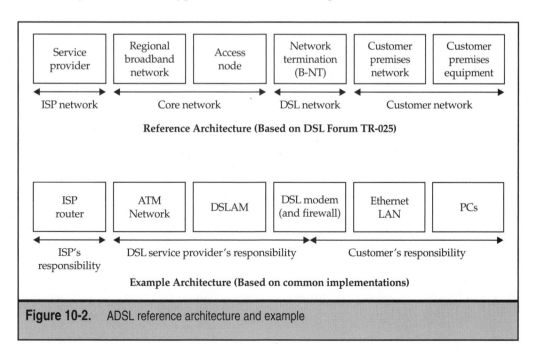

Figure 10-2.   ADSL reference architecture and example

The following are the six functional building blocks of a network with ADSL links:

▼ **Service provider** This is usually taken to mean the ISP's network, and so is really the Internet itself. Strictly speaking, this service provider network could represent specialized services (such as streaming digital movies and the like) right from the DSL service provider's network servers with no links to the Internet, but in actual practice this is the Internet, pure and simple.

■ **Regional broadband network** This network links the DSL service provider's DSLAMs to the service provider (ISP) network. The "network" could be as simple as a series of point-to-point leased lines from ISP router to DSLAM, but is usually much more elaborate. The *regional* nature of the network reflects the local exchange carrier (LEC) emphasis on ADSL: there are no nation-wide networks of DSLAM. The *broadband* nature of the network means that there is enough downstream bandwidth to support simple video (above 400 Kbps downstream).

■ **Access node** This is the DSLAM. The regional broadband network along with the DSLAM on one end and the ISP router or switch on the other make up the *core network* of the ADSL reference architecture because it is here that all of the user traffic from all the DSLAMs and DSL links comes together.

■ **Network termination (B-NT)** Strictly speaking, this is a *broadband* network termination device (video support again) and so a B-NT. This is the DSL modem, and the portion of the network from DSLAM to B-NT makes up the *DSL network* portion of the architecture.

■ **Customer premises network** This is how the DSL modem connects to one or PCs on the customer premises. Customer premises networks are explored more fully in Chapter 11.

▲ **Customer premises equipment** Usually just one or more PCs. The premises network and CPE make up the *customer network* in the reference architecture. If the DSL modem is an internal PC board, this essentially forms a *null customer network* arrangement. This configuration is rare today and not explored further.

The six components therefore make up four network segments: ISP network, core network, DSL network, and customer network. As mentioned, each network segment can send and receive information based on one of the three major forms of DSL transport: bits, packets, or cells. Conversion from one type to another for connections with the adjacent segments are the responsibility of the equipment at each end of the segment, of course. So if the core network is ATM, the DSLAM and ISP router must have at least one interface each that is capable of sending and receiving ATM cells. The cells might have IP packets inside, but these packets are "invisible" to the core ATM network.

Also shown in the figure are the usual network types used in ADSL implementations and deployments. This portion of the figure just gives more reality to the abstractions of the reference architecture. Typically, the ISP's router links to the DSL service provider's

DSLAMs over an ATM network usually run by the DSL service provider. The DSL service provider also handles the ADSL link side of the DSL modem, but the customer is responsible for any networking needed between DSL modem (which might include simple firewall functions) and the PC. These areas of responsibility are also shown in the figure.

However, just saying that there can be different network types on each segment of the DSL network is not that same as understanding what can and has been done on these segments in ADSL implementations. The real concern to the user is not how the core network gets traffic to and from the DSLAM. The user is much more concerned about the form of the traffic entering and leaving the DSL modem and customer premises network, which is most often some form of simple Ethernet LAN. The rest of this chapter more or less assumes that the core network between DSLAM and ISP is an ATM network carrying IP packets and explores the implications of how best to get this ATM cell content to the DSL modem and beyond, through the Ethernet LAN to the user's PC(s).

In real ADSL networks, there are really four methods that have been tried (and continue to be explored and developed) when it comes to linking user PCs to DSLAMs and beyond. These four major methods are shown in Figure 10-3 and examined in detail in the following sections.

Note that each method has drawbacks. This makes the choice of network type a nontrivial task for DSL service providers and equipment vendors. In what follows, keep in mind that if one method ever was shown to be far better than the others, the other three would most likely quickly vanish, at least for new deployments.

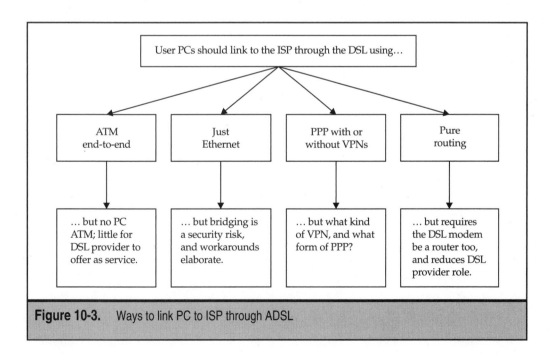

Figure 10-3.   Ways to link PC to ISP through ADSL

ATM End-to-End

It is very common in DSL service offerings to have an ATM core network linking the DSLAMs to ISP routers. The ATM core is a *virtual circuit* network, and only one link to the ATM network from each DSLAM and ISP router is needed to allow full mesh connectivity to all devices. The router could have many non-ATM interfaces in addition to the ATM interface for DSLAM connectivity. In ATM, virtual circuits are numbered by a *virtual path identifier* (VPI) and *virtual channel identifier* (VCI) in the ATM cell header. So ATM virtual circuit 0:37 could connect a router to a DSLAM on the north side of town, while ATM virtual circuit 0:92 could connect the router to a DSLAM on the south side of town, but over the same physical link to the ATM network (this link is called the ATM user–network interface, or UNI).

But what about the DSL link from DSLAM to remote DSL modem on the customer premises? Should the DSL link carry ATM cells as well? Why not? After all, G.lite requires the use of ATM cells on the DSL link, and full ADSL certainly allows, if not encourages, the use of ATM right to the DSL modem.

However, there is no ATM beyond the DSL modem. The DSL modem must convert the ATM cells to something else. Usually, the DSL modem will convert the IP packets inside the ATM cells to a stream of IP packets inside Ethernet frames for transmission over the customer's Ethernet LAN. Most DSL modems have only an Ethernet LAN connector for the user to plug into anyway.

Without user PC support of ATM, the ATM end-to-end option is just not viable. The customer premises network is the customer's responsibility, and no one could make the customer build an expensive ATM network when Ethernet and IP are so inexpensive, well-understood, and often built-in.

Not only that, but having ATM end-to-end, once the primary vision of ADSL service providers, now leaves little for the DSL service provider to do at all. Recall that even if ATM cells flow from PC to ISP router end-to-end, the main purpose of ATM is to deliver IP packets from router to PC and back. If all that flows between user and ISP is ATM cells, the DSL service provider cannot even provide simple services, such as a user's own Web page or simple security with login ID and password. The network between user DSL modem and ISP router sees nothing except a stream of ATM cells. Nothing is or can be done to the IP packets flowing on the network.

This type of ADSL network might not seem to be of much use, but there are times when such a "transparent" network can be useful. In fact, this type of ADSL network is referred to as a *transparent core ATM network* in the standard's documents. This type of ADSL network is shown in Figure 10-4.

The user's DSL modem now takes IP packets from the Ethernet LAN (not shown in the figure) and places them inside point-to-point (PPP) protocol frames. PPP is a standard frame structure normally used on the Internet to transport IP packets, but PPP can be used for other packets types as well. These PPP frames are given an ATM encapsulation trailer formatted according to ATM Adaptation Layer type 5 (AAL5) rules and then sent as a stream of ATM cells inside the ADSL frames and superframes. The DSLAM simply passes the cells—unchanged—through the core ATM network to the router with an ATM interface at the other end of the core network. There the IP packets can be extracted,

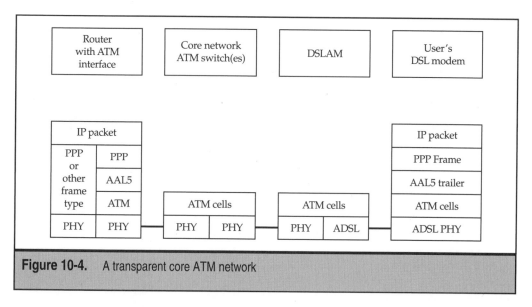

Figure 10-4.   A transparent core ATM network

placed in PPP one of the other frame types used by IP, and sent onto an IP network. The same process works in the reverse direction as well.

Transparent core ATM networks make a lot of sense in one key ADSL service-offering environment: remote access to a corporate LAN. In this situation, all the corporate organization wants is for the small office or home office (SOHO) worker to reach the corporate LAN. The company handles all login and access issues. All the company needs to tell the DSL service provider to do is to set up an ATM virtual circuit (VPI/VCI) from DSLAM to router (in this case, the router is not an ISP router, but the company's router).

As already pointed out, in many cases the IP packets sent back and forth from router to DSL modem are sent inside ATM cells. But why? What is it (or was it) about ATM cells that made them so attractive for ADSL use in particular and all DSLs in general? Why require ATM cells in G.lite? Because the legacy of ATM cell use and DSLs is one that will be present in DSL for some time to come, this is a good place to examine the issues regarding ATM use with ADSL.

Why ATM for ADSL?

When full ADSL is used with the remote splitter, use of ATM cells is the most common method of transport for the traffic most often of interest to users of ADSL for high-speed Internet access: IP packets. When G.lite is used without the need for the remote splitter, and simple filters are the rule, use of ATM inside the ADSL frames is mandatory. Naturally, the main use of G.lite, as with full ADSL, is for high-speed Internet access using a stream of IP packets. The Internet understands only IP packets, and the vast majority of user PCs can only use IP packets for communication anyway. So how did ATM achieve its position as the preferred method of transport for ADSL links? If ADSL is mainly used for high-speed Internet access from PC to ISP, and both understand only IP packets, why was ATM chosen as the preferred ADSL transport method in the first place? This section is intended to answer these questions.

It is always best to start at the beginning. However, this section cannot begin to describe all of the details of ATM, which are far beyond the scope of this book. But in order to appreciate the intentions of the standards makers with regard to ATM use with ADSL, at least some of the operational aspects of ATM must be presented.

Some of the basics of ATM have already been mentioned in several sections of this book. ATM is based on the idea of the fixed-length cell rather than a variable length frame or packet as the method of data unit transport. All ATM cells are exactly 53 octets (bytes) long, with a 5-octet header and 48-byte *payload,* or information area. This contrasts with the variable-length frames used in (for example) Ethernet, which range from a minimum of 64 octets to a maximum of 1,518 octets, or the variable length packets used in IP, which can be as small as 20 octets or as large as 65,535 octets (64K bytes).

The two main reasons for the invention of ATM cells were simple at heart: to make it easier and faster for network equipment to buffer and process data units and to allow both delay-sensitive and bulk-transport traffic to share a link with limited bandwidth without affecting the proper *quality of service* (QOS) required by both types of traffic. There are actually four major types of traffic in ATM, but these simple dual delay-sensitive and bulk-transport traffic classes are enough to understand the intent of ATM. Today, QOS needs are often just divided into delay-sensitive voice, which requires low and stable delays to sound good, and bandwidth-hungry data, which requires as much bandwidth as possible to transfer information efficiently. Video falls into a gray QOS area with twin needs of low and stable delays (for the audio) and yet more bandwidth than voice (for the visual).

At the time ATM was invented in the late 1980s, network equipment that dealt with variable length data units was slowed by the need to find proper sized buffers to hold the arriving frames before their total length was known and the need to manage memory buffers as the memory became more and more fragmented by variable length data units. It is just easier and more efficient to pack the trunk of an automobile with equal and small golf balls than a mixture of tennis balls, softballs, basketballs, and every size of ball in between. When bandwidth on a link is limited, a bulk file transfer that used huge frames for efficiency would hold up delay-sensitive voice traffic from time to time, and variable delays distort voice. The situation was similar to a stream of commuters hurrying to work in automobiles being held up at a railroad crossing by a 200–boxcar, slow freight train. When ATM cells were used, the boxcars and automobiles could take turns passing through the crossing. The trade-off is in terms of cell overhead (the "cell tax") that adds a 5-octet header to every 48 octets of user information. (Think of this as the price of adding a small engine to each and every boxcar instead of having a few diesel locomotives pulling all 200 at once.) This reason for using ATM is shown in Figure 10-5.

In the figure, voice and data user devices share a common link to a service provider and also share customer premises equipment (CPE). Digitized and packetized voice generates small, but delay-sensitive, information units. Information units carrying data tend to be as large as possible to cut down on transfer times and overhead percentage. If even a few voice packets pile up behind a large data frame, the low delay requirement for voice is not met. And if sometimes the voice sails out without a data frame in the way, the delay is too variable to guarantee good voice quality. ATM can easily intermix the voice and

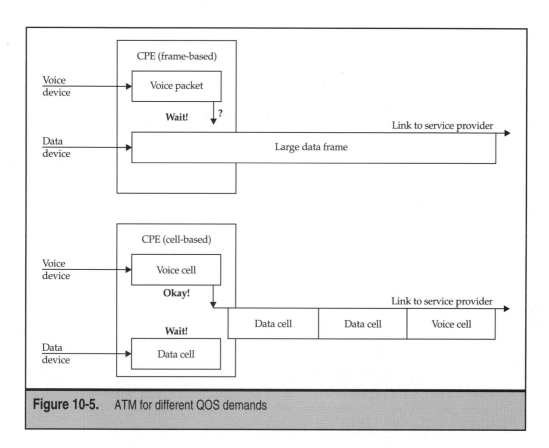

Figure 10-5.   ATM for different QOS demands

data traffic and so give the voice a low and stable delay while still passing data at a high rate. The cell-long wait that the data cell must now endure when ATM is used is deemed to be an acceptable price to pay for allowing voice to share the link.

These reasons for ATM use might sound compelling. What could be wrong with wanting efficient internal processing of information by network nodes and sharing the restricted bandwidth available equitably between applications with differing QOS demands? Nothing at all. It is just that both initial reasons are no longer valid in most situations and in most places on networks today. So the ATM cell tax is more of a burden than ever.

Consider network node processing first. Hardware architectures—in terms of speed and memory capacity—improved so rapidly throughout the 1990s that buffer management and internal processing speed were no longer the critical factors they once were. IP routers handling variable length Ethernet frames carrying IP packets were invented that could route at *wire speeds,* meaning that packets came in and went out as fast as the links could carry them. Plenty of memory was available for the packets and frames that had to wait at an output link, and in any case, ATM switches got congested by too many cells as well. Many of these fast router architectures functioned *internally* by making cell-like units out of the arriving IP packets, some of which looked much like ATM cells, right down to the header. But no matter what the packet or frame looked like inside the box, few vendors saw any reason to send and receive cells *externally* on the links between network nodes. Besides, most of these internal cell structures were vendor-specific and so could not just be shipped off over a link to a piece of equipment from another vendor.

Second, the increased use of high-capacity fiber effectively removed the need to worry about how slowly a 1,500 octet frame moved out over a link. ATM worked best in a *restricted bandwidth* environment when different QOS parameters were needed to service different traffic streams. Given enough bandwidth, the need to make many small cells in the first place tended to slow down rather than speed up operations on the link. This was a little like replacing the slow freight train with a speedy bullet train at the railroad crossing. Zoom! The train was gone and the automobiles flowed again right away. Several studies even seemed to show that at speeds above 600 Mbps or so, cells just got in the way of themselves because chipsets that made and unmade standard ATM cells that quickly were just not efficient.

Some ATM proponents saw these trends as an aberration. Sooner or later, hardware would be swamped by traffic and ATM cells would save the day. And bandwidth would never be plentiful enough to satisfy everyone. It seems for the moment that the ATM proponents were right on one count, but wrong on the other. The ATM proponents were wrong about hardware: Advanced architecture can route variable length IP packets as fast, if not faster, than an ATM switch (thanks to the internal use of cells, the ATM answer to that would go). But the ATM proponents were right about one thing: ATM still makes sense when bandwidth is limited. And there is no place where bandwidth is more limited today than on a local loop with a useful analog passband of 4 KHz.

So ATM was chosen as the ADSL preferred data unit mainly to support differing QOS requirements on the restricted bandwidth local loop. This made perfect sense, but there is one major problem with the continued use of ATM for ADSL links.

The problem is that in almost every single case, ADSL carries only IP packets. The IP packets themselves are often carrying not only data, but voice (in the form of voice over IP, or VoIP) and video (in the form of highly compressed and poor-looking video from a Web site, but video nonetheless). However, IP packets are all just "data" to ATM. So ATM exists to give proper QOS to voice (and video) along with data, but only so long as the voice (and video) are inside information units that can be distinguished from "ordinary" data-carrying IP packets. Given the structure of the Internet and Web today, this is all but impossible. IP packets—whether they're carrying voice, video, or data—all look the same to ATM and are treated the same by an ATM network.

The ATM-for-ADSL vision and current reality are shown in Figure 10-6. ATM can handle digital voice in the form of *voice and telephony over ATM* (VTOA) at the user end and on the network with an ATM QOS category known as constant bit rate (CBR). ATM can handle digital video in the form of compressed Motion Picture Experts Group (MPEG) format at the user end and on the network with an ATM QOS category known as *variable bit rate real-time* (VBR-rt). And ATM can handle IP packet data in the form of ATM Adaptation Layer type 5 (AAL5) encapsulation at the user end and on the network with

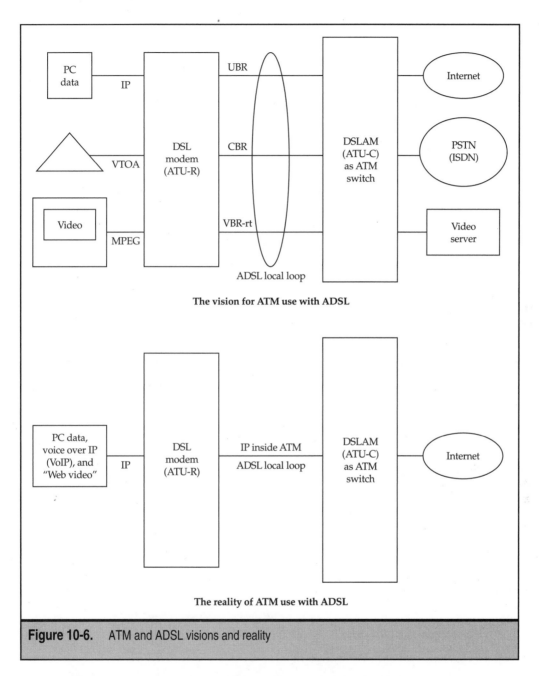

Figure 10-6.   ATM and ADSL visions and reality

an ATM QOS category known as *unspecified bit rate* (UBR). The DSLAM is also an ATM switch in this case, sending IP to the Internet, VTOA to an ISDN switch, and fetching video from a special ATM video server.

But note that the user devices required are a PC for the data, a digital VTOA tele-phone, and an MPEG television set. The PC is there already, of course, but there are cur-rently few digital televisions that can be hooked up to a telephone local loop, and even fewer VTOA-compliant telephones. The reality, as also shown in the figure, is that people will continue to use their analog telephone with ADSL (not shown in the figure, which shows only digital services) and use their PC to access the Internet over the ADSL link for data and video, at least for "Web video." The digital television shows that families watch for daily entertainment will link for the time being to digital cable TV and digital broad-cast networks, not to the Internet over ADSL. Because there is IP at each end of this real-ity, why is there ATM in the middle?

To be fair, there are several ways that ATM networks can be made more "IP aware." But making an expensive ATM switch as good as a fast IP router strikes many as an exer-cise in futility. Why not just get rid of the ATM and use IP everywhere? For now, the only reason (beyond the simplistic "the G.lite standard requires it") seems to be that ATM still makes some sense on the restricted bandwidth local loop, but only if truly separate (non-IP and/or non-Internet-based) voice and video services flourish in the near future. Both possibilities seem remote, however. ATM remains in most cases a solution without a problem.

Ethernet to the DSLAM

If ATM does not have much of a future outside of the core network linking DSLAMs to ISP routers, then why not get rid of ATM altogether in the ADSL architecture? Why not do what many cable modems do and just send IP packets inside of Ethernet frames be-tween DSL modem and DSLAM? The DSL modem most often sends and receives only Ethernet frames on the user (LAN) side anyway. Even if ATM must be used on the DSL link, why not use Ethernet instead of PPP to carry the IP packets?

Perhaps even the core network between DSLAM and router could dispense with ATM, as attractive as the ATM virtual circuits are for mesh connectivity. There are also virtual LANs (VLANs) and other technologies that could potentially link DSLAMs to ISP routers without the need for many point-to-point links.

This scheme seems to make perfect sense. This architecture essentially makes an *Ethernet bridge* out of the DSLAM on the DSL link side. A somewhat simplified view of this "Ethernet to the DSLAM" architecture is shown in Figure 10-7. Note that now the ADSL frames and superframes carry ATM cells containing not PPP, but Ethernet to and from the DSLAM.

Although PPP and ATM cells are still used from DSLAM to ISP router, on the cus-tomer's premises only IP packets inside of Ethernet frames are used. This makes a great deal of sense, because the user PC usually has an Ethernet NIC interface to link to the DSL modem's Ethernet interface (10Base-T, to be precise). PPP is a wide area network (WAN) protocol, not a LAN protocol, so PPP cannot be used directly on an Ethernet LAN.

So what's wrong with this ADSL and Ethernet architecture? The major drawback has to do mainly with security. A minor drawback is that Ethernet still leaves no way for the DSL service provider to authenticate the user with a login and password combination on

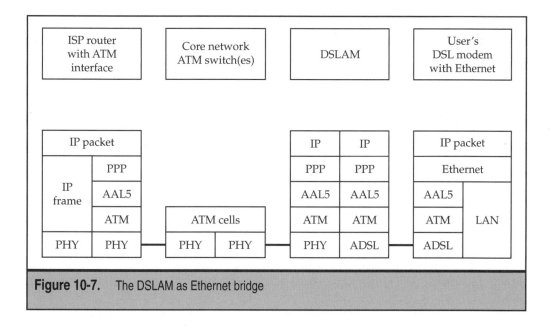

Figure 10-7. The DSLAM as Ethernet bridge

the user side of the DSLAM. Some form of PPP-based authentication (such as that provided by a *remote access dial-in user system* [RADIUS] server) can be used on the *back end* of the DSLAM, but this structure still relies on the ISP for authentication.

The security risk is a much more serious flaw when Ethernet frames are sent to the DSLAM. The risk is due to the fact that Ethernet frame use effectively makes the DSLAM into an Ethernet bridge on the ADSL link side of the DSLAM. The flaw is in the way that an Ethernet bridge handles the IP addresses they are assigned. What follows uses only enough IP address terminology to understand how the DSLAM Ethernet bridge introduces security issues. This is not intended to be a full discussion of IP addressing and DSLAMs.

IP addresses are currently (in IP version 4) assigned to end-user devices as 32-bit sequences expressed as four decimal numbers separated by dots. IP packets have both a source and destination IP address in the packet header. IP addresses consist of both a *network portion* and a *host portion*. Together, both portions add up to 32 bits, but (for example) 21 bits might be the network portion of the IP address, and the other 11 bits might be the host portion of the IP address. So a user's DSL modem or PC might have IP address 192.168.16.1/32, where the */32* is used to indicate that all 32 bits are needed to properly address this device (both network and host portion) in particular out of all other devices on the Internet.

But the network portion of the IP address plays an important role in the router. This portion of the IP address indicates the IP *network* to which the user device is connected. Routers that are not directly connected to the end-user device need not know exactly which end-user device is to receive the IP packet. It is enough just for these routers to know which *network* the IP user device is attached to. In this case, only the DSLAM needs to know which ADSL link is mapped to which IP address. All other routers need only know the IP network portion of the address. So on the router side of the interface, the DSLAM presents 192.168.16/21 to the rest of the routers on the Internet.

However, the rule regarding how IP addresses are used means that all of these 192.168.16.0/21 IP addresses must be assigned to the same *LAN segment* and bridged together at the Ethernet frames level by the DSLAM. This situation is shown in Figure 10-8. Up to 2,046 users (the first and last possible IP addresses cannot be assigned to users) share the 192.168.16.0/21 address space on the DSLAM. All other Internet routers need only have one table entry to route packets to this destination address.

The problem is that when Ethernet frames are bridged through a device, any Ethernet frame content is sent everywhere. Ethernet NICs are supposed to ignore any frames not specifically for them, but it is always possible for someone who knows what to do (that is, a hacker) to "listen in" on any traffic sent to and from the DSLAM. For example, all three DSL modems in the figure will receive a copy of any and all Ethernet frames sent to and from the DSLAM. So a neighbor could easily receive and interpret the packets inside these frames. This is a potentially embarrassing situation if the packet content is a complaint about the neighbor, and potentially even more damaging if the packet contains credit card or other financial information that could be used for gain by an unscrupulous neighbor.

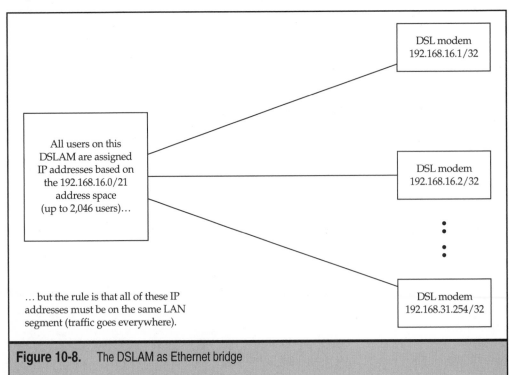

Figure 10-8.   The DSLAM as Ethernet bridge

Part of the problem here is that Ethernet is just not a WAN protocol, so there is no need for Ethernet for rely on the same authentication methods as a true WAN protocol like PP does. Now, there are ways to add security to the DSLAM-to-DSL modem environment to make the neighbors more secure. Such workarounds are done in cable modem environments all the time. These workarounds complicate the DSL architecture and tend to become more elaborate in the point-to-point DSL link environment than on the shared-access cable modem world. So the trend in DSL services is to look for other architectures to include PPP not only from DSLAM-to-DSL modem, but right through the DSL modem and onto the customer premises Ethernet LAN.

PPP With or Without VPNs

A third possible way to get information in the form of IP packets from user PC through a DSL modem through a DSLAM and on to an ISP router is to re-introduce PPP into the ar chitecture on the premises. Recall that PPP was an important part of the overall ADSL architecture when ATM cells were used for most of the segments, but PPP was not used on the Ethernet LAN between user PC and DSL modem. It is hard to use both PPP and Ethernet at the same time, because both are frames structures meant to carry IP packets directly.

It is hard to use PPP and Ethernet at the same time, but not impossible. In fact, ADSL gave birth to a new version of PPP just for this situation. This is called PPP over Ethernet (PPPoE). PPPoE use is quite common among DSL service providers because it allows for the use of Ethernet as a LAN from DSL modem to user PC while at the same time carrying PPP frames inside the Ethernet frames. This in turn both allows the DSL service provider to validate users with an authentication scheme such as RADIUS, and at the same time lets the DSL service provider run the DSLAM as a pure access multiplexer and not as a form of Ethernet bridge.

One of the signs that PPPoE is in use is that users must log in to the DSL service provider's network with a valid user ID and password. In many cases, it is the DSL modem that handles these login chores, especially if the DSL modem is providing some additional security in the form of IP network address translation (NAT) or the like.

PPPoE does exactly what it sounds like: it places the PPP frames containing IP packets arriving over the DSL link into Ethernet frames for transport over the Ethernet LAN to the user's PC. PPPoE use in the user PC requires the user's PC be equipped with the PPPoE protocol software, but this is just included with the user's DSL modem package. PPPoE was originally planned as a "temporary" solution until more sophisticated solutions came along and were available on the end user PC. But like many "temporary" workarounds, PPPoE has become so entrenched in DSL implementations that it is somewhat of a surprise when PPPoE is *not* in use.

What about situations where PPP is used with ATM on the DSLAM and in the ATM core network? Isn't this the same situation as the transparent ATM core network where only the ISP saw the PPP frames and could validate users? Ordinarily, the answer would be yes, the PPPoE does little to help make IP packets visible on the DSLAM and core network. But usually PPPoE is used along with a *broadband access server* that sits between

DSLAM and the ISP router. The broadband access server can now see not only the PPP frames inside the ATM cells, but the IP packets as well. DSL service providers with a broadband access service can not only validate users with RADIUS, they can also offer users their own Web pages, storage for backups, and so on.

The architecture using a broadband access service is also a valid DSL Forum configuration. Figure 10-9 shows what this architecture might look like, placing the broadband access server between DSLAM and ISP router.

Although the figure shows the broadband access server at the ISP end of the core network, this function can be done anywhere between the DSLAM and ISP router.

ADSL and Virtual Private Networks

So far, this section has only introduced PPPoE and the broadband access server as alternatives to simple ATM and/or Ethernet ADSL architectures. But once PPP is in place, it is possible to extend the architecture with the addition of features of what is called a *virtual private network* (VPN).

VPNs are important considerations in ADSL because once PPP enters the picture with PPPoE, it is possible for the DSL service provider to "see" the contents of the ATM cells as they make their way through the DSL network. In other words, the ATM cell contents are no longer "transparent" as they were when only ATM was in use between DSLAM and IP router.

Not that anyone would accuse a DSL service provider of compromising users' privacy. The risk is that hackers might be able to get into the DSLAM or broadband access server and hack into the IP packets that are now available whenever the ATM cell stream is converted to something else. (Conceivably, hackers could get into an ATM cell stream, but almost all hacking today is confined to the IP packet layer of a network.)

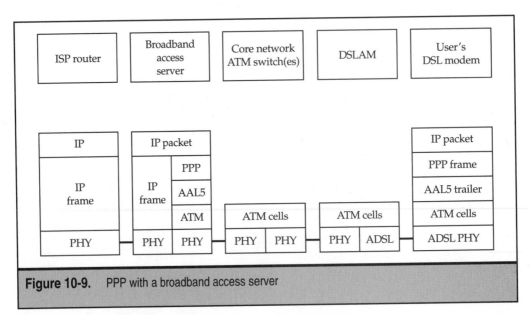

Figure 10-9.   PPP with a broadband access server

A VPN is a method of adding security to a public network to make it appear as secure as a private network would be. Recall that transparent ATM core networks were useful so that remote SOHO corporate users could access corporate LAN resources over the public DSL network using ATM virtual circuits as securely as if point-to-point private lines were used. Some of this security is potentially lost when PPP with a broadband access server is used. So PPP with a VPN is one way to try and restore this potentially lost security (there are other ways not considered here).

This section is not the place to consider all of the details of VPNs with regard to ADSL. It is enough to note that most PPP VPN schemes include some form of tunneling in some segments of the PC to ISP router network, encryption, or both. Tunneling is just the practice of placing one IP packet or frame inside another, thus effectively "hiding" the content of the inner IP packet. The inner IP packet is almost always encrypted as well, such as when secure IP (IPSec) is used.

Some of the ways to add VPN functions to ADSL are the following:

▼ **L2TP or PPTP**   The Layer 2 Tunneling Protocol (L2TP) and Point-to-Point Tunneling Protocol (PPTP) are ways to take IP packets inside of PPP frames and encrypt them before placing the now secure content into *another* IP packet (with another IP address) for transport over some portion of the core network (ATM or otherwise).

■ **IPSec**   Secure IP is a standard way to encapsulate and encrypt IP packets inside themselves. There are both significant differences and significant similarities between L2TP/PPTP and IPSec architectures and requirements.

▲ **MPLS**   *Multiprotocol Label Switching* (MPLS) is a way to replace the core ATM network with a purely IP routing architecture yet at the same time have the same concept of virtual circuits that ATM networks provide.

Pure Routing

The final alternative ADSL architecture on the segments of the network from user PC to ISP router is to do away with ATM entirely and move to an ADSL environment where there is only routing and there are only routers handling the flow of IP packets.

But doesn't the use of IP on the core network mean that there are no virtual circuits for the IP packets to follow as when ATM switches are used to link DSLAMs to ISP routers? Not necessarily. Most of the features available from ATM virtual circuits can now be added to a pure IP router environment with MPLS, introduced in the previous section. MPLS routers encapsulate IP packets not inside fixed length cells switched by ATM switches, but inside variable length frames (such as PPP frames) using a special MPLS header containing a *label* in front of the IP packet that is effectively a virtual circuit identifier. A full discussion of MPLS is not possible or necessary here. It is enough to note that the MPLS architecture looks much like the ATM core network, but with IP routers and MPLS replacing the ATM switches and ATM layers.

MPLS is a relatively new IP technology. It will be a while (but perhaps not too long) before MPLS core networks replace or enhance pure ATM core networks to link DSLAM to ISPs.

The Real World

This chapter spent a lot of time detailing four major ways to get IP packets across all the segments of an ADSL network. To review, they were the following:

▼ **ATM end-to-end**   Fulfills the original ADSL vision, but users will not have ATM on the premises.

■ **Just Ethernet**   Raises real security issues.

■ **PPP with or without VPNs**   Most viable current architecture using PPPoE.

▲ **Pure routing (using MPLS)**   Seems to be a nice long-term solution for a world without ATM.

Out of all four architectures, it is the third one with PPPoE that comes closest to what service providers do with ADSL today. This real-world architecture is shown in Figure 10-10.

This figure is busy, but very important when it comes to understanding how ADSL networks are built today. At the right of the figure, three common home ADSL arrangements are shown. First, in full ADSL, a splitter is used to split the signals to the existing home voice wiring and to a wire running to the DSL modem (this new wire must be installed by the service provider or customer). Alternatively, although shown on the same line, the filter used with G.lite allows the phone to be used without too much noise from the DSL line data (there is some hum or buzz on the line). Second, a home might have a full LAN connecting many PCs. Some DSL modems have multiple ports like an Ethernet hub, while others require a hub or router (for security) on the premises. Finally, a single PC can use the line with a simple crossover 10Base-T cable. The protocol stacks below each device shows the flow of information through it. The PCs all use PPPoE in this example.

The DSL modem takes the Ethernet frame and adds some headers (called the *Logical Link Control/Subnetwork Access Protocol* (LLC/SNAP) header and the ATM Adaptation Layer 5 (AAL5) header, and then sends the information out as a stream of ATM cells over an ADSL physical layer. The DSLAM aggregates traffic from many loops, changes the ATM connection identifiers, and sends the cells through an ATM switch network. The DSLAM also can split off the voice traffic and send it to the PSTN voice switch, but this service separation is often done before the traffic hits the DSLAM.

The ATM switching network in the figure uses SONET OC-3c links running at 155.52 Mbps to send traffic to a *gateway router*, although it is really a bridge because it looks only at the Ethernet frames. The function of this device is to form a gateway between the DSL service provider's ATM network backbone to the DSLAMs and the ATM network connecting to the router of the DSL customer's chosen ISP. Often these ATM networks are one and the same, but regulation in the United States still differentiates between DSL service provider and ISPs.

At the gateway router, which could be run by the DSL service provider or by the ISP, customer traffic is gathered from any DSLAM on the ATM network, the PPPoE is

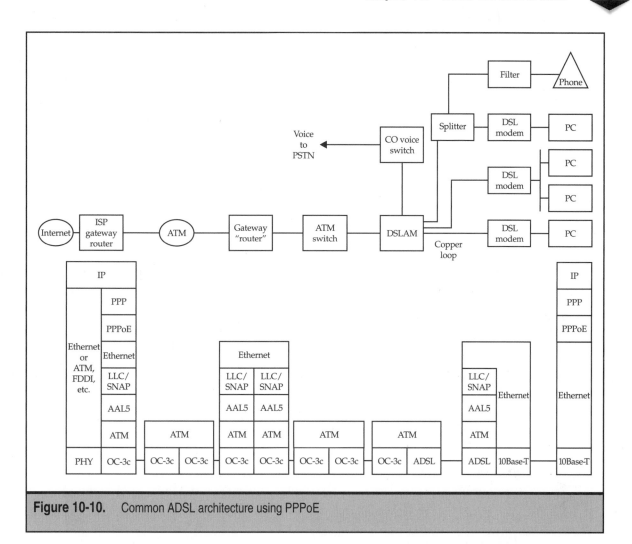

Figure 10-10. Common ADSL architecture using PPPoE

processed, and the IP packet sent onto the Internet inside an Ethernet frame, ATM cell stream, and something else. This is the logical endpoint of the DSL link.

This chapter has detailed what happens inside the ADSL frames as they flow from ISP to customer PC. However, many of the details of the customer premises network (such as the Ethernet LAN) linking the DSL modem to one or more PCs were left out. The next chapter looks at these premises arrangements in more detail and fully explores the issue of voice over DSL (VoDSL) as opposed to "sliding" analog voice under the ADSL bandwidth used for IP packet transport.

CHAPTER 11

ADSL in Action

Thus far this book has emphasized the networking aspects of the DSL family in general and ADSL in particular. The time has come however to shift the focus back onto the customer. That is, what does ADSL mean to a prospective user on a service provided by ADSL? After all, ADSL itself does nothing except shuttle bits back and forth. What can ADSL do for a user? What might a user have to do to prepare their home or business for ADSL?

Suppose a local service provider, be it a cable TV company or the local telephone service provider, announces the availability of ADSL in a given area. Interested users contact the service provider to find out what they can do with it. Many only want faster Internet access. These customers would be extremely happy because almost all ADSL services start with basic Internet access at 64 Kbps upstream and anywhere from 256 Kbps to 1.5 Mbps downstream (most upstream speeds are in the range of 422 Kbps to 768 Kbps).

But what else can be done with this bandwidth? Right now, ADSL services form a faster bit pipe through the ADSL service provider to the Internet ISP. But to some it seems silly to expend all this time and effort on ADSL just to have it form a fast bit pipe through the DSL service provider onto the Internet. Not that Internet access will not be the main attraction of ADSL, but the local ADSL service provider can add value to the ADSL link by providing local servers for local users. These servers can still be based on IP, of course, and may still be attached to the rest of the Internet, but the key here is that messy Internet bottlenecks can be avoided because the server is right at the end of the big ADSL bit pipe.

With this in mind, the ADSL Forum outlined a series of speeds at which ADSL should run upstream and downstream to provide a variety of services. The ADSL Forum is now the DSL Forum, but in this chapter the term ADSL Forum is used because all of these issues were addressed in the days when the organization was still known as the ADSL Forum. Remember that all of these services, with few exceptions, could still be provided directly from the global public Internet; it may be simply more efficient to get them from the local service provider.

ADSL TARGET SERVICE SPEEDS: VIDEO

As was mentioned previously, many discussions of ADSL network services revolve around references to high-speed Internet access and vague assurances of unspecified broadband services gathered under the "video" umbrella. As important as Internet access is, it might be more important that ADSL perform more than one neat trick to assure itself a place in the hearts and minds of potential customers. Therefore, video services might assume a more important role in ADSL as time goes on. All are characterized by a need to transfer full motion streaming video to the end-user.

ADSL promises enough bandwidth (and perhaps sufficiently stable delays and low bit error rates) to deliver many types of video services to the home. In an effort to be more helpful and specific, the ADSL Forum has documented a number of possible video services that can be provided with ADSL and has even included the suggested downstream and upstream bandwidths that these services would need. It should be noted that this particular document has been replaced by newer specifications, but the basic speed ranges are still sound. However, the emphasis here is on the characteristic speeds associated with the services, and not so much on how the services will actually be delivered on an ADSL network. These are listed in Table 11-1.

Video services such as the following will function well with 1.5 Mbps downstream and a modest 64 Kbps upstream:

▼   Home shopping

■   Information services, such as a person announcing library hours

▲   Computer gaming—usually used to mean a type of game system played over the network, such as Monopoly or Doom, but sometimes used as a polite term for lotteries and gambling

Others, such as videoconferencing and traditional video games, may require a range of speeds, depending on the video quality and detail required. Still others, such as distance learning, near video-on-demand (movies start every 15 minutes), and movies-on-demand will require downstream bandwidths as high as 3 Mbps. Broadcast-quality TV images, given state-of-the-art digital compression techniques, will still require 6 to 8 Mbps. Bit rates above 1.5 Mbps might be hard to come by in ADSL systems.

Application	Downstream	Upstream
Broadcast TV	6–8 Mbps	64 Kbps
Movies on demand	1.5–3 Mbps	64 Kbps
Near video on demand	1.5–3 Mbps	64 Kbps
Distance learning	1.5–3 Mbps	64–384 Kbps
Home shopping	1.5 Mbps	64 Kbps
Information services	1.5 Mbps	64 Kbps
Computer gaming	1.5 Mbps	64 Kbps
Video conferencing	384 Kbps–1.5 Mbps	384 Kbps–1.5 Mbps
Video games	64 Kbps–2.8 Mbps	64 Kbps

Source: ADSL Forum

Table 11-1.   ADSL Target Service Speeds for Video

Only a couple of the services, distance learning and videoconferencing, would require an upstream bandwidth of more than 64 Kbps according to the ADSL Forum. In many implementations, videoconferencing would require symmetric bandwidth, up to a full 1.5 Mbps in both directions. Such video applications enter the realm of *High bit-rate DSL* (HDSL) or *Symmetric DSL* (SDSL), rather than pure *Asymmetric DSL* (ADSL).

How could an ADSL service provider use the information in Table 11-1 to provide services to prospective customers? Well, at the very least, there could be a LAN located right next to the DSLAM. The LAN could have routers and servers attached to it, of course-perhaps several of each. Local libraries, government offices, and other local organizations could have prerecorded video clips of information about hours, meetings, or almost anything else on this server.

Retail stores at the local mall could have a separate server for a home shopping service, all based on video. The customer could "walk" through the "virtual mall," enter a store, browse around, even "talk" to a video sales associate. People tend to sneer at such lazy and easily pleased shoppers, but this type of service has real attraction for retailers. Most retailers do almost half of their annual business between Thanksgiving and Christmas, regardless of religious mix in the area. In fact, some only stay in business from one holiday season to the next from the profits of that brief period. How damaging to their revenues is a November ice storm or December blizzard? When people do not come out to shop, retailers suffer a disaster. Now, customers—no matter how much they prefer to socialize at the mall with real humans—do not have to brave 12 inches of snow to pick out a hideous tie for a relative. The presence of bookstores on the Internet has already revolutionized the retail book industry. It is not unusual for people to browse the bookstore as before, but order the books on the Internet. Stores can charge a shipping fee, but people seem to prefer paying that over state sales tax.

The same thinking applies to local schools. School closings due to weather are normally thought to be a symptom of life in the Northeastern United States. However, when a freak snowfall strikes Georgia, the resulting paralysis is much worse than in the snow-savvy areas of the country. After all, how many snow plows are there in the Sun Belt? Distance learning with the help of video can keep things going when everything else has stopped.

A corporate video conferencing service could help with work-at-home employees, or *telecommuters.* Their effectiveness is often limited because they are in contact with the office through e-mail and faxes, but are cut off from human contact and often vital social interactions. Video conferencing services help telecommuters to participate more fully, even from home.

Movies could be provided by a local video server also, but few plan to offer or even deem feasible offering cable TV–like services with ADSL. Even if half of the cable TV customers in a given area switched to ADSL TV, the revenue stream would probably not be great enough to justify the expense of bandwidth- and gigabyte-gobbling digital video.

This list could easily be extended, but the point is made: ADSL needs services to be useful. The service could simply be Internet access, and, in fact, most if not all of these services detailed could easily be provided on the Internet itself. The attraction of the services to the local ADSL provider and customer is that the information they represent could easily be localized for more immediate relevance. This approach has already been tried with smaller, local ISPs, with some success. All of the services could still be IP-based, and Internet access through a router would still be available.

ADSL TARGET SERVICE SPEEDS: MISCELLANEOUS

The initial ADSL Forum series of target speeds supported a wide range of video services, but the target speeds did not exclude other types of services, and miscellaneous services, such as imaging and "legacy" services, are included. Legacy services provide what the customer is accustomed to, and the others add to service capabilities. These types of services are shown in Table 11-2.

Data communications covers Internet access, remote LAN access (from a home office to a business, perhaps), and distance learning (non-video-based). These should function well downstream with anywhere from 64 Kbps to a full 1.5 Mbps available. Internet and LAN access should have 10 percent or more of the downstream available bandwidth (for example, more than 150 Kbps upstream at 1.5 Mbps downstream). Distance learning should have marginally more, topping out at 384 Kbps upstream.

Service	Application	Downstream	Upstream
Data communications	Internet access	64 Kbps–1.5 Mbps	>10% of downstream
	Remote LAN access	64 Kbps–1.5 Mbps	>10% of downstream
	Distance learning	64 Kbps–1.5 Mbps	64–384 Kbps
Image-based	Home shopping	64 Kbps–1.5 Mbps	64 Kbps
	Information service	64 Kbps–1.5 Mbps	64 Kbps
Legacy services	POTS	4 KHz	4 KHz
	ISDN	160 Kbps	160 Kbps

Source: ADSL Forum

Table 11-2.   ADSL Target Speeds for Miscellaneous Services

Image-based services, such as home shopping from an online catalog or information services in the form of a graphical page showing business hours, should have anywhere from 64 Kbps to 1.5 Mbps available downstream and a modest 64 Kbps upstream because user commands should be terse in this environment.

Legacy services include plain old telephone service (POTS) and ISDN (not broadband ISDN, but narrowband ISDN). ISDN needs 160 Kbps in both directions (144 Kbps for Basic Rate Interface (BRI), plus some overhead), and POTS requires only 4 KHz analog for support on an ADSL link.

Some of the services, such as distance learning or home shopping, are the same in concept as the video equivalents, just without the video component. The others, such as remote LAN access, are familiar to anyone working in an organization of almost any size. The nice thing about these services is that people do not have to go to the office to get them.

ADSL TARGET SPEEDS AND DISTANCES

A frustrating fact of life is that ADSL is evolving so rapidly that hard and fast speeds and distances are difficult to translate into absolutes. ADSL equipment that provided 1.5 Mbps downstream at 12,000 feet now runs at 3.0 Mbps, and it stretches to 15,000 feet and beyond in its new product packaging.

Nevertheless, the ADSL forum has put together a number of documented speeds and distances that makers of ADSL equipment should "target" in their products. Keep in mind that even these speeds and distance are a moving target (pun intended), but they are still helpful.

Table 11-3 shows the target speeds and the distances at which they should be achieved, for both 24 AWG (American wire gauge) and 26 AWG copper loops (the most common gauges used, although they are frequently mixed).

Downstream Speed	24 AWG	26AWG
1.544 Mbps (T1)	18,000 feet	15,000 feet
2.048 Mbps (E1)	16,000 feet	12,000 feet
3.088 Mbps (2 × T1)	?	?
4.096 Mbps (2 × E1)	?	?
4.632 Mbps (3 × T1)	14,000 feet	12,000 feet
6.312 Mbps (T-2)	12,000 feet	9,000 feet
8.448 Mbps (upper limit)	9,000 feet	?
Source: ADSL Forum		

Table 11-3. ADSL Target Speeds and Distances

Note that with the exception of a few landmarks that stand out, many of the distances are not yet well-known enough, in terms of equipment trials and real-world deployments, so a ? marks the unknown. However, a rough estimate involving a form of interpolation (3 Mbps should be achievable roughly halfway between 14,000 and 18,000 feet) might be a start for fixing these distances.

The 18,000 feet on 24 AWG loops (15,000 on 26 AWG) for 1.5 Mbps seems to be a given. In other words, all ADSL equipment vendors seem to be comfortable with offering this speed at this distance. Beyond 18,000 feet, loading coils and line extenders enter the equation. The T-2 speed of 6.312 Mbps is another benchmark speed as well, pegged at 12,000 feet and 9,000 feet for 24 and 26 AWG loops, respectively. Note that dropping the speed to 4.632 Mbps (three times the T1 rate of 1.5 Mbps) only marginally increases the distance achievable (from 12,000 to 14,000 feet, and 9,000 to 12,000 feet).

The upper limit of 8.448 Mbps defined for ADSL is based on the current technology line code speed limitations and a wish to clearly define speed limits. Aside from this, however, nothing prevents an equipment provider from trying to exceed these target speeds and distances. A lively competition will likely result from this.

Note that the table mentions nothing about impairments due to the possible presence of bridged taps or mixed wiring gauges. In the real world, at 12,000 feet, the most common maximum downstream speed for full ADSL is between 4 and 8 Mbps, although with G.lite the limit is 1.5 Mbps due to the way G.lite uses DMT bins. Upstream maximum speeds range from 422 Kbps to 768 Kbps, and this usually depends on the amount of power that the DSL modem vendor uses in their product (the greater the speed, the pricier the unit). At 15,000 feet, the most common maximum is 2.2 Mbps for full ADSL and 1.5 Mbps for G.lite, with the same 422–768 Kbps upstream speed as before. Although ADSL is defined for local loops up to 18,000 feet long, it is sometimes hard to find an ADSL service provider that will commit to providing any form of ADSL service beyond 15,000 feet, and sometimes even beyond 14,000 feet.

ADSL IN THE HOUSE!

The hypothetical ADSL service provider has a DSLAM and perhaps even a group of servers and routers on a LAN to provide all or some of the services shown in the tables in the previous sections. All that is needed is a pool of customers eager to hook their local loops up to the DSLAM instead of the PSTN switch (of course, the ADSL splitter or filter in G.lite still handles the analog voice). But wait a minute. What about the DSL modem (ATU-R)? What about the remote splitter in full ADSL and the interface at the customer side of the link? Where do they come from? How are they packaged? Is there anything else that needs to be done to prepare a home for ADSL?

First, some background to the issues raised here. Under the 1984 Divestiture agreement, all telephones and premises wiring in the United States are owned by and are under the control of the customer (technically, the owner of the premises, but this is usually the customer anyway). Customers are free to buy and use any FCC-approved telephony device and may hire anyone they want (even the telephone company) to run new wiring or maintain old wiring. Customers can install their own wiring and often do, but few are able to comply with all relevant electrical or power standards, or are even aware of them.

The boundary between what is controlled by the customer (the devices and premises wiring) and what is controlled by the network (the telephone company) is established by a demarcation point, or just *demarc*. This demarc is typically a small box mounted in the basement or garage in residential environments. In newer housing developments, the demarc is located outside of the premises. In ADSL networks, once the local loop is digitized, this becomes the *network interface device* (NID).

In many other countries around the world, the wiring and telephony devices continue to be owned and controlled by the service provider. Deregulation movements in many countries will change this eventually, but this section emphasizes the current situation in the United States and few other places.

After a firm NID or demarc has been established, the location and arrangement of the full ADSL POTS splitter to provide continued analog telephone support, as well as the DSL modem, become an issue. What should be the relationship between the full ADSL splitter and the ATU-R, and how should the devices that connect to the ATU-R be wired? Fortunately, the ADSL Forum also addressed this issue. There can be no universal or "correct" way to do this for the simple reason that the ADSL devices, the existing telephone(s), and the premises wiring are beyond the demarc and beyond the control of the network—an ADSL network or otherwise. ADSL service providers have as little control over the ATU-R as telephone companies have over modem or telephone handset appearance, location, and function.

Nevertheless, the ATU-R equipment vendors have a lot of control over the situation. If a particular configuration is not offered by any vendor, it will not matter whether customers want it or not. This does not mean that all ATU-R packages will be identical. There is always variety for the sake of product distinction. However, ATU-R and required splitter can be arranged in products in several fairly standard ways, each with its own implications for the customer, especially where wiring is concerned. This section examines the issues involved with ADSL installations at the customer premises. Please note that G.lite does away with the issues regarding the position of the splitter on the premises because there is no remote splitter in G.lite. Also, G.lite requires no new wiring except to a location that has no previous modem connectivity (G.lite with microfilters will support analog voice). In other words, G.lite makes most if not all of these premises issues about splitter packages inoperative.

INSTALLING FULL ADSL: SPLITTER AT THE DEMARCATION?

The simplest way to accomplish a full ADSL installation with splitter is shown in Figure 11-1. The customer acquires an ATU-R in different ways (purchase, lease, or straight give-away). In any case, the customer makes the decision. In this scenario, the splitter is built into the

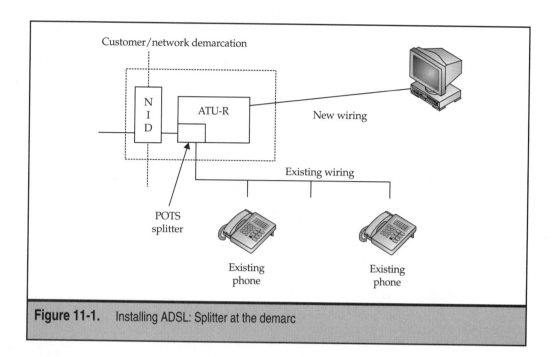

Figure 11-1. Installing ADSL: Splitter at the demarc

ATU-R, which is mounted as close to the demarc as possible. At this time, the demarc becomes a NID. Existing premises wiring for existing telephone(s) is then simply unplugged from the NID (the demarc) and plugged into the ATU-R splitter, RJ-11 connector to RJ-11 connector. (RJ-11 is the standard analog connector that is built into the splitter at the ATU-R). The ATU-R is plugged into the NID—so much the better for analog telephone support.

However, because the ADSL devices such as PCs or cable TV set-top boxes are likely to be far from the NID, new wiring must be run to support the PC or set-top box that wants to access the new ADSL services. While not an insurmountable problem, the lack of control that the service provider can exercise in this situation could be troublesome. If the customer wants to install their own wiring, the job might not be done properly. It seems reasonable, however, that most customers will defer to the advice of the service provider ("let us do it—for a small price").

Benefits of this arrangement include minimal risk of appliance or POTS coupled noise; plus, it is the best solution if the service provider "owns" the ATU-R. Drawbacks include lack of the ready alternating current (AC) power needed for the ATU-R at most demarc locations (garages, basements, and the outside of homes), as well as the presence of environmental extremes (heat and cold, wet and dry) in most of these locations. Small business environments should be better in this regard.

Another potential drawback is the presence of the splitter and ATU-R in the same device. This circumstance maximizes the possibility of interference from POTS ringing and other high-voltage signals on the ADSL bit stream. In fact, putting the splitter under the same covers is possibly the worst place to put the splitter for these reasons. Nevertheless, most ATU-R equipment vendors have packaged the splitter and ATU-R for convenience and simplicity of design. Interference between POTS and ADSL has so far not been a problem or concern.

INSTALLING FULL ADSL: SPLITTER IN THE PC BOARD?

Again and again, regardless of where the full ADSL splitter is located, the overwhelming concern is the amount of new premises wiring required. As mentioned previously, the control that a service provider may exercise over the type and workmanship of new premises wiring installed is exactly none whatsoever. Although standards exist for electrical and power requirements covering premises telecommunications wiring, inspections are rare and violations common. With any new full ADSL installations, it might be a good idea to try to minimize the new wiring required.

It is anticipated that most full ADSL service offerings will feature high-speed Internet access, almost exclusively. This being the case, it might be a good idea to package the ATU-R used with full ADSL to closely resemble a modem. Most PC users are familiar and comfortable with modems and may already have a telephone line at the desktop location. This ADSL arrangement is shown in Figure 11-2.

When the ATU-R and splitter are packaged as an internal modem board or external modem device, the wiring task might be easier. Only one new cable run—from the NID to

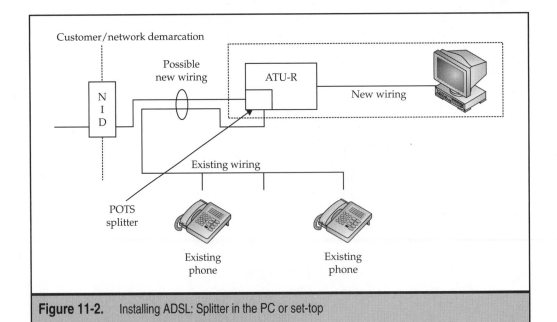

Figure 11-2.   Installing ADSL: Splitter in the PC or set-top

the ATU-R—might be needed to connect the splitter in the ATU-R to the existing POTS. If a second line has been used for Internet access along with a line for a telephone next to the telephone handset before the ADSL install, this wiring can be used to link the ATU-R splitter to the existing analog wiring without any new wiring at all. Even so, the wiring at the NID would have to be rearranged, as shown in the figure.

Naturally, other new wiring runs to other ADSL devices (PCs) would still need to be installed. The runs should be much shorter, however, as the figure illustrates. Additionally, no wiring is needed to connect to the device that holds the integrated ATU-R. Because few ADSL service providers or equipment vendors will support more than a single upstream and downstream ADSL channel, this approach makes a lot of sense.

The benefits of this arrangement include the minimum requirement for new wiring and the need only for new POTS wire (very tolerant of "nonstandard" installations). Drawbacks include the fact that continued POTS service now depends on the presence of the ATU-R, which is similar to a modem. In other words, unplugging the POTS wiring from the "modem" would interrupt analog voice service. A special wall connector (known as RJ-31X in the United States) might help solve the problem, which is not really an ADSL problem at all. A more serious drawback is that the long parallel run of POTS and ADSL signals from NID to splitter, as seen in the figure, cross-couples all of the POTS noise onto the ADSL link. POTS noises include ringing, pulse dialing, hook flashes, and so on. The ADSL Forum reported that just a few feet in parallel have been shown to cause ADSL errors. Interleaving within the ADSL frame and higher layer error control in the transfer protocols such as IP or ATM will help, but this arrangement could impact perceived service quality.

All in all, this scenario has serious technical problems—most critical is the coupling problem on the parallel wiring run.

INSTALLING FULL ADSL:
LOW PASS/HIGH PASS FILTERS?

In any ADSL installation, there are two problems with combining the POTS splitter to provide continuing support for existing analog telephones inside the ATU-R (Remote) housing itself. The first is that this arrangement is sure to maximize the risk of coupling POTS noise onto the ADSL wiring, to the detriment of ADSL signal quality. The second is that this arrangement forms a single point-of-failure for both new ADSL services and analog voice services.

This being the case, it might be better to separate the POTS splitter from the ATU-R altogether. In almost all current full ADSL installations, this is what is done.

Figure 11-3 shows how analog and digital signals would both exit the NID. A small *low pass filter* (LPF) is mounted near the NID. The low pass filter is needed because POTS support is provided by the continued presence of analog voice baseband signals at the low end of the frequency range (4 KHz is needed for analog voice support). The filter, which needs no AC power, passes these voice signals to existing telephones over existing wiring.

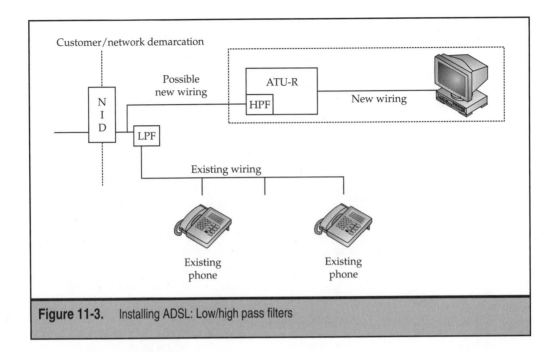

Figure 11-3.   Installing ADSL: Low/high pass filters

The ATU-R now contains a *high pass filter* (HPF) that filters out the low frequency analog voice signals, allowing only ADSL signals to enter the ATU-R—this is basically just another form of splitter. Now the ATU-R can be packaged exactly like an analog modem and can be located at the most convenient place on the premises. Wiring to the ATU-R might be new or existing if the PC previously used a second line for Internet access.

The benefits of this arrangement include short new wiring distances and elimination of parallel POTS–ADSL wiring that could cause signal coupling and interference. Of course, if an existing telephone is adjacent to the PC, this benefit could disappear. Also, the LPF can actually be installed before the ATU-R, perhaps months earlier in preparation of an ADSL service offering, and the splitter is now independent of the functioning or presence of the ATU-R (that is, no inadvertent disconnects of voice service). The ATU-R can now be purchased, installed, and employed in precisely the same way as a modem.

The drawback is that all splitters are intended to function with a specific ATU-R from the same vendor. Another potential problem is in countries where lines need an active POTS splitter. In this case, the ATU-R (or some other network component) would have to provide power to the splitter over the ADSL wiring.

INSTALLING FULL ADSL:
LOW PASS FILTERS FOR ALL PHONES?

Figure 11-4 shows a distributed split POTS splitter arrangement. Although this arrangement could be used for full ADSL, this architecture essentially is exactly what G.lite does.

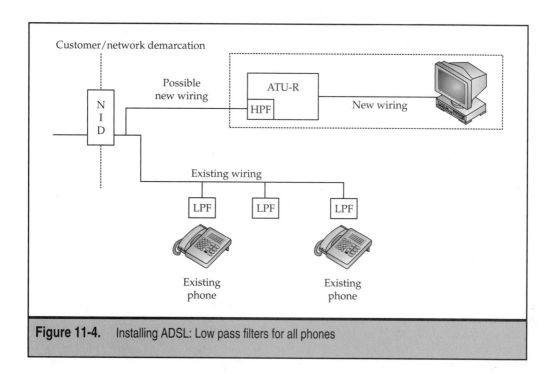

Figure 11-4. Installing ADSL: Low pass filters for all phones

Note the absence of any device that could be labeled a "splitter." Each telephone and jack now has a low pass filter or microfilter installed, possibly as a faceplate-mounted device. The low pass filter is needed because POTS support is provided by the continued presence of analog voice baseband signals at the low end of the frequency range (4 KHz). The filters, which need no AC power, passes these voice signals to existing telephones over existing wiring.

The drawback here is the same as before—all splitters are intended to function with a specific ATU-R from the same vendor. Another potential problem is that removing any low pass filter, even from unterminated (phoneless) jacks, might affect the performance of the ADSL link. Also, when four or more microfilters are used on the same line, analog voice often becomes muffled.

The preceding ADSL splitter and DSL modem (ATU-R) discussion has detailed all possible packages for full ADSL and G.lite. The last two arrangements for full ADSL and G.lite are the most common today. Generally, no matter which particular architecture is used, the POTS splitter should be isolated from the ATU-R. This eliminates a single-point-of-failure scenario and minimizes the chance for POTS-to-ADSL signal coupling. Also, it is important that the wiring inside premises essentially belongs to and is controlled by the customer in the United States and some other countries. Therefore, it can be quite hard to control and maintain the quality of the wiring, especially in cases where the wiring was installed by nonprofessionals (perhaps even the customer) with little regard or respect for standards. Indeed, they might not even be aware of the relevant standards.

This being the case, installations that minimize the requirement for new inside wiring are always the preferred approach. Additionally, this will also minimize the cost to the customer if they want to have the service provider install the new wiring.

Keep in mind that the full ADSL architecture supports a variety of *premises distribution networks* (PDN), not just point-to-point wiring. A 10Base-T LAN or wireless premises network based on a technology such as Bluetooth is possible, along with other more exotic arrangements. However, for most ADSL installs, a single new Category 5 (Cat-5) wire that runs from the DSL modem to the family PC should be all that is necessary. This cable is usually included with the DSL modem. More elaborate configurations for the customer distribution network when multiple PCs are in use are considered next.

ADSL PREMISES NETWORK ISSUES

Linking a DSL modem (ATU-R) to a single PC is usually a simple affair that most customers can manage by themselves. In cases where the customer needs assistance, usually a single call to an 800 number will suffice to step the customer through the procedure, assuming the DSLAM and ISP access has been set up ahead of time.

In most cases, the installation involves no more than plugging in the DSL modem to an AC power outlet, using a wire supplied with the DSL modem to link the DSL modem to the wall jack, and using a LAN crossover cable (also supplied) to connect the DSL modem RJ-45 LAN jack to a PC Ethernet NIC (the NIC is often supplied by the DSL service provider as well). When PPPoE or something other than pure bridged Ethernet is used by the DSL service provider, some software needs to be loaded onto the PC as well. All of the scenarios in the previous section showed a single PC linked to the DSL modem.

However, because more PCs are sold today as second PCs rather than first PCs, and most homes have more than one person needing access to a computer and the Internet at the same time, it is much more common to find DSL services with more than one PC on the link. This requires a true customer premises distribution network to connect all the PCs to the DSL modem. This section examines all of the more popular ways to link multiple PCs to a single DSL modem.

First of all, it is important to note that this is not and cannot be an exhaustive survey of all possible DSL modem and home networking packages. Many DSL modems today are not just a DSL modem, and many DSL modems include a LAN hub, routing, and some security functions. But this section will at least be useful in distinguishing the major forms of DSL premises arrangements for multiple PCs.

It is assumed in this section that the DSL modem is not simply a DSL modem and nothing else. The DSL modems discussed in this section are assumed to have a simple routing function built in. That is, like all IP routers, the DSL modem uses the DSL service provider's assigned IP address on the ADSL link itself, but uses a separate IP address space (usually an IP address space reserved for LANs and other private networks) on the customer distribution network. It should be noted that this "inner and outer" IP addressing used by a router is not quite the same as a feature called *network address translation* (NAT). NAT is often used for security purposes today, but NAT was originally developed just as a way to conserve public IP addresses by allowing for the use of private IP addresses on isolated segments of the Internet. So NAT is needed if the IP address space used on the customer's premises network is private, and this internal address space usually is. This is because private IP addresses can never be used on the global public Internet. So not only does a DSL modem route between ADSL link and inner customer network, the DSL modem must translate between public IP addresses assigned by the DSL service provider and used on the ADSL link and the private IP addresses used on the customer's premises network and typically assigned by the DSL modem itself.

It is also assumed in this section that the premises distribution network is not run through the single PC linked to the DSL modem. Such a scenario is possible and is shown in Figure 11-5. Most operating systems allow for such "workgroup" networking, but the burden of DSL link access is placed on the single PC with direct connectivity to the DSL modem. And if this PC is not on, there is no DSL access. One method using this configuration employed a premises networking scheme called HomePNA (Home Private Network Architecture) to link other PCs to the PC with the ADSL link. This scenario is not discussed further.

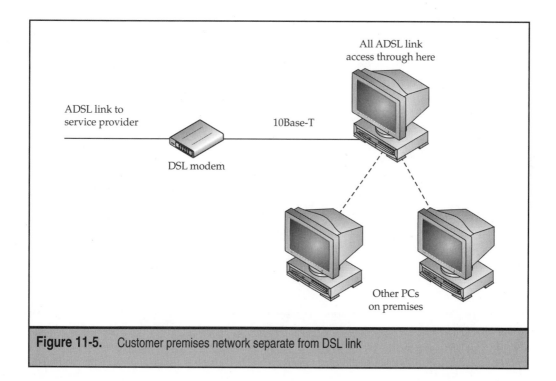

Figure 11-5.   Customer premises network separate from DSL link

For the rest of this chapter, it is assumed that the DSL modem is a station on the premises network, and not merely a device linked to and accessed by a single PC. This is shown in Figure 11-6. This arrangement is certainly the most common position for the DSL modem today.

Note that the figure adds inner and outer IP address spaces to the architecture. Typically, the ADSL service provider assigns only one public IP address on the link, and this address is for the DSL modem. Some ADSL service providers will allows users to purchase more public IP addresses, but this is not common. (The IP address from the 172.16.0.0/16 IP address space shown on the ADSL link in the figure is actually *also* a private IP address, and used here for security reasons). But even when one IP address is used, it is common for the DSL router to both assign private IP addresses with the 10.0.0.0 address range for the customer premises network and to perform NAT on the traffic between them (NAT must be done when private IP addresses are used on the premises network). Another popular private IP address space to use with DSL routers is the 192.168.1.0/24 address space.

What follows says nothing about the protocol stacks used on the premises network between DSL modem and user PCs. The possible encapsulations for IP packets between DSL modems and PCs were detailed in the previous chapter. It is enough here just to mention that there are now three types of encapsulation possibilities that IP traffic might use from DSL modem to PC. The pure Ethernet bridging done through the DSLAM is not

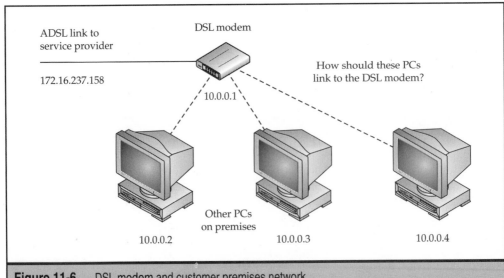

Figure 11-6.   DSL modem and customer premises network

an issue from the premises perspective once the DSL modem is a router, as is assumed here. The three possible encapsulations are as follows:

▼ **ATM cells**   Virtually unheard of; few PCs understand ATM cells, and 10Base-T Ethernet is not a native ATM cell transport.

■ **PPPoE and other types of tunnels**   PPPoE and even more elaborate forms of tunnels can still be used as long as the PCs and DSL modem all know it.

▲ **Pure routing**   Simple IP packets in frames sent to the DSL modem without PPPoE or other forms of tunneling.

All of the arrangements described in the rest of this section can use either the PPPoE (or related tunneling) and the routing encapsulations. ATM cell use requires specialized hardware, software, and networks not considered here.

ADSL PREMISES NETWORKS

This section explores three ways that multiple PCs can link to a DSL modem:

▼ 10Base-T Ethernet LAN

■ Wireless (IEEE 802.11b)

▲ Bluetooth

It is assumed that the DSL modem is also an IP router and has only a 10Base-T Ethernet port (some models can run at 100 Mbps, but this is a minor variation).

10Base-T Ethernet LAN

There are other ways to link multiple PCs to a DSL modem, but no details for these alternatives are presented. In most cases, the method closely mirrors one of these overall architectures. For example, some DSL modems come with a *Universal Serial Bus* (USB) port instead of or as well as an Ethernet 10Base-T port. Just as 10Base-T does, the USB network requires a hub (in this case a USB hub), cabling (USB cables), and the proper PC interface (most PCs now have built-in USB ports). So the details of USB-based customer networks follow 10Base-T LAN concerns closely enough that a separate section on USB networks is not necessary.

When any device has a 10 Mbps Ethernet connector on it that is capable of supported 10Base-T wiring, the natural temptation is to build an Ethernet LAN on the premises to reach multiple devices. Most PCs either come with an Ethernet NIC, or one can be added for only a few dollars more when the PC is ordered. PCs without a NIC (and there still are many) can have a NIC added for as little as $20 (although a few more dollars spent will often avoid compatibility issues, particularly in the Linux environment).

A 10Base-T LAN requires a hub. Simple four-port 10Base-T hubs cost about $40, and sometimes even less. A DSL modem linking three PCs through a 10Base-T hub is shown in Figure 11-7.

The hub is an AC-powered device and must be plugged in. The cable linking the DSL modem to the hub is no longer the simple crossover cable used when the DSL modem links directly to a single PC NIC. In most cases, Cat-5 or better cables must be purchased separately, and these cables can cost many times more than the NICs and hub they connect.

A common variation on this basic theme is when the DSL modem is just that: a DSL modem with a simple 10Base-T port. The DSL modem in this case has no sophisticated router or NAT capabilities. So a "home network" router/NAT/switch/LAN-hub device now adds the intelligence needed for the DSL modem arrangement. This variation just supplies in the "hub" any features needed that are lacking in the DSL modem.

The biggest advantage to these related arrangements is their simplicity. The customer distribution network is now just a common Ethernet LAN. Routers and PCs with NICs more or less expect to use this type of network. Typically the hub is quite close to the DSL modem and at least one of the PCs.

The biggest problem to this approach is installation. Cables can be ordered with connectors in preselected lengths, as long as the length is known. But cable with connectors are hard to snake through walls and between floors, and simply nailing them to the wall is unattractive in the living areas. Unconnectorized cabling is much less expensive, but special tools and expertise are required to add the connectors after the cable is run. This expertise is often beyond the capabilities of the homeowner. But for small offices serviced by ADSL, or homes with enough expertise available, this approach is popular and attractive.

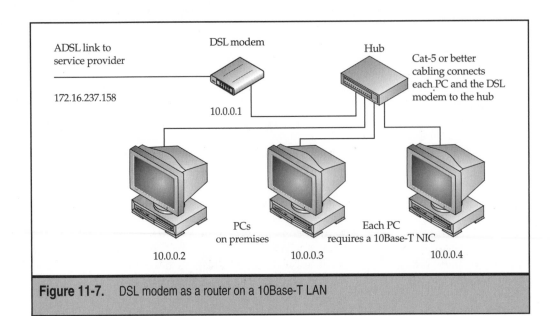

Figure 11-7. DSL modem as a router on a 10Base-T LAN

However, even after installation, this home LAN arrangement has one major draw-back. This is the problem of adds, moves, and changes. Suppose only two PC locations are initially wired. If a third PC is added, a third cable must be run. And what if a PC is moved to another room? Then the cable must be rerun, or abandoned and new cable installed.

If mobility is a concern, some form of wireless network is a good form for the customer premises network to take. Wireless networks take all of the work out of adds, moves, and changes, but can be more expensive than wired LANs. This is not always true, of course: it all depends on the lengths of Cat-5 cable required and whether preconnectorized cables are used.

Wireless (IEEE 802.11b)

Not too long ago, many forms of wireless networking supported Ethernet speeds. Almost all of these were vendor-specific, because the official IEEE wireless LAN (IEEE 802.11) until recently ran at only a fraction of Ethernet's 10 Mbps. But the recent IEEE 802.11b standard offers true 10 Mbps wireless Ethernet, and most vendor-specific products have made or are making the shift to 802.11b compliance (or compatibility).

There are two kinds of IEEE 802.11b wireless LANs: *ad-hoc* and *access point*. The ad-hoc type is not particularly suitable for DSL premises applications and is not considered further. More permanent wireless LANs, such as DSL premises networks, should use another type of wireless LAN using the concept of an *access point*, or *base station*. In this type of wireless LAN, the access point acts as a wired LAN hub. The hub provides the connectivity among the wireless computers.

Two types of access points are defined in IEEE 802.11b: dedicated hardware access points (sometimes seen as HAPs) and software access points (sometimes SAPs). Generally, hardware access points offer more features and better overall performance, but this can vary from product to product. Software access points use special software that runs on a computer equipped with both wireless NIC, as in an ad-hoc wireless LAN and the usual wired NIC (most likely Ethernet). Because this section considers only premises networks where the DSL modem is stationed on the LAN, only the hardware access point configuration is shown in Figure 11-8.

In theory, any hardware or software from any vendor that adheres to the IEEE 802.11b standard can interoperate. However, the standard itself is relatively new, so verification testing is still a requirement. This is especially true because the original IEEE 802.11 standard specified two different methods for wireless communications. These made up the *air interface* of the wireless devices. IEEE 802.11 wireless LANs could use either *frequency hopping* (FH) or *direct sequence spread spectrum* (DSSS) as an air interface. These are distinct methods and cannot interoperate. The newer IEEE 802.11b uses DSSS exclusively (with minor variations allowed).

Care is also necessary when it comes to operational speeds. The original IEEE 802.11 specification ran at 1 and 2 Mbps, much slower than the basic 10 Mbps Ethernet speed. IEEE 802.11b runs at either 5.5 Mbps or 11 Mbps (full Ethernet plus some overhead). Many vendors, in the interests of backward compatibility, support all speeds. But some vendors support only the higher speeds, and of course older equipment cannot run faster

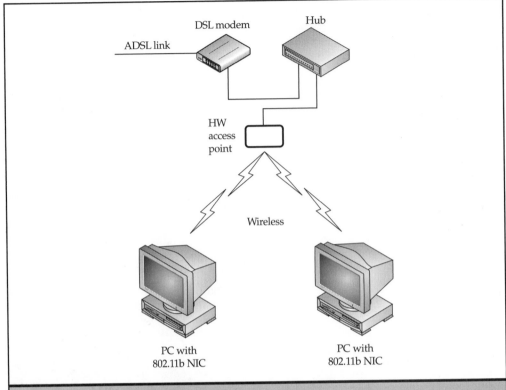

Figure 11-8. IEEE 802.11b hardware access point as a LAN hub in ADSL

than 2 Mbps at all. And FH-type IEEE 802.11 equipment is essentially obsolete. There are also vendors who insist on "extending" a standard like IEEE 802.11b with their own unique features. Needless to say, this practice also complicates interoperability and might even prevent any form of meaningful interoperability among different vendor's IEEE 802.11b products.

When it comes to range, each access point has a finite range beyond which a client computer cannot communicate with the access point. The actual distance will vary depending on the environment. Vendors are fond of stating the maximum feasible distance, of course, and usually provide both indoor (better) and outdoor (worse) ranges to give a reasonable indication of performance. It is important to note that at the distance limits, connectivity might be maintained, but performance will drop—often dramatically—as the connection quality drops and the network tries to compensate, usually by slowing the air link down.

Typical indoor ranges are between 150 to 300 feet (about 75 to 100 meters), but can be much shorter in some older concrete and steel buildings. Support for longer distance is possible, but only at greatly reduced throughput rates. Outdoor ranges are sometimes quoted at 1,000 feet (more than 300 meters), but this depends on the environment, as always. IEEE 802.11b LANs operate in the 2.4 GHz ISM (industrial, scientific, and medical) frequency range, so no license is required.

Security is always a concern in wireless networking, and wireless LANs are no exception. An intruder does not even need physical access to the wired LAN in order to gain access to privileged information. But IEEE 802.11 information cannot even be received, much less decoded, by simple sniffers and scanners. Special equipment can be used to intercept and decode wireless LAN information, although this special equipment can be quite expensive. Moreover, all IEEE 802.11 wireless LANs have a function called *wireless equivalent privacy* (WEP). This is a form of encryption that provides privacy at least as good as that achieved on a wired LAN. Using WEP in a wireless LAN is always a good idea. However, it should be noted that the encryption security included in IEEE 802.11b has been criticized as much too simplistic for most network situations. It should also be noted that all normal PPPoE, tunneling, and other virtual private network (VPN) procedures will function in exactly the same way over a wireless LAN as they do over a traditional wired LAN.

Security for customer premises networks using ADSL is discussed more fully in the next section of this chapter.

Bluetooth

The Bluetooth wireless technology was discussed in Chapter 5, where it was pointed out that it is more suitable for premises use than as an alternative to DSL. This section considers Bluetooth as a wireless alternative to IEEE 802.11b for a DSL customer premises network.

Bluetooth devices have three power classes: 100 milliwatt, 2.5 milliwatt, and 1 milliwatt. The 100 milliwatt device range is 100 meters (a little more than 300 feet), the 2.5 milliwatt range is 10 meters (about 30 feet), and the 1 milliwatt range is only 10 centimeters (a few inches). The optimal throughput for Bluetooth devices is 1 Mbps. However, there is always some error-correcting overhead to add in and the need for retransmissions with some protocols and with high bit error rates. Throughput in the range of 700 to 800 Kbps should not be unreasonable to expect. Bluetooth operates in the same range as IEEE 802.11 wireless LANs, so the two technologies can and do interfere with each other, degrading the performance of both. So while the speed of Bluetooth should be enough to keep up with many ADSL links, especially G.lite, interference with a neighbor's IEEE 802.11 wireless LAN is a distinct possibility.

Bluetooth supports both voice and data traffic, but this is not an issue when Bluetooth is used to links PCs to DSL modems. Some of the frequency channels that Bluetooth uses to hop around on are also shared by microwave ovens. Any data just vanishes when Bluetooth hops onto a noisy channel. Bluetooth security is a combination of link-level and encryption security. The security is device-based and does not require the user to do anything at all to invoke the security. Bluetooth has three levels of security:

▼ **Level 1**   No security at all.

■ **Level 2**   Service-level-enforced security. This security is established after the Bluetooth channel connection has been negotiated.

▲ **Level 3**   Link-level-enforced security. This requires the user to enter passwords and so on for services.

Level 2 security is recommended for most instances. Level 3 is the most secure, but requires full encryption and constant user intervention. Level 3 security is not transparent, and so makes Bluetooth more complex to use with DSL. Whatever the security level, Bluetooth devices are always vulnerable to denial of service attacks from flooding the channels with interference. There is nothing like IEEE 802.11b's WEP defined for Bluetooth.

ADSL SECURITY ISSUES

In July of 2001, many users of Qwest's ADSL service received a strange message on their PC screens. The message said that the private IP address (such as 10.0.0.2) handed out to the PC by the DSL modem using the Dynamic Host Configuration Protocol (DHCP) had "expired," and the DSL modem could not "renew the lease." What this essentially meant was that the user could not use the ADSL link, even though it was up and running.

Why? This was an instance of the Code Red *distributed denial of service* (DDoS) attack. Older DSL modems deployed by Qwest and other DSL service providers were at risk, and the DDoS attacker basically took over some 240,000 DSL modems that July. The DSL modems were not even the targets of the attack, which was aimed at Web sites.

In the case of Code Red, the DSL modems were more or less just innocent bystanders. But hacker attempts to break in to home PCs and LANs through DSL modems have increased in recent years. Corporate sites have hardened their security measures in many cases, and hackers out for an easy success have been turning to ADSL users (and, to be fair, cable modem users) because home security tends to be lax (or non-existent) compared to normal corporate procedures.

DSL and cable modem links are "always on." Even when PPPoE is not used or the user is not logged in, the link from DSL modem to DSLAM is still up and running. In this case, "always on" means "always at risk."

This section does not outline all of the ways that user information can be at risk when DSL is used. The last chapter mentioned Ethernet bridging in the DSLAM as one such security risk, and the threat was taken seriously enough that most DSL modems today are routers, not bridges. But as an example, Figure 11-9 shows how a hacker might gain access to a user PC through ADSL.

The file sharing referred to in the figure is Windows file sharing. In some types of Windows systems, it is easy to decouple the file sharing from IP, but in others (especially the newer forms of Windows) it is not even possible. Linux machines running SAMBA

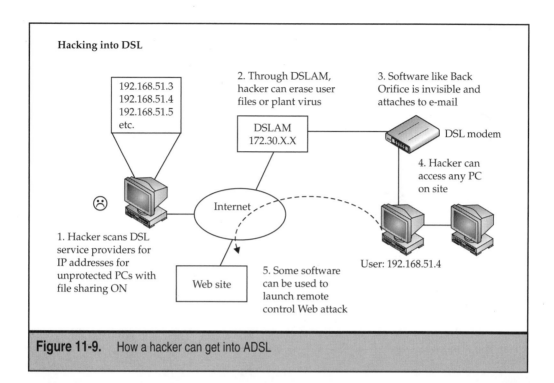

Hacking into DSL

192.168.51.3
192.168.51.4
192.168.51.5
etc.

2. Through DSLAM,
hacker can erase user
files or plant virus

3. Software like Back
Orifice is invisible and
attaches to e-mail

DSLAM
172.30.X.X

DSL modem

4. Hacker can
access any PC
on site

Internet

1. Hacker scans DSL
service providers for
IP addresses for
unprotected PCs with
file sharing ON

Web site

5. Some software
can be used to
launch remote
control Web attack

User: 192.168.51.4

Figure 11-9.   How a hacker can get into ADSL

are also vulnerable to this form of attack. These vulnerabilities make the following discussion of personal firewalls so important.

Note that this form of attack on ADSL relies on the user PC having a public IP address that the hacker can determine. Again, making the DSL modem into a router helps prevent this type of abuse, and using NAT in the DSL modem with a private IP address space on the premises network helps even more.

But what if the DSL modem does not offer router and/or NAT protection? Older DSL modems were simple bridges, and are vulnerable to many forms of hacker attack. And not many DSL service providers are in the habit of upgrading their customer's DSL modems for any reason (in fact, it is unheard of). Turning off file sharing is always a good idea, but is not always possible in some situations.

A more comprehensive approach to securing the premises PCs is to use some type of special hardware security device between the network and the DSL modem and/or some form of security screening software on each PC. More sensitive users will use both, while more complacent users will rely on one or the other (and other users might not rely on either until it is too late).

It has been noted that many current DSL modems include such "security" features as NAT (user packets never have the host's IP address on packets sent out over the Internet), DHCP (the DSL modem's IP address on the public Internet can change), and even some simple firewall filtering rules (packets that are known to be harmful can be blocked by the

DSL modem). The problem with relying on the DSL modem to be a security device is that, as Code Red has shown, the DSL modems *themselves* can be a risk. For instance, the DHCP client lease often acquires the same IP address over and over again. And DSL modems are primarily designed to pass packets quickly, not pass packets securely.

This is not the place to try and compile a comprehensive list of hardware and software security products. Such a list quickly becomes outdated, and product details and comparisons are readily available from other sources. Here is an alphabetical list of some hardware security devices that can be used with ADSL:

▼ D-Link DSL/Cable Residential Gateway

■ Linksys EtherFast Cable/DSL Router

■ Linksys Instant GigaDrive

■ Macsense Xrouter

■ NEXLAND ISB2LAN

■ Ramp Networks WebRamp 700s

■ UMAX Ugate-3000 Cable/ADSL Modem Sharing Gateway Router

▲ ZyXEL Prestige P310 WAN Router

In most cases, these devices take the place of the 10Base-T hub and connect directly to the DSL modem (or can even *replace* the DSL modem). In other cases, the device is placed between the DSL modem and the LAN hub, which is retained. Some products can be used in both ways. Most also feature firewall filtering, support for PPPoE, tunneling such as IPSec or PPTP VPN, a DHCP client for the ADSL link, a server function for the attached user PCs, as well as NAT (this can complement the DSL modem's NAT feature).

The high-end devices are not only Ethernet hub, but fast Ethernet (100 Mbps) switch. Most have a simple Web-browser management interface for simple user configuration. Prices start at about $100 or sometimes even less.

Software security products run on each PC (in some cases, one PC can provide protection for all PCs on a premises LAN). Here is an alphabetical list of some security software packages that can be used with (or without!) ADSL:

▼ BlackICE Defender (from NetworkICE)

■ CyberArmor (from InfoExpress)

■ ConSeal PC Firewall (from C&C Software)

■ Internet Firewall 2000 (IFW2000) for PCs (from Digital Robotics)

■ Personal Firewall and Internet Security Suite (from Symantec)

■ Tiny Software Personal Firewall

▲ ZoneAlarm (from ZoneLabs)

When all is said and done, it does not so much matter which hardware and/or software security package is used with ADSL. What matters most is that *some* form of additional security is used beyond the DSL modem.

VOICE OVER DSL

At least one of the hardware security products mentioned in the preceding section comes with a seemingly odd feature. This is an ordinary analog voice telephony jack that is right next to the expected RJ-45 jacks for use with Ethernet. The user can plug an ordinary telephone into this port on the device. What is that voice port doing there? Isn't analog voice already supported on the ADSL line under the ADSL frequencies? Yes. This port is not for voice *under* DSL. It is for one form of *voice over DSL* (VoDSL).

As might be implied in the previous sentence, there are two forms of VoDSL, one using VoIP and the other using telephony over ATM. This section closes this chapter with a look at just what VoDSL is, and how VoDSL can be used to support voice when VoIP or ATM is used on the DSL link.

VoDSL does not have to be a form of VoIP. VoDSL can be used with both packet-mode and ATM-mode DSL links. VoDSL can be delivered with a form of ATM designed especially for the needs of delay-sensitive applications (like voice) known as ATM Adaptation Layer type 2 (AAL2). Recall that even when ATM is used on the ADSL link to carry IP packets, a form of ATM encapsulation known as AAL5 is used. AAL5 addresses the needs of "data" applications such as IP when the main concern is not a low and stable delay, but minimal overhead for transported packets. But this section will consider any voice delivered using digital techniques instead of analog compatibility over a DSL link to be VoDSL, even VoIP.

So what is the difference between the VoIP form of VoDSL and the ATM-based version of VoDSL? VoIP uses packetized and digitized voice to reach not a traditional PSTN voice switch (and using the DSLAM and ATM network to get to the voice switch), but VoIP uses packetized and digitized voice to reach an *Internet-attached* voice switch through the Internet. And that makes all the difference. Whether the DSL link is running packet-mode (IP packets) or ATM-mode (ATM cells), the official VoDSL architecture links to voice gateways over an ATM core network, while the *Internet-based* VoIP "form" of VoDSL links to voice gateways on the Internet.

If this difference sounds confusing or trivial, perhaps a few visual representations of ATM-based VoDSL and VoDSL with VoIP will help. VoIP does not rely on any DSL to be useful. VoIP can be used with any Internet connection, dial-up or leased private line. VoIP can save users money on long-distance calls, especially to countries outside of the United States. VoIP over ADSL makes perfect sense if Internet users want to continue (or start) using VoIP over the Internet on their ADSL link.

So why isn't VoDSL the same as VoIP? Mainly because VoDSL cannot and should not *rely* on IP connections to the Internet to be useful. Some DSLs need VoDSL if they are to support voice at all. ADSL is not in that category, but HDSL and HDSL2 are, and so are many types of SDSL (the position of VDSL will be considered later). Recall that local loops cannot really be analog and digital at the same time. So if the line code used on the local loop is some form of digital line code such as PAM, there is really no place for the analog voice band at 0–4 KHz. With the line codes used in HDSL, HDSL2, and most forms of SDSL, once the local loop supports the DSL, there is no support for analog voice. This was not such a problem when HDSL, HDSL2, and SDSL were mainly targeted at businesses, because many businesses had digital voice lines separate from their data lines already. But for residential services, or small offices, only one local loop should be used for both voice and data. So if the line code is digital, then some form of voice over DSL must be standardized and used. And because the PSTN voice switch is right next to the DSLAM (in most cases) in the DSL service provider's central office, there is no need to use the Internet to link DSL voice users to the PSTN. That is why VoDSL differs from VoIP.

The DSL Forum's reference model for VoDSL is shown in Figure 11-10. The biggest change from analog voice support is the introduction of the premises' *integrated access device* (IAD) and the presence beyond the DSLAM of the *voice gateway.*

In the figure, the DSL modem has become the IAD, and now has a number of voice ports along with the RJ-45 port used for the premises link to the PC(s). Any PCs still send their IP packets through the DSL modem and DSLAM onto the core network to an IP router, as shown in the figure. And just as before, the IP packets can then make their way through the router onto the Internet and/or the some corporate site. So far, nothing much has changed from the PC perspective.

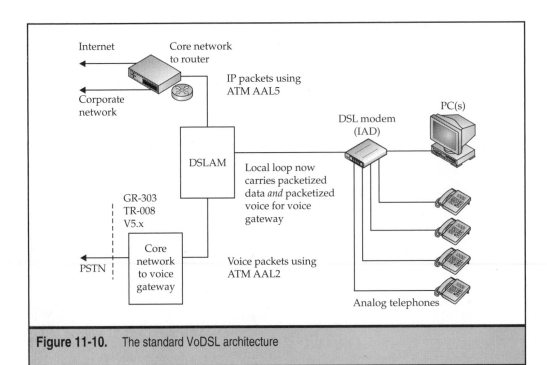

Figure 11-10.   The standard VoDSL architecture

The change is with regard to the presence of the telephones on the IAD voice ports. These telephones can be digital telephones such as those used in offices or with ISDN, but they are most likely normal analog telephones in a residential environment. The choice to support analog and/or digital telephones is up to the manufacturer of the IAD. If analog telephones are supported, the conversion to digital voice is done in the IAD. Now, the official network model for VoDSL from the DSL Forum shows individual telephones for each IAD port (the model also shows four ports on the IAD). Residential users might be better served by a single port that could accommodate all of the telephones on the current premises wiring, but this architecture not only applies to ADSL, but also any other DSL, especially DSLs like HDSL2 and SDSL that do not support analog voice "under" the DSL frequencies. For the time being, residential users with ADSL would be best served by retaining the analog services with a splitter or microfilter arrangement because the end result is the same with the IAD. Note that in the VoDSL model, there is no splitter or microfilters anywhere to be seen, even on the DSLAM. VoDSL is for digitized and packetized voice on the local loop. Period.

The end result for the telephones is the same, however. That is, the voice still ends up on the PSTN through the central office switch (not shown in the figure). But instead of a splitter being used in or near the DSLAM to send analog voice to the voice switch, the DSLAM must selectively send voice packets over the same core network used to link the DSLAMs to routers. But these packets go to the new *voice gateway* while the DSLAM still sends "data" packets through the core network and on to the router (the DSLAM link to the voice gateway might also be very short if the DSL service provider is also a telephone company). In the official VoDSL model, this is easy for the DSLAM to do because IP packets arrive inside ATM cells using AAL5 over the DSL link on one virtual circuit from the customer, and the voice packets arrive inside ATM cells using AAL2 over the DSL link on *another* virtual circuit (it is an ironclad rule in ATM that two different AAL types can never be used on the same virtual circuit). So the AAL5 virtual circuit (perhaps VPI:VCI 0:37) ATM cells go onto the core network as before and the AAL2 virtual circuit (perhaps VPI:VCI 0:38) ATM cells go to the voice gateway.

In the VoDSL network model, what the IAD does is called the *customer premises interworking function* (CP-IWF, because it now must *interwork* with voice as well as data) and what the voice gateway does is called the central office interworking function (CO-IWF). But no matter what it is called, the voice gateway (or CO-IWF) must convert between ATM cells carrying AAL2 voice and the type of digital voice understood by the central office switch attached to the PSTN. This is how people without DSL can still call folks with IADs, and vice versa.

The main set of rules for accomplishing this conversion are set out in Telcordia specification GR-303. An older related specification called TR-008 is also often used, and sometimes only GR-303 is listed. These specifications were initially developed for digital loop carrier (DLC) connections to the central office switch, so the end result is that voice traffic to and from the PSTN looks like a series of T1s carrying voice. Outside of the United States, the ITU-T V5.1 or V5.2 specification is used to accomplish the same thing with E1s rather than T1s.

In summary, the official form of VoDSL is a way to allow telephony service providers to link DSL users up to the voice switch even though the DSL method used does not support analog voice directly. VoDSL is all but required in HDSL, HDSL2, and most SDSL, at least if voice services are also supported by the local telephony and DSL service provider. When used with these forms of DSL, VoDSL is often called *derived voice*.

This form of DSL makes a lot of sense if the DSL service provider is a telephone company. The voice has to make its way onto and off of the PSTN somehow. Usually the DSLAM is right near the central office switch anyway. And ATM supports voice quite well. Because ATM is usually used on most DSL links (and required on many), it makes sense to use ATM cells for voice as well.

So much for ATM-based VoDSL. How does VoIP-based VoDSL differ? In ATM-based VoDSL, the digitized and packetized voice is sent over the DSL link as ATM cells with AAL2. In VoIP-based VoDSL, the IAD sends the voice inside IP packets that are sent in turn inside ATM cells (if used) using AAL5. With VoIP, the voice and data cells all look the same, so the DSLAM cannot send the voice IP packets on to the voice gateway as before. Only a router can select the VoIP packets for sending to a voice gateway, this time over the Internet or with a direct link from the router. An Internet connection to the voice gateway is shown in Figure 11-11.

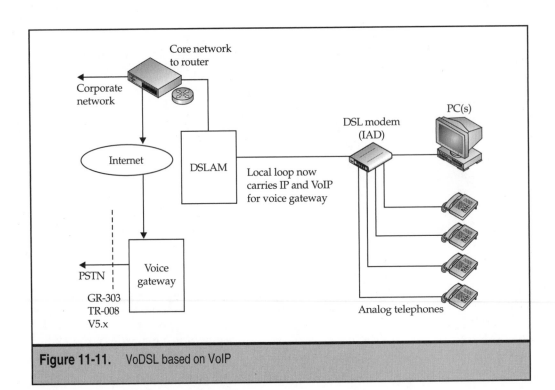

Figure 11-11. VoDSL based on VoIP

Note that the end result of both ATM-based VoDSL and VoIP-based VoDSL are exactly the same. Sooner or later a voice gateway with links to the PSTN is encountered, or else "off net" calls are impossible to make or receive. The difference is whether ATM AAL2 is used or not. If the core network is ATM, it is the router that must "find" the voice packets. It is possible that the DSLAM could also be a router, and in this role send the voice and data IP packets directly to router and voice gateway, but this possibility is not shown in the figure.

CHAPTER 12

The Other Side of ADSL:
The DSLAM

A s important as the ATU-R (the DSL modem) and ATU-C (in the central office) are for the operation of an ADSL link, the Digital Subscriber Line Access Multiplexer (DSLAM) is just as important. Without the DSLAM, the bits have nowhere to go once they reach the ATU-C. The DSLAM multiplexes and demultiplexes the traffic to and from the ADSL links on the customer side of the network (the DSLs) and attaches the ATU-Cs with the service side of the network (the Internet, video servers in some schemes, and so forth).

This chapter focuses on the DSLAM and related issues such as DSLAM configuration and placement. The emphasis is on the network side of the ADSL link in all cases. Other DSL technologies are included because the use and function of a DSLAM is in no way limited to ADSL.

On that note, it is important to realize that DSLAMs are used with more than just ADSL. In the ADSL network architecture, the device that interfaces with the ATU-Cs in the serving office technically is the ADSL access node. Now, an ADSL access node is a very good example of a DSLAM, but the DSLAM is a more generic device and is not just tied to ADSL. In other words, all ADSL access nodes are forms of DSLAMs, but not all DSLAMs are ADSL access nodes.

DSLAMs have their own architectures apart from ADSL. Additionally, DSLAMs support a variety of DSLs, not just ADSL. A DSLAM might support one or more of the following DSLs: *High bit-rate* DSL (HDSL), *Symmetric DSL* (SDSL), *Asymmetric DSL* (ADSL), and so on—this is just on the DSL side. The ATU-C splitter can be built into the DSLAM itself, just as the ATU-R splitter can be housed in the ATU-R. Of course, the DSLAM side splitter can be implemented in a variety of ways, just as the ATU-R splitter was in the previous chapter.

On the service side, the DSLAMs can interface with ATM switches, TCP/IP routers, *switched digital video* (SDV) servers, LANs, and many other possibilities. If this seems like a lot of functionality and, thus, room for a wide range of competitors and product capabilities, it is.

In fact, the battle for the DSLAM and services side of the ADSL network architecture is just heating up. Premises ADSL equipment must be relatively inexpensive, and so profit margins from this equipment might be low. Perhaps added revenue from active DSLAM markets could help profits.

DSLAM CONCEPTUAL ARCHITECTURE

Figure 12-1 shows the basic conceptual architecture of a DSL access multiplexer. Note that the architecture is not tied to any type of DSL, nor is it tied to any particular product or service. The DSLAM details all depend on the support that the vendor builds into a particular product. In one of its simplest forms, the DSLAM might be an ADSL access node and support ADSL packet mode links and Internet access through a TCP/IP router exclusively.

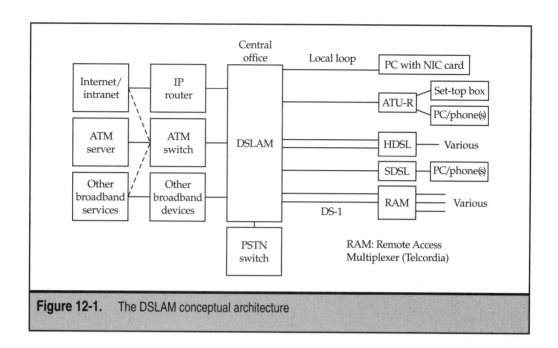

Figure 12-1.   The DSLAM conceptual architecture

The figure shows both the central location of the DSLAM within the ADSL network, but also the versatility of the DSLAM when it comes to service and line configuration support. Some of the roles of the DSLAM are not obvious.

The DSLAM usually sits in the central office (CO). Other interconnection possibilities exist with remote or virtual co-location arrangements, some of which will be discussed later in this chapter, but most DSLAMs are inside a telephone company central office. In some competitive environments, the DSLAM could be owned and operated by a service provider other than the incumbent local exchange carrier (LEC). In any case, the DSLAM interfaces with the PSTN switch if it must continue to provide analog voice services, which it typically will.

On the customer side, the DSLAM can support any one of a number of DSL techniques and loop arrangements. The local loop can terminate directly at an internal PC DSL modem for ADSL, or at a separately packaged DSL modem for a number of ADSL devices. High bit-rate DSL (HDSL) with two pairs of wires (or HDSL2 on one pair) can support a variety of devices. However, this would require a T1- or E1-compatible interface on the premises—not always a given in residential environments. Symmetric DSL (SDSL) can also be supported, even with POTS in some cases, in the same fashion as ADSL. Although not shown in the figure, ISDL is a distinct possibility for DSLAM support also. Telcordia has defined a *remote access multiplexer* (RAM) configuration with ordinary DS-1 support on two pairs of wire for small businesses. One or all of these DSL technologies may be supported by the DSLAM. However, it should be kept in mind that T1/E1 units in a DSLAM will be extremely rare.

On the service side in the figure, the DSLAM might support IP routers, ATM switches, or even other broadband devices to access services on other platforms such as SDV servers (which could also be on the ATM network). Again, support may be for one or all. Usually, the IP router would be used for Internet or corporate intranet access, and the ATM switch for ATM-based servers. Alternatively, the DSLAM architecture allows for all servers and services to be accessed through an ATM switching network. Typically, on the core network side tying a number of DSLAMs together, the interface to the network side of the DSLAM would be a either an ATM transport such as SONET/SDH or a 100 Mbps 100Base-T LAN interface.

Please realize that the basic DSLAM is not necessarily a switch or router. It is in more ways than one just a type of multiplexer. That is, the DSLAM combines streams of bits in the channels coming upstream from homes and small offices and splits up a large bit stream coming downstream from the IP or ATM network. The DSLAM could split this bit stream based totally on channels, as a multiplexer always does. In this case, the link from the DSLAM to some other piece of equipment, whatever it may be, must be able to carry the sum total of the traffic arriving from customers all at once.

This aggregation of traffic based on the sum total of the input bit rates is commonly known as *time division multiplexing* (TDM). However, other types of multiplexing are used in networks that send and receive digital information. Often considered a step above straight TDM is a technique called *statistical time division multiplexing*, or statistical TDM, or just "stat muxing." Statistical multiplexing takes advantage of the fact that many applications that generate these bits are bursty applications, as has been mentioned many times. Lulls often exist in the stream of bits, during which other applications can share the same bandwidth. This is the basis of the whole concept of packet switching in the first place. In some sense, packet switching in statistical TDM applied not on a link-by-link basis, but on a network-wide basis; this is an oversimplification of the situation, but it is a useful analogy.

What does all this have to do with the DSLAM? A DSLAM today will rarely use simple TDM. The DSLAM usually includes features like tagging of priority traffic, traffic shaping to smooth out packet or cell bursts, and even cross-connect features. This makes the DSLAM more expensive, but more effective. Here is why.

Suppose a DSLAM services 64 ADSL links in a communications rack, 8 per shelf, and 8 shelves are in the rack. This is by no means an uncommon number for a small DSLAM, although some modern DSLAMs can support 256 ADSL customers per DSLAM. Naturally, multiple racks can be used to service more ADSL customers, although this adds to the heat given off inside the central office as well as power requirements, and requires more floor space. Now, suppose that each ADSL link operates at a modest 1.5 Mbps downstream and 128 Kbps upstream. Again, these numbers are quite representative, although ADSL is capable of much more. If the DSLAM were a simple TDM device, how fast would the links coming out of and going into the back of the DSLAM have to be to handle this traffic load?

Consider upstream first in this example. If there are 64 inputs to the DSLAM (the upstream user traffic) running at 128 Kbps each, the link from the DSLAM to anything else must be able to handle 64×128 Kbps = 8,192 Kbps, or 8.192 Mbps. The easiest way to do this is with a series of T1 links to separate serial ports on the router (six T1s will do), but consider the situation downstream. Each ADSL link operates at 1.5 Mbps downstream. The same 64 outputs must operate at 64×1.5 Mbps = about 96 Mbps into the back of the DSLAM from the router. It is not feasible to put 64 T1 links onto the router, and even two T3s at 45 Mbps falls short of 64 T1s (a T3 can only carry 28 T1s, or 56 total for two T3 links). Now, as it turns out, given the asymmetrical nature of ADSL, the upstream traffic aggregate of about 8 Mbps can be carried on the same T3s (however many) used for downstream transport. This is quite wasteful, however, because a lot of upstream bandwidth is being unused.

However, if the DSLAM is just a passive TDM device, little else can be done to carry the bits through the DSLAM to and from the ADSL links. Perhaps some amount of intelligence could be added to the DSLAM to allow the DSLAM to statistically multiplex the traffic from and especially to the ADSL links themselves. Recall that the AS and LS channels in bit synchronous ADSL contain idle bits, as well as information bits. Perhaps the DSLAM could filter out the idle bits on each ADSL link and send only live data bits to the router. However, there are other ways that a DSLAM can make better use of the bandwidth between the back of the DSLAM and whatever other switches or routers used to access the ADSL services provided. Suppose that the DSLAM and the router communicate not over a T-carrier serial link at all. Suppose both the router and DSLAM are linked by an Ethernet LAN running at 100 Mbps. Technically, this is a 100Base-T LAN, but the point is the speed, not the architecture. However, the DSLAM cannot simply dump bits onto the Ethernet LAN. The DSLAM network interface card in the DSLAM must send and receive Ethernet frames, just like any other LAN-attached device, and so bits coming in from the ADSL links must be packaged as a series of Ethernet frames and sent to the router. The router must respond in kind, but this is not a difficult situation at all.

What is difficult is determining the content of the frames. By definition, the content of a frame is a *packet*. The DSLAM must be able to send packets inside frames across the Ethernet LAN. No one should be surprised that one of the allowable ADSL link operational modes uses IP packets inside of Ethernet frames (with or without PPPoE). The whole idea is to make the DSLAM into a more active network device, and not just a simple multiplex. ATM mode on the ADSL links is also supported, with an ATM transport such as SONET/SDH on the core network side of the DSLAM. On the return path to the DSLAM, the router sends IP packets as well. It is important to realize that if the packets sent to and from the DSLAM are also sent to and from the customer on the ADSL links, this is nothing more than ADSL packetmode in action. Note also that this implementation of a DSLAM has already moved far beyond the simple TDM device at the beginning of this section. Why bother? The whole point is to make the DSLAM more versatile and efficient. But why stop here? Why not put some packet switching or routing intelligence directly in the DSLAM? This idea is explored more fully in the section "The Expanding Role of the DSLAM."

THE "TYPICAL" DSLAM

This chapter has mentioned general DSLAM architectures and capabilities several times. However, the information might have seemed vague and unsatisfactory; there is a reason for this. The basic internal DSLAM form and function is not covered by any set of ADSL or other DSL standards documents. The fundamental concept of the DSLAM as a housing for multiple ATU-Cs, HTU-Cs, or something else is a given in all cases. Beyond that, what the DSLAM does and how it does it is completely up to the DSLAM equipment vendor. There are some general rules for DSLAM placement and the relationship to the ATU-C splitters, but these details are discussed later in "DSLAM Configuration Possibilities."

This does not mean that the entire field of DSLAM operation is a free-for-all. Most DSLAM products all support some common features quite consistently, but DSLAM products vary with other features, in terms of numbers and speeds. This section attempts to sort out just what features would be found on a "typical" DSLAM, if there is such a thing.

Recall that the DSLAM occupies the key position in the entire ADSL architecture. All traffic to and from the users passes through the DSLAM. All traffic to and from the servers on the network behind the DSLAM must pass through the DSLAM also. The DSLAM is typically located in or near the main distribution frame (also called the wire center) of the central office because of the need to have convenient access to the local loops. This location is common when the service provider is a LEC or CLEC, but other types of ADSL service providers are possible. All that is needed is access to some local loops leading to customers, a DSLAM, and some services to offer.

It might be useful to divide the entire ADSL network architecture into three parts, especially from the DSLAM perspective. This model has been common in many DSLAM vendors' documentation. In this model, the ATU-Rs, or other remote DSL devices such as an HTU-R, form the *service user* (SU) portion of the network. The ATU-C, HTU-C, or other interfaces in the DSLAM form the *network access provider* (NAP) portion of the network. The access network itself and the network or networks on which the services may be found are part of the *network service provider* (NSP) portion of the network. The central role of the DSLAM as NAP is to form the connection between service user and service provider. This model of the DSLAM's role is shown in Figure 12-2. This is just a somewhat simpler version of the ADSL architectures given in many of the previous chapters.

In the figure, the DSLAM does more than just house the ATU-Cs. In its most general form, the DSLAM also includes the proper type of connections to what is called in the figure the *access network* (or core network) to the service network provider(s). The type of access network supported depends on the type of connectivity offered by the DSLAM vendor, of course.

A DSLAM services local loops in a *card* that contains the software (microcode) and hardware for the proper type of DSL supported and the proper line code. The cards are gathered into *shelves*, which usually share a power supply and may have their own network management capability. The shelves are housed in rack, which is the basic unit of the DSLAM. That is, the size of the rack determines the total capacity of the DSLAM. Almost all DSLAMs will do ADSL with either CAP, DMT, or both. In addition, many other DSLAMs will support

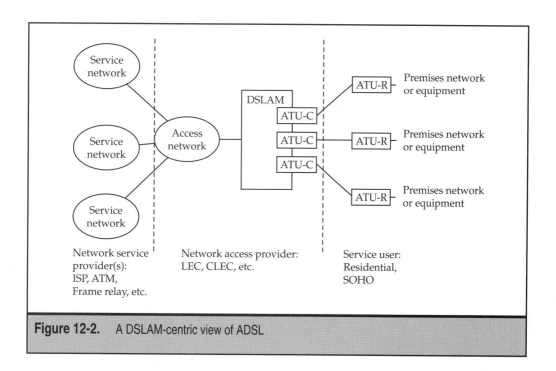

Figure 12-2.   A DSLAM-centric view of ADSL

cards for HDSL, IDSL, SDSL, VDSL, and many variations on HDSL especially. When using these other xDSLs, the DSLAM will generally employ QAM or 2B1Q line code.

The access network connection type will most likely be an ATM SONET/SDH connection or 100 Mbps Ethernet (100Base-T). However, some DSLAMs will support the *high-speed serial interface* (HSSI) connections common in frame relay networks or some form of T-carrier (useful for pure bit synch DSL modes), most likely T1, T3, or both. The DSLAM itself might also perform some bridging and routing functions. The DSLAM might be able to send and receive ATM cells or IP packets in most cases. Sometimes, other protocols might be supported, such as Novell's *proprietary packet protocol* (IPX), the Internet's *point-to-point protocol* (PPP), *frame relay* (FR), an older LAN protocol called NetBIOS, and IBM's *synchronous data link control* (SDLC) protocol for SNA networks. More details on this bridging and routing with ATM and FR are explored in the next section.

As far as network management goes, almost all DSLAMs support the widely supported *Simple Network Management Protocol* (SNMP), first used on the Internet but now seen everywhere. In addition, some support the international standard *Common Management Interface Protocol* (CMIP). Also, some DSLAMs do not employ SNMP directly, but allow the DSLAM to be indirectly managed by means of a proxy agent, usually a PC that understands SNMP and whatever network management protocol the DSLAM uses. Proxy agents are common when a product's network management software is proprietary and not standard, such as SNMP or CMIP. Proxy agents are generally frowned upon except when absolutely necessary because of their perceived inefficiencies and added complexity.

The number of loops serviced by a single rack of DSLAM shelves vary from fewer than 100 to more than 1,000. The pricing varies accordingly, depending on more than just numbers of local loops supported, such as network management capabilities.

THE EXPANDING ROLE OF THE DSLAM

The simple time division multiplexing DSLAM has already given way to units that not only aggregate the traffic from many DSL links, but also concentrate the traffic with simple techniques, such as filtering out idle bit patterns. Obviously, there is lot more functionality that could be built into a DSLAM to expand its already central role in ADSL.

Many DSLAM products come with 100Base-T ports. These DSLAMs can be linked by means of a 100Base-T LAN hub to a router. The router links to the Internet or some other IP-based network, but this scenario requires the service provider to purchase not only the DSLAM, but also the 100Base-T hub and router. This increases the complexity of the configuration, not to mention the total cost before a single ADSL user can be offered Internet connectivity.

It is therefore attractive to service providers to consider buying a DSLAM with IP routing already built-in inside the DSLAM. This cuts down on the total cost of the configuration, eliminates the hub entirely, and is a more compact device. In either case, the ADSL links are presumably running as packet mode ADSL because the packets have to go to and come from somewhere outside of the DSLAM. The differences between these two configurations is shown in Figure 12-3.

This is not a perfect solution. Router purchasers have historically been more comfortable with separate routers than with routers built into other products. For example, early 10Base-T hub vendors put a board in their hub that could do everything a separate router box could do. Potential buyers did not like the idea. They preferred routers separate from everything else. Apparently, they felt that a separate router reduced the complexity of the

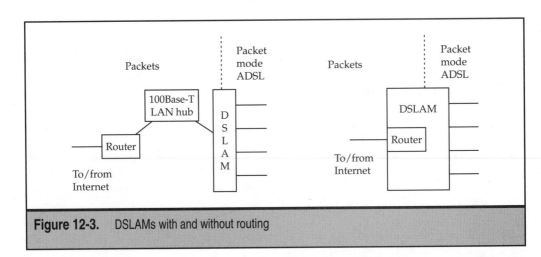

Figure 12-3.   DSLAMs with and without routing

hub and made troubleshooting the entire configuration easier because the router could be isolated from the hub. Perhaps the same thing will happen with DSLAMs, and the router will always remain separate from the DSLAM.

Several of the ADSL link models send and receive ATM cells, not packets. And not surprisingly, many DSLAM vendors support interfaces capable of linking to an ATM switch port. From there, traffic would make its way onto a core broadband ATM network either where the servers are located, or (more likely) to an ISP router with an ATM interface. The evolutionary path for the ATM switching could be the same as for the router. That is, the ATM switching function could be incorporated into the DSLAM itself. All of the same advantages and disadvantages of the DSLAM-based router apply to the ATM switch scenario. (Even if ATM is used on the ADSL link, the DSLAM could still have a router or 100Base-T interface.)

As ADSL—and, in truth, all of the DSLs—matures, the switching and routing capabilities of the DSLAM will continue to expand. Even if the potential buyers of DSLAMs reject the integrated switch/router configuration, perhaps the DSLAM could still supply some form of switching or routing on the user side of the DSLAM. This additional switching or routing need not necessarily be limited to IP packets or ATM cells.

This needs a few words of explanation. Consider a service provider with a frame relay service backbone. Frame relay is a fast packet switching network that uses frame relay switches as network nodes and something called *frame relay access devices* (FRADs) as the user interface to the network. FRADs may be separate devices all on their own, much like a LAN hub, or incorporated into routers as FRAD software modules. Such routers are said to be capable of supporting the "FRAD function" on a given serial port.

Access to a frame relay network is usually over a 64 Kbps digital link or even a full 1.5 Mbps T1. When used with frame relay, the link carries frame relay frames (naturally) formatted according to the link access procedure for frame relay (LAP-F) protocol. This link must be provisioned by the local service provider, often at great cost. ADSL can offer a better way to provision frame relay access. Figure 12-4 shows how.

The reason for putting frame relay switching in the DSLAM is both to multiplex and concentrate it for transport over the access network, most likely to a corporate site where frame relay is used as the normal WAN technology. This corporate access frame relay network is not uncommon, and the service network of interest in this case is a series of frame relay service providers, of course. Note that in this scenario, ADSL can be used to provide frame relay access through the LEC to many other frame relay IXC or private corporate networks, not just one.

The ATU-R is now either a module in a multiprotocol router implementing a FRAD function or in a separate FRAD device altogether. The whole point of enhancing the DSLAM beyond the simple time division multiplexing of frame relay *permanent virtual circuits* (PVCs) is to make more efficient use of the bandwidth available on the access network. Of course, one characteristic of doing this with ADSL is that the bandwidth is inherently asymmetrical. If symmetrical bandwidth is needed, as would be the case if the SOHO user had a LAN with its own servers on the premises, the same DSLAM can support users with HDSL or HDSL2, in which case the ATUs become HTU-Cs and HTU-Rs. In fact, almost any DSL can be used, including SDSL and IDSL.

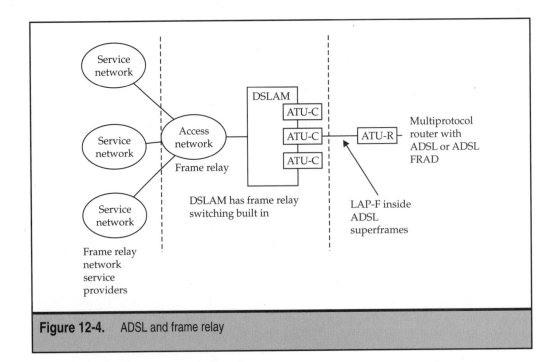

Figure 12-4. ADSL and frame relay

Another way that an "expanded role" DSLAM can benefit customers and service providers alike is in the area of a service that is usually called "fractional T1," but that is technically Nx64 service. With Nx64 service, the link runs at a multiple of 64 Kbps (the DS-0 rate of a T1), where in the United States the value can range from N=2 (128 Kbps) up to N=24 (which is basically a full DS-1 running at 1.536 Mbps). Fractional T1 services are popular with users, but not so popular with local exchange carriers. Users like having a place to go beyond 64 Kbps that is not a full 1.536 Mbps, but rather can grow there as bandwidth needs and the value of N increase over time.

The reason the LECs are not overly fond of fractional T1 is that there is no real way to provide a piece of a T1 to a single location. Two pairs of wire carry all 24 channels of a full T1. Yet the customer pays for only 384 Kbps (N=6), 768 Kbps (N=12), or some other piece. This provisioning of 24 channels where only a few are actually used is not very efficient. What the LEC will typically do, therefore, is immediately to take the 64 Kbps channels from a number of customers and concentrate them at the wire center in a digital cross connect (DCS). The trouble is that this concentration must be done in contiguous channels. That is, the N channels must stay together through the wire center DCS. This is shown in Figure 12-5.

Why should this be a problem? Because the customer knows that fractional T1 enables them to go from 384 Kbps to 768 Kbps essentially by changing from N=6 to N=12. Suppose Customer #1 in the figure is so successful at whatever the 384 Kbps link is used for, that

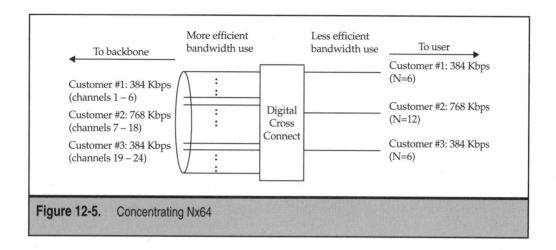

To backbone — More efficient bandwidth use

Customer #1: 384 Kbps
(channels 1 – 6)

Customer #2: 768 Kbps
(channels 7 – 18)

Customer #3: 384 Kbps
(channels 19 – 24)

Digital Cross Connect

Less efficient bandwidth use — To user

Customer #1: 384 Kbps
(N=6)

Customer #2: 768 Kbps
(N=12)

Customer #3: 384 Kbps
(N=6)

Figure 12-5.   Concentrating Nx64

the customer wants to upgrade their fractional T1 from N=6 to N=12. The capacity is there, of course, because the link to the customer is still effectively a full T1 anyway. The service provider would like to have the increased revenue, of course. Unfortunately, the only way to go from N=6 to N=12 fractional T1 service is to give Customer #1 contiguous channels 1 through 12 through the DCS. As the figure shows, this is impossible because those channels are currently used by Customer #2. The customer will still get 768 Kbps eventually, but only after a great deal of reconfiguration and effort.

How can DSL and the DSLAM help? The DSLAM can more intelligently concentrate the Nx64 traffic arriving from the customers and rearrange the channels easily. In this case, HDSL or HDSL2 would be more appropriate because circuits are inherently symmetrical in ADSL. And now the service can be provisioned on one pair with HDSL2 instead of two pairs as in a "real" T1. Therefore, provisioning fractional T1 with a DSLAM and DSL makes a lot of sense.

So the DSLAM can contain a routing function, a frame relay switching function, and even a cross-connect function for delivering fractional T1 services. These functions extend the capabilities of the DSLAM and help to support various modes of ADSL, such as packet mode, but the DSLAM can also be used effectively to deliver ATM services over the ADSL network.

ATM cells are transported on a variety of physical links, such as T-carrier and SONET. The transport of cells on these diverse physical link types is handled by the ATM *Transmission Convergence* (TC) sublayer of the ATM Physical Media layer. The use of the term "convergence" reflects that a variety of the media types must now transport ATM cells inside their respective transmission frame structures instead of their normal circuit-oriented bit patterns.

A potential complication is the presence in ADSL of the "fast" and "interleaved" buffers. Only support for the interleaved buffer is required in ADSL.

The fast and interleaved buffers have been mentioned before. This is a good place to review their operation and the role they play in ADSL. The interleaved buffer is a way of adding extra protection against transmission errors by using a technique called *interleaving*. This was deemed valuable for data services. The disadvantage is that it adds time delay (latency) to the information stream, as much as 20 milliseconds. Other services, such as video, cannot tolerate this added delay. However, video is much less sensitive to transmission errors.

The fast and interleaved buffers in the ADSL modems are associated with each service type. An ADSL system can carry multiple services on one line, and the modem will divert the bits to the appropriate fast of interleaved buffers.

Figure 12-6 shows a possible way of supporting ATM services over ADSL with the DSLAM again playing the central role.

In the figure, the ATM transmission convergence is shown for both the ADSL fast and interleaved buffers. Only ATM cells are sent on the ADSL link. No explicit relationship exists between the ATM cells and the ADSL frame structure. The DSLAM performs all ATM layer functions, which includes the key switching aspect of ATM. The presence of the ATM network in the figure is not really an exclusive arrangement. ATM allows for both service and network interworking with other networks types such as IP and frame relay.

On the customer premises, the ATM cells can originate and terminate in an "ATM LAN" running at 25 Mbps, but it will be far more common to have an Ethernet LAN on the premises, in which case the "higher layer functions" in the figure are definitely needed. In the Ethernet case, the premises equipment will need an ATM *segmentation and reassembly* (SAR) sublayer to convert the Ethernet frames to ATM cells and vice versa.

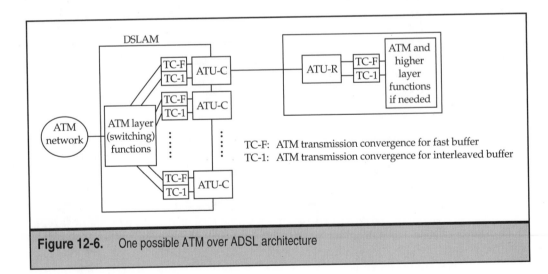

Figure 12-6.   *One possible ATM over ADSL architecture*

DO-IT-YOURSELF ADSL

This entire chapter has thus far assumed that the DSLAM is located in the local service provider's central office and that the DSL service provider is a state-certified LEC or CLEC. This section explains that this is not always the case. In fact, the presence of the DSLAM in places other than the LEC central office was at one time evidence of a growing "non-telephone company movement" toward ADSL, whether the local service provider embraces ADSL of not.

At first glance, this statement seems astonishing. How could someone obtain ADSL services if their local telephone company does not provide it? The key is understanding that in the deregulated local telecommunications marketplace, it is no longer a given that the LEC controls all access to the local loop. Local loop access has officially been "unbundled" to a large degree, and even where it has not, organizations, such as ISPs and groups of users, have been very inventive at gaining access to the local loop in ways the LECs never dreamed of. Of course, a wholesale turn away from "data CLEC" ADSL services has taken place recently, complete with bankruptcies and mergers. But many things tend to run in cycles, so the issues behind such "do it yourself" ADSL are still worth exploring.

After someone unbundles access to the local loop through a long-term lease agreement, all that is technically needed to provide ADSL service is to put an ATU-R at one end and a DSLAM with a group of ATU-Cs at the other. The DSLAM can add some of the capabilities outlined previously, such as ATM switching or IP routing.

In many cases, the LECs are less than pleased at such movements, and who can blame them? They are accustomed to having absolute control over their facilities. The LECs built the physical plant of which the local loops are part at great financial cost and risk. If any other entity is allowed to use the local loop, the LECs want to make sure that they are adequately compensated for it.

In spite of the LEC's concern and sometimes outright opposition, ADSL services actually appeared without the participation of the LEC. Some organizations, such as ISPs and even groups of ordinary citizens, can bring ADSL—or any DSL for that matter—to their homes without LEC involvement by using two main methods: this section calls them the "alarm circuit approach" and the "grass roots" approach.

There is little doubt that ADSL implementations just feature fast Internet access as their premier service. Of course, this type of service will only allow the channels provisioned on the ADSL link to connect two things together: the user's PC and the ISP site's servers/routers. None of the envisioned sophisticated ATM switching and broadband services are available in this simple model, but maybe this is all right. Throughout this book, ADSL (and even the other DSLs) have been positioned mainly as a way of getting packet-oriented Internet traffic off the circuit-oriented PSTN. So what if access is limited to the Internet? The whole world is out there on the Internet already.

This Internet-only scenario does not rule out ADSL use for telecommuters or others. A corporate intranet can just as easily be accessed through the proper ISP as anyway else. The overwhelming concern is security, and the security can be provided in different ways at a variety of steps between user and server, as already mentioned in the previous chapter. However, this application of ADSL emphasizes the use of ADSL by an ISP for simple Internet access, although ADSL is obviously not limited to this fast Internet service only.

Here is how *alarm circuit* ADSL came to exist. An area's ISP wants to offer ADSL services to customers in that area. In addition to the obvious revenue opportunities for the ISP, there are other incentives to do so. Many ISPs may already exist in the area, and the ISP in question might want to distinguish itself from the others. The LEC may be making noises about becoming an ISP themselves (or may already have done this), and the ISP is seeking to keep ahead of the LEC. And so on.

Of course, providing ADSL service does consist of a little more than having a pool of customers with ATU-Rs and a compatible DSLAM. As has already been discussed, the DSLAM can easily interface with an ISP's internal server/router setup and take users onto the Internet and Web. The ATU-R is customer equipment in the United States and beyond the control of the LEC, so this is not a problem. However, the DSLAM is usually located in the LEC's office space. What about the situation where the DSLAM is owned and operated by the ISP and not the LEC?

In this case, the rules under the Telecommunications Act of 1996 (TA96) take effect. Regardless about how one feels about the overall effectiveness or shortcomings of TA96, a few things in it are fairly explicit. The FCC and state regulators have established rules regarding the co-location of other service provider's equipment within a LEC serving office. This applies to all switching locations, but the LEC location is clearly the most important. If an incumbent LEC desires to do certain things under TA96 (such as offer long distance calling), they had to offer co-location to competitors. The ISP could, therefore, conceivably pay rent to the LEC and place their DSLAM right inside the switching office, within easy cabling distance of the wire center.

Co-location does have a couple of drawbacks. First, a lot of potential competitors possess a lot of different equipment today. Floor space is at a premium and may not be available. Second, the rent can be very high. Which companies will want to add revenues to their biggest competitor? Third, it is not really a good idea to put such an essential piece of equipment in somebody else's floor space. Access alone could be a problem, and often is.

To address these issues, the FCC and many state regulators made co-location easier to do in practice.

It is now allowed for access to an incumbent LEC's equipment when the competitor's equipment is not inside, but is in close proximity to the LEC's local exchange. Usually, this translates to "across the street" but not always or exclusively. The actual cable distance depends on what is actually being done. For some voice services, such as competitive voice mailboxes, a simple channelized voice trunk to the competitor location and back is sufficient. If 600 feet of T3 coaxial cable for a tail-end could not stretch, fiber optic cable could extend the reach to a mile or so, in some cases. It all depends on the physical limit for the port type in the LEC's equipment being accessed, whether switch, wire center, or something else.

The nice thing about this type of co-location is that it got around the disadvantages of "regular" co-location. Floor space did not matter, rent was just a connection fee (which still might be substantial), and the equipment was often now in the competitor's own office space. Sometimes the ISP equipment was in a "broomcloset" of a company that specialized in central office proximity, but the result was the same.

With this newer co-location, an ISP with owned or rented space close enough to the LEC office obtains a DSLAM from one of a variety of vendors, pays to interconnect with the LEC's wire center or main distribution frame (a not inconsiderable fee, and coupled with many related charges), and distributes compatible ATU-Rs to their customers in the area. The only missing piece is the local loop from LEC to customer—but this is a big piece.

It is possible to try to buy access to the actual local loop running from LEC to customer. In the normal ADSL configuration, the splitter in the ATU-C and ATU-R keeps the analog voice service going into the voice switch. However, this requires a complex cabling arrangement between ISP and LEC because the DSLAM (and presumably the ATU-C splitters) is now typically across the street from the LEC. This is shown in Figure 12-7.

In the figure, voice and data from a customer splitter and ATU-R enter the wire center. The loop is cross-connected to a cable (perhaps a T1 or T3) leading across the street to the virtually co-located ISP. There, the analog voice is split off at the ATU-C in the DSLAM. The voice is sent back to the LEC via more cables to the PSTN switch ports. The data continues into the ISP network and equipment. The whole system is quite complex and

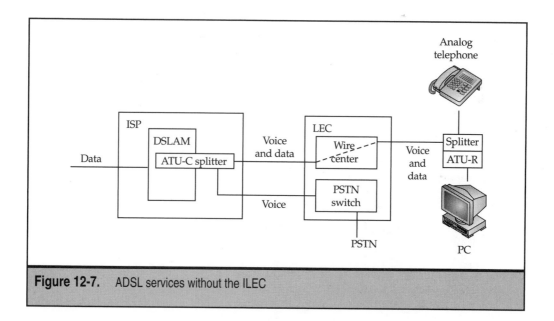

Figure 12-7.   ADSL services without the ILEC

expensive to set up. Also, the cable distance to the ISP counts toward the 12 or 18 kft limit of the DSL method, and a reasonable wire center signal loss must be factored in, as well. Of course, the cable distance back to the LEC must not exceed the supported distance for the switch port. In addition, connector types must match, and so on. All in all, a less than ideal situation.

The whole point of the ISP (or "data CLEC") supporting ADSL, however, is to provide faster Internet access, not voice service, and many people already have second lines used almost exclusively for Internet access, so maybe it is not essential to support the analog voice on the ADSL link at all. In fact, in most cases, all the ISP wants to buy is the actual copper wire and nothing else—not access to the LEC switch, not access to the customer's main telephone, nothing but a copper wire pair from wire center to residence.

Unfortunately, this is sometimes easier said than done. The competitors that are usually allowed to co-locate, virtually or otherwise, typically have to be certified by the state they are located in, which can be a complex and expensive process. Also, the LECs have never totally unbundled the local loop from a number of other things that must be bought along with the wire. This is not universally true, but typically is. For instance, the philosophy is that if a competitor wants access to the local loop, the competition is going to offer voice services also, and so the competitor must also buy access to the LEC switch port, and access to directory services, and maybe operator services. And what about reports on usage and billing services? In a totally voice circuit network, all of these things make sense. This is the whole bundling approach: sell competitors what they need to compete. When the situation concerns packets and not circuits, however, unbundling makes even more sense. Let competitors pick and choose what to buy in almost a totally a la carte fashion; this is sometimes done, but rarely. Bundling is more typical, but it is important to note that allowing *any* competitive access to the PSTN is technically known as *unbundling*. The actual issue here is one of degree and extent. Already the topic of court cases, the basic philosophy of TA96 was that if the LEC demonstrates enough competition in local services, the LEC is permitted to offer long distance services.

So what is an ISP to do if faced with unreasonable (from their perspective) unbundling charges to access the local loop? Why not put in a totally new line and use it exclusively for ISP access? That way, there is no consideration of analog voice support at all, but what about the unbundling? Easy. Just buy a circuit that is tariffed to be the minimal type of service on a pair of wires in the first place. Alarm circuits!

This concept needs some background to explain. Most business burglar alarms are run over simple local loops. No switching is needed, nor dial tone, nor directory assistance, and so forth. All that is needed is a low-quality pair of wires from customer to burglar alarm company. When something trips the alarm at the customer premises, a signal is sent over the wire and triggers something at the alarm company's premises, such as a siren and a flashing red light on a console. The alarm service may be a private security firm or the local police. No matter, the concept is the same.

Alarm circuits can be very low-quality. No voice bandwidth is needed, although it is usually there. The alarm circuit can be a point-to-point leased line, although some have dial tone on them. In any case, the point is that the alarm circuit is functional, inexpensive (about $25 per month), and usually carries none of the "extra" unbundling features the ISP could do without.

Several ISPs and data CLECs have tried this approach already. Order a bunch of alarm circuits for their customers, have them terminated at the ISP site, give ATU-Rs to each user, and away they go. However, the LECs have recently begun to fight back over this use of what are usually called *local-area data service* (LADS) lines. Other names include *dry pairs, burglar alarm wire, modified metallic circuit with no ring down generators*, and *voice grade 36 circuits*.

The ISPs argue the restricted local loop access and unwanted unbundling features make this method the only efficient way to get ADSL services to their customers. The LECs argue that this use of the lines circumvents regulations and floods lines intended for low bit rates with high speed data. Some LECs have even stopped selling LADS lines to ISPs (or indeed to anyone at all).

Both sides were at one point playing tough. The ISPs threatened legal action over "anti-competitive" practices when LECs claim there were no more wire pairs available to a given ISP location. Such a claim on the part of a LEC was a little strange when dial-up service was replaced with ADSL, but the recently terminated dial-up service pairs seemed to mysteriously disappear. The LECs were correct in saying that they do not have tariff approval to offer these lines for ADSL, and it was also true that running ADSL signals in cable bundles can disrupt other services in the same bundle. This usually happens not when the ADSL signals in the downstream direction were running from the same origin to the same destination, but just the opposite. In other words, if ADSL "downstream" from the PSTN switch was opposed to an ADSL "downstream" from the ISP in the same bundle, the crosstalk and interference might knock out both lines.

This should not happen often. Few complaints have been made about the use of ADSL over LADS lines and other burglar alarm-type services. But the LECs were both concerned and nervous. Some of the LECs have even been accused of placing or keeping loading coils on these circuits to prevent their use for ADSL. Today, the collapse of the whole data CLEC and "do it yourself" ISP ADSL movement has put a damper on this debate. The ILECs have won the ADSL war.

The biggest problem with the alarm circuit approach was always that it did not scale well. That is, it was impractical and uneconomic for widespread services. The ISPs and data CLECs were still paying retail and not wholesale costs for the lines, for instance. The "grass roots" approach was really just another variation on the alarm circuit theme. In this scenario, local users themselves got together to purchase access to their local loops. The same unbundling issues applied, but with significant differences. Some state regulators had been more sympathetic to users buying their own loops than LEC competitors buying access to the loops. The LECs themselves seem less concerned about citizens buying the loops than competitive LECs or ISPs. And even in cases where certification is required, the process for such "community groups" was often quick and inexpensive.

Some states have set initial costs for direct user access to the local loop very low, even lower than the alarm circuit price of about $25 per month. The LECs are claiming that such low rates are below the costs of providing the loop, but there is little hard data to go on, and so the same battle lines exist, but often on a much smaller and less intense scale.

In the *grass roots* approach, users serviced by the same LEC switch organize and approached the LEC about buying access to their local loop—or even buying them outright. They could then rent office space according to the virtual co-location rules, site their DSLAM as before, and tie the DSLAM into a recognized ISP. Alternatively, they could become their own ISP themselves. In this case, the voice trunks back to the PSTN switch might be absolutely necessary if continued voice support was needed. Then again, the Internet could supply all such needs on its own.

In any case, the grass roots approach appealed to concentrations of groups of highly technical users around the country. Most already had their own LANs at home. They did not hesitate to use new technology to help their situation. Usually, they wanted faster access not only to the Internet and Web, but to their own corporate Intranet when telecommuting. Today, like "alarm circuit" ADSL, this form of ADSL service has all but disappeared. People wait for their incumbent LEC to support ADSL, or they look to cable modems or other forms of high-speed Internet access.

DSLAMS AND SONET/SDH RINGS

Most service providers offering in DSL services today are local telephone companies or LECs. After all, the LECs have control of the analog local loops that ADSL enhances for high-speed digital access.

Many service providers also have another distinctive technology available to them, the *Synchronous Optical Network / Synchronous Digital Hierarchy* (SONET/SDH) ring, which transfers standard transmission frames on fiber optic cable rings. Outside of the United States, the term used is SDH, but the result is the same. Whenever a service provider mentions the "backbone" links in their network today, this usually refers to the SONET/SDH rings. The rings help to prevent service outages because if a signal traveling one way around the ring is interrupted by a cable break or failure—which was the cause of 75 percent of service outages in the 1990s—the signals reroute themselves the other way around the ring. This rerouting occurs in about 60 milliseconds, which is faster than the blink of a eye (which happens in about 100 milliseconds). SONET/SDH rings also offer enormous fiber bandwidths for aggregate traffic between switching offices, well into the multiple gigabit-per-second ranges (a Gbps equals 1,000 Mbps). Perhaps there is a way to interface ADSL—or any DSL technology—with SONET/SDH rings, and so maximize the benefits of both. In fact, such a method is shown is Figure 12-8. In the figure, only the SONET term is used, as it is in the rest of this section. But SDH applies as well.

The central feature in the figure is the SONET ring itself, which may span many miles and link switching offices to special *remote digital terminal* (RDT, or just RT) devices as part of a *digital loop carrier* (DLC) architecture, serving what are called *carrier serving areas* (CSAs). CSAs are typically employed in office parks and housing developments to minimize the need to run several point-to-point wires over several miles to a serving office. Of course, the whole CSA concept and terminology is also employed in all local loop situations today, whether an RDT is present or not. The ring provides protection switching, as well. The RDT may be a small cabinet of modest size and cost.

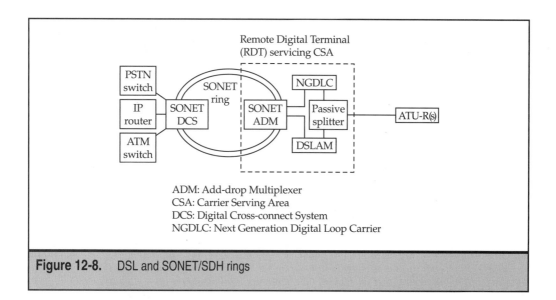

ADM: Add-drop Multiplexer
CSA: Carrier Serving Area
DCS: Digital Cross-connect System
NGDLC: Next Generation Digital Loop Carrier

Figure 12-8.   DSL and SONET/SDH rings

Inside the RDT, a SONET *add/drop multiplexer* (ADM) interfaces with the ring. Although the figure shows only one such device, there would in reality be many RDTs on a ring. Note the location of the DSLAM in this architecture: inside the RDT. What is it doing there, of all places?

As more and more local loops are deployed at the end of DLCs, some movement of the DSLAM to the remote location must occur. More details about the exact form of the DSLAM in the DLC cabinet will be discussed in the last section of this chapter. Here it is just a "DSLAM."

As shown in the figure, a passive splitter is located at the end of the local loop. Now, the analog voice still enters the *next generation DLC* (NGDLC) on the SONET rings, but the ADSL digital information now passes through the distributed DSLAM onto the SONET ring on its own. The SONET ring contains enough bandwidth to accommodate both numerous analog voice channels as DS-0s running at 64 Kbps (the digitization is a function of the NGDLC) and just as many ADSL channels running at higher and asymmetrical speeds. No matter, the flexible configuration capabilities of SONET rings allows for this.

Simply put, a NGDLC is a DLC optimized for use on a SONET ring. A DLC is frequently used today to support many analog local loops in a CSA, but is not appropriate for ADSL in many cases. The NGDLC will have more flexible and remote configuration capabilities, and have the traffic carrying capacity to tolerate the presence of ADSL on the loops. Most important of all, the NGDLC will have an interface designed for SONET ring connectivity. A passive splitter combines analog and digital signals to and from the ATU-Rs. All analog voice is digitized by the NGDLC for transport on the SONET ring.

In the switching office, a SONET *digital access cross connect system* (DACS, or just DCS) splits off the digitized voice traffic from the ADSL traffic. Voice still goes into the PSTN switch, and other traffic still finds its way into an IP router or ATM switch. In fact, the switches and routers could be located almost anywhere on the SONET ring where a DCS is employed, even in a competitor's service location.

This use of ADSL and SONET rings gives maximum service outage protection and carries enough traffic that thousands of ADSL downstream traffic flows could be supported at once. However, it may be a while before such architectures become common.

Issues still to be resolved are the following: Is the RDT fed by a SONET ring now, or is it just a point-to-point link? Is there space in the remote cabinet or area to hold the remote DSLAM? Will enough customers want the ADSL (or other DSL) service to justify the remote DSLAM expense?

A remote DSLAM can be made small enough today to fit into almost any DLC site. There are DLCs in small, pale green cabinets or huts (something called a *controlled environment vault* [CEV]) in urban areas or even in the basement of an office building. A key issue here is the availability of power, battery backup, heat dissipation, and other mundane operational factors. Most DLC sites have power and battery backup already. Space in a RDT cabinet is another concern, because they were never designed or intended to house extra equipment like ADSL.

The penetration of DLC and CSA arrangements has increased over the past 15 years or so. A lot of this occurred when the CSA rules were introduced in 1983. By limiting new loops to 12 kft with 22 or 24 gauge wire, and 9 kft with 26 gauge wire, and forbidding loading coils, the CSA rules virtually guaranteed the need for DLCs for new shopping malls, condominium and town house units, and office parks. Under CSA rules, bridged taps were allowed, but were limited to 2,500 feet total, with no single bridged tap exceeding 2,000 feet. The older rules continued to allow loading coils, of course, but with restricted distances that mandated new local exchanges if DLCs were not used with CSAs.

The CSA system has become quite popular around the world, and as DLC systems increase and SONET rings multiply, the remote DSLAM and NGDLC arrangement will become even more popular.

DSLAM CONFIGURATION POSSIBILITIES

So far, this chapter has given some of the details regarding a "typical" DSLAM and discussed on a general level some of the particulars about DSLAM variations and configurations. The time has come now to be more systematic in considering the specific configurations for the DSLAM in and out of the central office. This section examines the advantages and disadvantages of various central office, co-location, or remote co-location choices for the DSLAM, and the next section examines the position of DSLAMs with regard to DLC local loops. When a DLC is used, a home run to the central office DSLAM is not possible, and so any DSLAM functions must move to the remote DLC location.

First, a look at central office DSLAM arrangements. The general model for what a central office DSLAM is and does is shown in Figure 12-9. In a sense, this figure is just a closer look at the DSLAM itself.

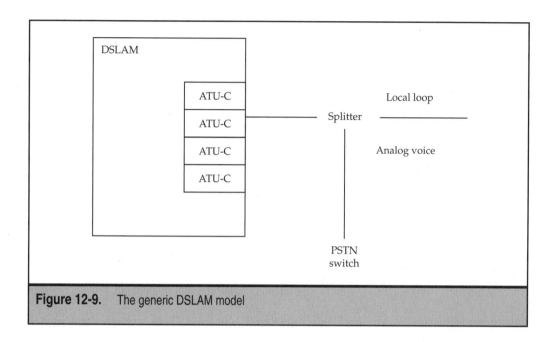

Figure 12-9.   The generic DSLAM model

Note that in the figure the ATU-C splitter function that sends analog voice to the PSTN switch and data on to the DSLAM is not in the DSLAM itself. As it turns out, just as putting the remote splitter in the DSL modem (ATU-R) itself turned out to minimize flexibility and maximize noise, placing the splitter inside the DSLAM with the ATU-R had exactly the same effect. So, it is better to take the splitter and separate the splitter from the DSLAM itself. The questions are how far should the splitter be from the DSLAM and what form should the splitter take?

Usually, access lines (local loops) entering the central office are organized at the *main distribution frame* (MDF) in the central office. The local loops enter the MDF typically as 100- or 200-pair unshielded but twisted cable binders containing multiple 50-pair binder groups. The purpose of the MDF is twofold. First, the MDF provides protection for the central office switch from lightning strikes on the local loop and other forms of "over voltage" protection. Second, the MDF provides a location for the *line terminal block* for punching down the pairs that are actually active and need connectivity to the voice switch. A *punch down* is simply a physical connection between the active local loop cable pairs and the PSTN voice switch ports. Terminal blocks are typically connected to the switch by 25- or 32-pair cables, although the punch downs are done pair by pair, of course.

When the splitter is physically located in the DSLAM, connectivity between the active ADSL users and the DSLAM (and then on to the voice switch for voice services) is accomplished as shown in Figure 12-10. The distance from the MDF to the DSLAM is seldom more than a few hundred feet, even if the DSLAM is in another building.

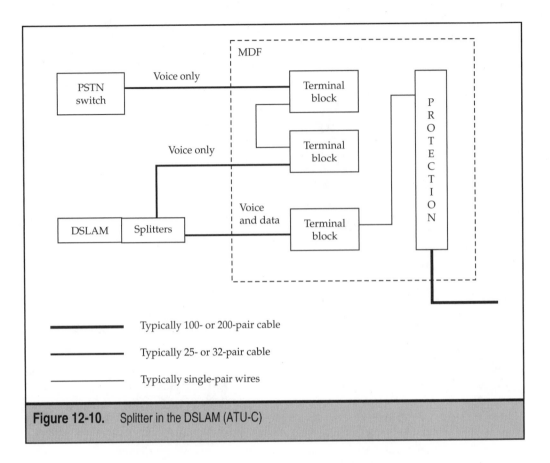

Figure 12-10.   Splitter in the DSLAM (ATU-C)

In the figure, the connections to the DSLAM splitters are also with 25- or 32-pair cables. But extra terminal blocks are required to cross connect pairs leading to the splitters with both voice and data on the ADSL links, and then returning carrying voice alone. The DSLAM carries the data to the core network, of course, but this data path from the DSLAM is not shown in the figure.

Ordinarily, the subscriber path for a customer is from local loop cable through protection to the terminal block connected directly to the PSTN switch. These voice-only customers are not shown in the figure. What is shown is that when a customer adds ADSL, the DSL service provider (the ILEC in this case) can simply punch down that new ADSL customer's pair to the "outbound" terminal block leading to the DSLAM splitter. Then they would also punch down a connection between the "inbound" terminal block from the DSLAM to the terminal block connected to the PSTN switch. Normally, this last punch down would lead to the same switch port as before, so the customer's telephone number would not change. This is the path shown in the figure. Single pairs used for this service function allow customers to be added one at a time.

The chief advantage to this configuration is that the DSLAM does not interfere with PSTN or MDF operation in terms of noise. Neither is the DSLAM very vulnerable to MDF or switch noise. Also, the combined splitter and ATU-C have fewer powering or form factor issues with regard to the splitter.

However, there are many disadvantages to this configuration. If the ATU-C in the DSLAM has to be replaced or removed, this will disrupt voice service, and this is always a concern for ILECs (voice is their main business and regulators keep a close eye of outages). Also, two cables are needed to link the DSLAM to the MDF, along with terminal blocks. The MDF has no access to the ADSL signal, of course. This is a drawback because many LECs rely on the MDF for low-level troubleshooting and testing (the MDF is the common place to store such testing equipment). Because the punch downs are totally manual, if a customer moves—even next door—the jumper cables between terminal blocks must be reconfigured to continue to provide voice service. But even worse from the standpoint of ADSL, the splitter is just too close to the DSLAM (technically, the low pass filter [LPF] for voice is too close to the high pass filter [HPF] for data). Whenever an analog phone line rings, the noise could disrupt service on many adjacent DSLAM ports.

Maybe the LPFs for voice could be removed from the DSLAM ports, but kept nearby, perhaps in the same rack to conserve floor space. This configuration is shown in Figure 12-11.

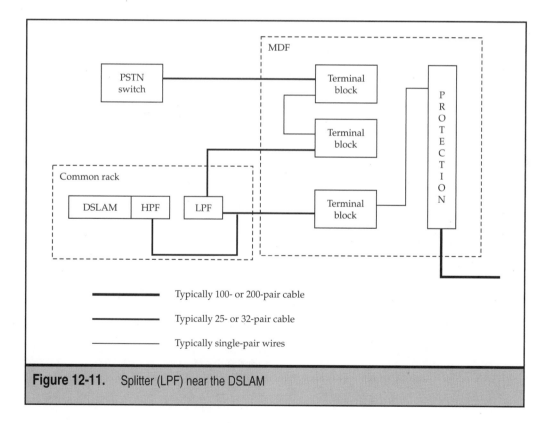

Figure 12-11. Splitter (LPF) near the DSLAM

The cables that carry voice and data and voice only are no longer labeled just to minimize clutter in the figure.

Note that more cable is needed just to connect the entering ADSL links to both LPF and HPF in the DSLAM. However, this configuration does have distinct advantages. Power and floor space for the splitter is still available and conserved. But now removing an ATU-C line card does not disrupt voice service at all (always an advantage for an ILEC). Also, ringing voltage is much less of a noise concern even with this minimal separation of LPF from DSLAM line card. And if a data CLEC or ISP uses this arrangement, the LPFs and associated cables are not even necessary because there is no voice on the pair.

On the other hand, more cables are needed, and the ADSL signal is still not accessible at the MDF for monitoring and testing. Customer moves also pose the same problem as before. In summary, the disadvantages are almost the same, but the benefits are many.

If some way could be found to give access to the ADSL signal in the MDF, this would be an even better configuration. Maybe the LPF, or even the HPF as well, could simply move to the MDF. This arrangement is shown in Figure 12-12.

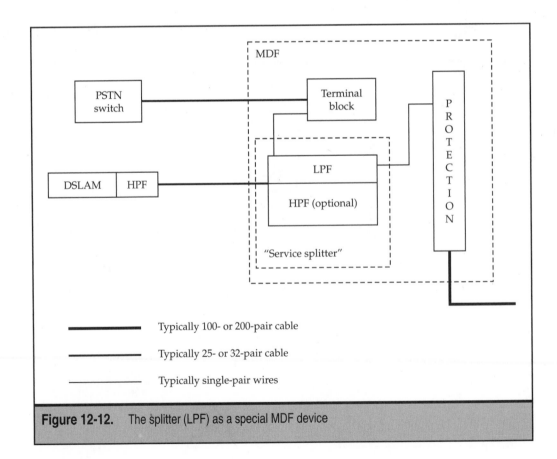

Typically 100- or 200-pair cable

Typically 25- or 32-pair cable

Typically single-pair wires

Figure 12-12.　The splitter (LPF) as a special MDF device

In the figure, there is no need for extra cables or manually punched down terminal blocks, because a "service splitter" can selectively pass only the data arriving on the ADSL links on to the DSLAM. The service splitter can also house the HPF for the DSLAM, making the ATU-C even simpler. When a customer gets ADSL service, the only actions necessary are to run the pair to and from the service splitter. Service splitters are often electronic devices that can run their own suite of tests on ADSL links, and monitor ADSL service as well. In their simplest forms, service splitters are just devices that house LPF units. Advantages are now many of the same as before, plus some new ones. Removing an ATU-C line card does not impact voice service, one cable is all that is needed to the DSLAM, and noise is reduced. In addition, the optional HPF gives full access to the ADSL link signal at the MDF, and maximizes ATU-C port density in the DSLAM (because the HPF is no longer needed there). And if only data services are provided, the LPF is not even needed on the pair.

The only disadvantage to this configuration is that the conditions—such as heat and power considerations—or floor space in the MDF might limit the number and size of service splitters deployed. In these cases, the service splitter might be located outside of, but still close to, the MDF. In these cases, the extra cables and terminal blocks are again needed, but the advantage of MDF visibility for ADSL might override this concern.

The only other DSLAM configuration to be explored is the case where the DSLAM is physically remote from the central office in a "remote co-location" site. This is shown in Figure 12-13 as "reverse ADSL" for data CLECs and ISPs.

Note that only the data is sent to the DSLAM. It makes little sense to send voice across the street and back again. One reason that ILECs tend not to get excited about this arrangement is that it adds no advantages to the previous configuration with the service splitter (and allows competitive ADSL services). The disadvantages have already been mentioned, but just to mention one of the biggest drawbacks, the total ADSL link distance now must add the extra footage to the total local loop length, and the link to the MDF has to paid for. Moreover, new 100- or 200-pair cables might have to be specially run to the site, and additional MDF wiring and protection blocks are needed.

All of these DSLAM configurations have been tried. There is no real right way or wrong way to configure a DSLAM and associated splitters. It all depends on the main goals of the DSL service provider, such as to minimize additional cabling requirements or gain MDF access to ADSL signals.

DLC AND DSLAMS

The earlier section "DSLAMs and SONET/SDH Rings" mentioned the position of ADSL and DSLAMs with regard to the DLC. In that section, the emphasis was on the SONET/SDH rings used to link the remote DSLAM and DLC to the central office. The time has come to be more specific about the form of the DSLAM used in the *remote terminal* (RT) of the DLC. It turns out that there are three major forms that a DSLAM, or DSLAM function, can take in a DLC cabinet. All of them share common concerns with regard to power requirements, heat given off, the possible need for battery backup, the room in the cabinet

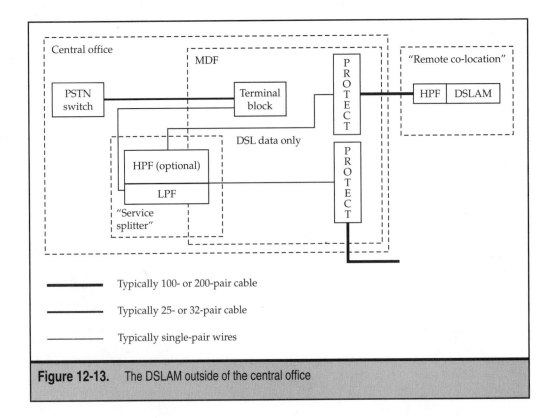

Figure 12-13. The DSLAM outside of the central office

required, the port density of the DSLAM, and related concerns about control and management. The three ways examined here are the following:

▼ A complete remote DSLAM

■ ADSL line cards (in two forms)

▲ The RAM DSLAM

Before detailing each, a quick review of DLC components is in order. Figure 12-14 shows the generic architecture of a modern DLC arrangement servicing a *carrier serving area* (CSA) cluster of customers, such as suburban housing developments, townhouse units, or apartment complexes. This figure shows only 24 analog telephones on the DLC, but DLC can service up to 1,000 or so customers depending on the size of the DLC cabinet and the type of link to the central office.

The DLC in the figure is serviced by a digital link up to 12,000 feet (up to 3.7 km) in length, which might be using either some form of T1/E1 links, HDSL/HDSL2, or fiber such as SONET/SDH. This link takes the digitized voice channels back to the *central office terminal* (COT, not shown) and on to the voice switch. The voice channels run at 64 Kbps, or sometimes 56 Kbps, and that is the problem when a DLC customer wants ADSL. As

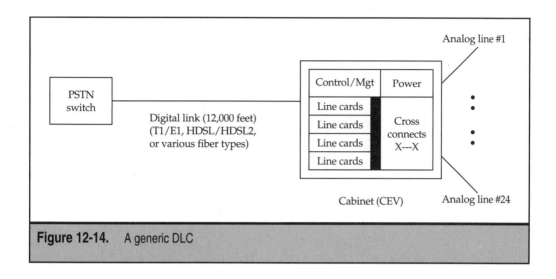

Figure 12-14.   A generic DLC

has been pointed out several times before, there is no way to make the DLC pass more than 64 Kbps per customer line.

Inside the cabinet, or *controlled environment vault* (CEV), is the DLC itself. There is a lot of variation among DLCs, because compatibility on the network side is needed only with the COT, which is always made by the same vendor as the DLC. But most DLCs will contain some control and management module, power supplies including batteries, the line cards to digitize the analog voice (or convert to a different digital form), voltage protection (not shown), and cross connects on small terminal blocks for the active pairs. In many cases, the cross connect terminal blocks are located outside of, but close to, the DLC cabinet itself. The line cards are connected to the digital link by a common backplane, shown in black in the figure.

Now, the easiest way to bring ADSL to customers on a DLC is to just put the DSLAM in the cabinet along with the DLC. This form of remote DSLAM in a cabinet is shown in Figure 12-15.

In the figure, the cross connect now sends the voice and ADSL data to the DSLAM splitter and then sends the returning voice on to the DLC for digitization or conversion. The data is taken back to the core network on separate links, because the voice occupies the same links as before. DSLAMs can be made small if necessary, and only a few ports might be required. Remote DSLAMs can service up to 600 to 1,000 users. Management can be done using the same procedures as with a central of DSLAM, just over the digital link to the DLC. There is also no concern about DSLAM and DLC compatibility, because the DSLAM just sends the voice right to the DLC, and the DLC notices no difference in the voice at all.

However, this can be a very expensive solution. In most cases there is no place to put the remote DSLAM in the DLC cabinet. Either a new cabinet is required, or the existing cabinet must be replaced. This might not even be possible if space is at a premium. Not

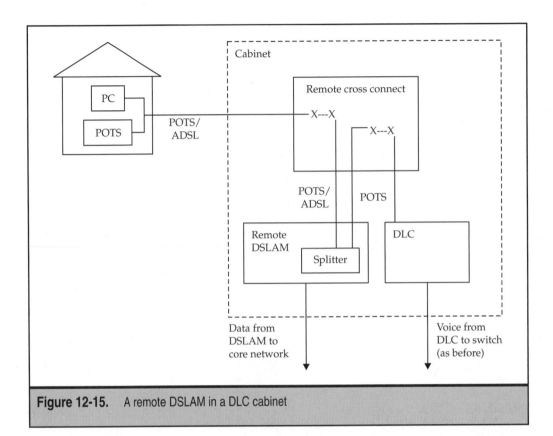

Figure 12-15.   A remote DSLAM in a DLC cabinet

only that, but the number of additional cross connections required could easily exceed the capacity needed in the DLC and/or cabinet. Most cross-connect blocks are intended only to service the number of pairs that the DLC supports, and there are only a few spares. Adding a remote DSLAM could not only require a larger cabinet or a new cabinet for the DSLAM, but the number of extra cross connects needed depends on the density of the ADSL subscribers, a number that has proven hard to predict with any confidence.

So remote DSLAMs are the easiest way to add ADSL to customers serviced on DLCs, but also require a large degree of funding up front and engineering guesswork.

The second way to service customers on a DLC that want ADSL is with an ADSL line card. The usual DLC line cards digitize only the 4 KHz voice passband and ignore all other frequencies. But ADSL line cards can pay attention to ADSL frequencies and generate proper ADSL line codes. There are two ways that ADSL ports can be added to the usual DLC voice ports. Either the ADSL ports can be placed on a separate line card for placement in empty DLC slots, or the ADSL ports can be added as an integrated voice and ADSL line card with a higher port density. Both forms are shown in Figure 12-16.

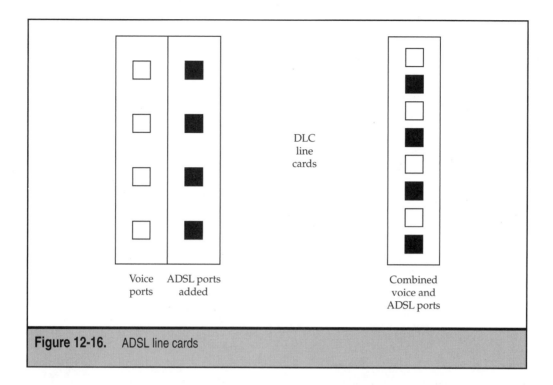

Figure 12-16. ADSL line cards

There are usually several empty line card slots in a DLC cabinet. If the ADSL sub-scriber density is low in the housing complex or apartments, this is an optimal solution. When the density is high enough, it makes sense to replace the voice line cards with inte-grated, high-density voice and data line cards as shown on the right of the figure. Some cross connecting might still be needed within the DLC cabinet, but this should be mini-mal, and separate links are still required to take the data onto the core network, but in theory no wholesale reengineering of the DLC cabinet space should be needed. To this major advantage could be added the minor point that when separate ADSL line cards are used, the empty DLC slots are there to be filled, and ADSL line cards will fill them.

However, the major disadvantages differ slightly between the two forms. If separate ADSL line cards are used, the problem is that many DLC cabinets are packed already, with little room to add anything at all. What good is a four- or eight-port ADSL line card if inside a DLC cabinet servicing 100 or possibly even 1,000 lines only two slots are avail-able? And if DLC slots are used by the ADSL line cards, this impacts the ability to expand the voice capacity of the DLC. But replacing voice line cards with integrated line cards might mean that in order to add one customer, the cost to the DSL service provider is the same as adding four or eight customers. Not only are the integrated line cards expensive, but they tend to require more power and give off much more heat than their voice-only counterparts. The battery backup power available can service the voice on the lines during a commercial power outage, but adding the ADSL to this requirement is asking a lot of the batteries. (Fortunately, most DSL service providers that are telephone companies are only required to battery back up voice services, not data. But how is this selective backup possible with integrated line cards and when VoIP or VoDSL is in use?)

Whichever form of ADSL line cards are used, DLC compatibility is a big issue. A slightly different form of line card might be needed for each DLC vendor, and there are many of those. The line card must match up with the customer's DSL modem as well, even if the DSL modem is supplied by the service provider (it almost always is). So the service provider must make sure that the DSL modem is compatible with both DSLAMs in the central office (for home run customers) and in the DLC. Also, the ADSL line cards do not and cannot perform all of the functions of a DSLAM, such as traffic aggregation or routing/switching. So usually even if ADSL line cards are used in the cabinet, the ADSL links serviced in this fashion still require a more or less complete DSLAM in the central office to provide more coordinated ADSL functions. Naturally, management is an issue as well.

So while ADSL line cards impact the physical aspects of the DLC cabinet less than remote DSLAMs, there are still issues with regard to compatibility, heat, power, and management.

The final method for adding ADSL services to a DLC line is to install what is called a *RAM-based DSLAM* into the cabinet. The major difference between a remote DSLAM and a RAM-based DSLAM is that the RAM-based DSLAM is designed to be integrated into the DLC cabinet more or less painlessly. RAM-based DSLAMs are often called "cigar boxes" or "pizza boxes" because only the essentials of DSLAM operation are retained in a package jammed with electronics. It is important to note that RAM-based devices have been used for years to add ISDN to DLC lines, so the concept is not new at all. The idea behind RAM-based DSLAMs is shown in Figure 12-17.

The major advantage of the RAM-based DSLAM is that, like remote DSLAMs, the RAM-based DSLAM is totally independent of the DLC and voice, so they will work with any vendor's DLC product. And like line card solutions, the integration of the small RAM-based DSLAM into the DLC means that little cross connecting is required. And even if the DLC is jammed full of DLC equipment, the small RAM-based DSLAM form makes it much easier to locate outside of the cabinet than a full remote DSLAM.

The biggest problem with RAM-based DSLAMs is scalability. RAM-based DSLAMs are small, but so is their ADSL link capacity. In some cases, it might be more cost effective to re-engineer the DLC cabinet than to worry about how to add another RAM-based DSLAM every other month. Also, like line cards, RAM-based DSLAMs might still need full DSLAMs in the central office. And there are always heat and power issues.

So RAM-based DSLAMs can be an attractive solution to the DLC issues in some cases, but not always.

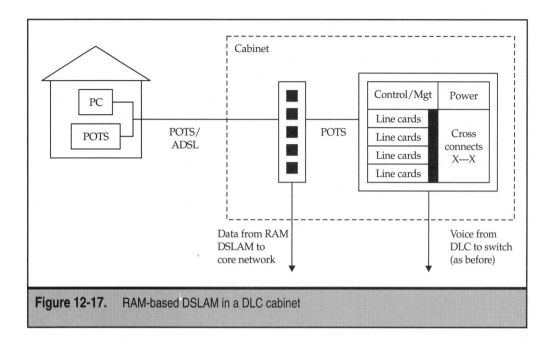

Figure 12-17.   RAM-based DSLAM in a DLC cabinet

Some studies claim that up to 50 percent or more of the DLC cabinets in the United States, especially older ones, simply cannot be upgraded to ADSL in any fashion at all. There might be a considerable wait for ADSL service as these older cabinets and DLCs are phased out and replaced.

CHAPTER 13

DSL Migration Scenarios

As has been pointed out numerous times in this book, DSL comes in many flavors. The emphasis for the last few chapters has been on ADSL, but that does not mean that ADSL is the end of user and service-provider migration strategies. All of the various forms, such as HDSL (High bit-rate), SDSL (Symmetric), and IDSL (based on ISDN-like links), could play a role in the total DSL picture. However, it is undeniably true that ADSL (Asymmetrical DSL) and VDSL (Very high data rate DSL) seem to occupy most of the interest of service providers and technical writers, especially when it comes to residential or SOHO applications. This is no accident. Most equipment vendors and service providers have focused on ADSL and VDSL for residential and small business environments for two reasons. First, ADSL appears to offer the broadest base of near-term services (fast Internet access, for example) to the widest possible range of local loop conditions. Second, VDSL appears to offer the bandwidth needed for almost any broadband service, including video. If this is the case, not only is ADSL the most promising member of the DSL family, but VDSL might well be the future of ADSL.

In fact, there are a number of possible migration paths that a customer might be offered, and a service provider might propose to furnish ADSL and VDSL. The ADSL Forum spent a lot of time and effort on this topic in their early days and mention should be made of their work. This chapter relies heavily on documentation from the ADSL Forum. Along these lines, several DSL migration paths have been outlined, as listed here:

▼ **From analog modems to ADSL**   Customers using analog modems migrate to ADSL services in order to get faster Internet access.

■ **From DLC to ADSL**   Customers currently serviced by a *digital loop carrier* (DLC) feeder (not necessarily fiber-based) migrate to ADSL services in order to get faster Internet access.

■ **From ISDN to ADSL**   Customers currently using ISDN services for Internet access migrate to ADSL services in order to get faster Internet access.

■ **From ADSL to next generation DLC (NGDLC)**   Customers currently using ADSL, in some form, migrate to next generation DLC systems that can provide a wider range of services to a broader range of customers in a more cost-effective manner.

▲ **From ADSL to VDSL**   Customers currently using ADSL (or NGDLC-based ADSL) migrate to VDSL, which offers the highest speeds and widest range of broadband services.

Many more migration paths than those listed here exist, but these are the ones most likely to be offered to the vast majority of DSL customers and are probably the most important. Each migration scenario has its own set of issues, benefits, and hazards for customers and service providers alike.

DSL MIGRATION ARCHITECTURE

A number of possible migration paths have been outlined to furnish ADSL and eventually VDSL to a wide variety of customers. Not all of them are discussed in this chapter, but the main and most important ones are all treated in detail. These DSL migration scenarios presented in the preceding section are shown in Figure 13-1 as a number of different paths through the total DSL migration architecture. Many service providers and equipment vendors envision migration from an analog plain old telephone service (POTS) world to ADSL, and then eventually on to a VDSL network.

DSL migration scenarios can include technologies such as HDSL, HDSL2, and others; however, the migration strategies shown are the DSL technologies and related migration strategies that the equipment vendors and service providers have emphasized most often to supply ADSL and VDSL services. The DSL migration scenarios shown in the figure do not mention HDSL2 or any other DSL technology besides ADSL and VDSL. However, almost any of the three paths shown in the figure could include other DSL family members. For example, the ISDN path could potentially include IDSL, and DLCs could employ HDSL or HDSL2. The key idea throughout this chapter is that the migration will probably move from an analog POTS world to ADSL services, and then ultimately on to a full VDSL network.

More information on the five paths indicated in the figure are listed here:

1. **From analog modems to ADSL**   The nice thing about customers using analog modems is that they are fairly familiar with the technology. Much of the ADSL marketing strategy emphasizes ADSL devices as just a new type of modem (the term *DSL modem* has been used frequently throughout this book). Because most ADSL services are expected to include nothing more than faster Internet access outside of the congested and slow PSTN, there is a natural attraction to customers and real incentive for service providers to perform this migration to ADSL.

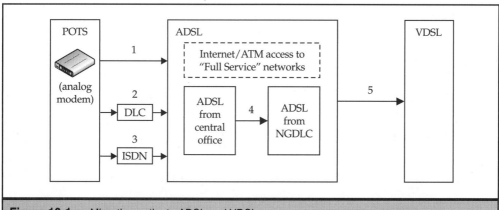

Figure 13-1.   Migration paths to ADSL and VDSL

2. **From DLC to ADSL**   ADSL makes maximum use of twisted-pair copper wire, but not everyone is at the end of a single-pair "home run" of copper wires. It has been mentioned several times that more customers are serviced by a DLC feeder (not necessarily fiber-based, but a digital feeder), especially when following the guidelines from the CSA concept, which tends to put a DLC at 12,000 feet (about 3.7 km) to service clusters of customers in the area. These users must be able to migrate to ADSL services for faster Internet access as well.

3. **From ISDN to ADSL**   It came as a shock to some telephone companies that customers wanted ISDN not to make telephone calls or to access telephone company switch-based ISDN services, such as they were. They just wanted a faster bit pipe to the Internet. Even when ISDN switches were used for simple point-to-point links at 64 or 128 Kbps for Internet access, the link still looks like ISDN more than anything else. This means that each Internet connection still has to pass through the ISDN voice switch, even though the voice switch adds exactly nothing to the Internet connection except possibly an easy way to switch ISPs by dialing (then again, this is currently difficult to do with ADSL, so switched services do have some advantages). So customers currently using ISDN services for no more than faster Internet access should be given a migration path to simple ADSL services that can provide faster Internet access with much less stress and strain on the voice network.

4. **From ADSL to NGDLC**   After customers in sufficient numbers get on ADSL, and the whole CSA concept becomes even more prevalent than it currently is as more and more developments and townhouse complexes are built farther and farther from city centers, it might be a good idea to migrate to a more distributed architecture. Now ADSL could be provided on SONET/SDH rings, DSLAMs could be distributed, as well, and the whole process of furnishing ADSL services would be more efficient as a result. In this scenario, ADSL is a prerequisite, but customers currently using ADSL in some form who are migrated to next generation DLC systems can potentially be provided with a wider range of services in addition to fast Internet access.

5. **From ADSL to VDSL**   ADSL is quite fast, no doubt about it. But ADSL will still struggle to supply the bandwidths needed to support true broadband applications, such as broadcast-quality video and the like. Customers currently using ADSL, either on home-run local loops or as NGDLC-based ADSL, will be migrated to VDSL at some point in the future. VDSL, of course, offers the highest speeds to support the widest range of broadband services.

Note that some of the paths involve conscious decisions on the part of customers. For instance, customers must choose when and if the transition from analog modem to ADSL should be made, but other migration paths are totally independent of and transparent to the user. Any decision on the part of the service provider to furnish ADSL services on DLC systems, for example, is not only invisible to the customer, but totally out of the customer's hands. Always remember that other possible migration paths exist, but are not shown in the figure.

This last point is especially true within the ADSL network itself. Note that within the ADSL network, the migration from simple, central office-based ADSL to providing ADSL services from NGDLC systems is essentially transparent to the users and requires no changes, or even awareness, on the customer's part.

The migration architecture also points out that the intent of ADSL is to provide either Internet (that is, Transport Control Protocol/Internet Protocol, or TCP/IP) or ATM network access to a "full-service network" that provides customers with all the services they need. Whether this network is the Internet with more broadband capability or an ATM network used as the core network to tie DSLAMs to service providers (or even an ATM network with ATM-based services) is not essential to ADSL itself. ADSL will perform in either situation.

MIGRATION URGENCIES AND PRIORITIES

More is needed than just an outline of possible migration paths to furnish ADSL and VDSL. All of the migration paths, even those not considered in this survey, can be characterized in terms of a starting network and a target network scenario. For instance, the first migration path listed, from analog modem to ADSL, could have "analog modem Internet access" as the starting network and "ADSL for Internet access" as the target network.

The five major migration paths considered in this chapter are shown in Table 13-1. Most of the information presented here is based on ADSL Forum documentation. Keep in mind that the starting and target networks shown in the table are not all that exist, but these seem to be the ones on which most service providers and equipment vendors are

Starting Network	Target Network	Urgency	Priority
Analog modem Internet access	ADSL for Internet access	Immediate	High
Analog modem on DCL	ADSL delivered service(s)	Immediate	Medium
ISDN for Internet access	ADSL for Internet access	Immediate	High
ADSL from local exchange	ADSL from NGDLC	Future	High
ADSL delivered service(s)	VDSL delivered service(s)	Future	High

Table 13-1.   DSL Migration Paths

currently concentrating. Each migration scenario is assigned an urgency and priority based on service provider and equipment vendor assessments of just how crucial each one is for the success of ADSL. The other migration scenarios are mainly concerned with variations of these major themes, and none of those omitted can be considered to have immediate urgency and high priorities.

Each migration scenario has been assigned an urgency and priority for standardization and implementation. These are detailed as follows, using the starting and target terminology instead of numbered transition paths:

▼   **From analog modem Internet access to ADSL for Internet access**   The urgency is immediate, the priority is high. The pressure placed on the PSTN switches and trunks from packets-on-circuits Internet access make this a critical migration scenario.

■   **From analog modem on DLC to ADSL delivered service(s)**   The urgency is immediate, the priority is medium. The special challenges of DLC systems (such as limited bandwidth, remote configuration concerns, and so forth) for ADSL make this an important item; the priority is only medium, though, given the current distribution of DLC systems and the workarounds available (for example, moving DLC ADSL customers onto home runs). This priority may increase in the future.

■   **From ISDN for Internet access to ADSL for Internet access**   The urgency is immediate, the priority is high. The concerns are basically the same as those for analog modem Internet access.

■   **From ADSL in the local exchange (LE, usually called the central office in the United States) to ADSL in the NGDLC equipment**   The urgency is future, the priority is high. NGDLC systems will not be constrained by bandwidth and may be integrated with SONET/SDH rings and even ATM switches. This will further decrease traffic aggregation woes at the central office (LE). However, NGDLC systems are not in the immediate future of ADSL just yet.

▲   **From ADSL delivered service(s) to VDSL delivered services**   The urgency is future, the priority is high. Most see VDSL as a natural evolutionary path for ADSL, but probably only until NGDLC fiber-based systems become much more common than they are today.

Each of these starting and target network migration scenarios has its own set of pros and cons for both the customers and service providers. These are considered in the sections that follow.

FROM ANALOG MODEM TO ADSL

The most common migration scenario for customers and service providers venturing into the wonderful world of ADSL is most likely to be from a starting network scenario of limited

Internet access speeds with 56K or slower analog modems, although 56K modems do share some asymmetrical features with ADSL, because 56K modems are still 33.6 Kbps upstream and therefore asymmetrical. The target network would be higher Internet access speeds with ADSL. Of course, the limited analog modem speeds offer a ready market for ADSL services in the first place, which should explain the anticipated popularity of this migration scenario. (Of course, analog modem users could also move to cable modems or some form of wireless access, but this chapter is about DSL migration.)

The whole point of this move by customers would be to speed up current long waits for Web graphics and files. It is not unusual today to wait 30 seconds or longer for a full Web page of writhing graphics like flaming corporate logos. File downloads at 33.6 Kbps can take about six minutes per megabyte on a good day and time, and twice as long during congested intervals. The shorter wait is the greatest benefit to ADSL Internet users, although it sometimes seems that ADSL congestion can be as bad as analog modem congestion.

The greatest benefit to the service provider is that it moves long-holding-time Web sessions off of the voice network, and this is true even if the service provider is not a telephone company. After all, even an ISP has to use the PSTN to reach their customers at the end of the local loop in the first place, and so some ISPs have been avid early adapters of ADSL. Of course, some ISPs have accused the incumbent LECs (ILECs) of blocking their attempts to deploy ADSL for the ISP's customers because the ILECs might plan to offer their own ADSL services at some point in the future. However, many of the delays with ADSL deployments can be traced to the simple growing pains of any new technology, and not necessarily the sinister actions of incumbent service providers (users in more rural parts of the United States might not agree).

If the ADSL service provider happens to be a telephone company (pretty much a given today), an added benefit of this ADSL migration scenario is that it does not require expensive ISDN upgrades to PSTN switches, which are needed to give customers faster Internet access without ADSL. ISDN upgrades can cost about $500,000, not to mention the need to coordinate and manage the upgrade process.

This migration plan also gives the telephone companies an answer to the possibility that competing cable modem services from the cable TV companies could become even more popular than they are now. This is not the place to rehash cable modems, but it should be pointed out that much ADSL activity seems to be a direct response to cable modem developments and service offerings. Ironically, many telephone companies, especially "long distance" telephone companies, considered buying and even did buy cable TV companies to deliver higher bandwidth services than the copper loop could deliver, but many telephone companies that have experimented with becoming cable TV companies, and even offering voice services on these cable networks, seem to be abandoning their early efforts. It seems that customers like watching TV on cable, but talking and accessing the Internet through a separate telephone network.

Finally, all of this makes even better sense as more and more telephone companies become ISPs, as is the case in the current deregulated environment in the United States.

FROM DLC TO ADSL

Many local loops in the United States and around the world are currently serviced with some kind of pairgain system, which multiplex many telephone customers onto a single carrier system (thus, gaining pairs that can be used for other purposes). The vast majority of these pairgain systems are DLC systems, as were detailed in a previous chapter. As mentioned in that chapter, DLCs pose problems for even simple ADSL deployment scenarios.

Most troublesome is the fact that current DLCs will not pass 1.1 MHz ADSL bandwidths, so the service provider end of the ADSL loop is not at the central office or local exchange, but at the DLC remote digital terminal. This restriction in bandwidth occurs because the most common DLC systems, such as SLC-96 (Subscriber Loop Carrier carrying 96 analog signals on a group of four T1s), simply digitize the analog voice in the 4 KHz baseband range and filter out the other frequencies. The voice passband is digitized at 64 Kbps (or 56 Kbps), essentially limiting any ADSL deployed at the end of a DLC system to this speed downstream (and upstream as well, but the downstream speed in ADSL is much more critical, of course).

The previous chapter outlined several ways that the DSLAM or some DSLAM functions could be moved to the DLC cabinet. However, this section assumes that wholesale movement of DSLAMs to DLC is not planned, yet the DSL service provider still wishes to offer ADSL services to customers on DLCs. Perhaps some form of voice over DSL (VoDSL) can help. In such DLC-to-ADSL migration scenarios, it might be possible to derive a 64 Kbps channel from the total ADSL bandwidth, which means that the analog voice passband is digitized at the DLC and can still enter the DLC equipment as before. In this case, the ADSL signals above and beyond the analog voice channel must find their way to the local exchange some other way, probably on new wire pairs (or fiber optic links) installed expressly for this purpose. This would preserve the DLC line for voice service, but calls into question the entire ADSL strategy of employing and preserving existing copper pairs. Also, it obviously requires some way to digitize the analog voice on the customer premises, perhaps with an *integrated access device* (IAD) built into the DSL modem. But then the question is how users can hook all their telephones up to the IAD. All in all, a messy situation.

Alternatively, and more simply, it might be possible to shift non-ADSL customers onto the DLC and move ADSL customers onto home run links. In other words, when a customer serviced on a DLC system wants to use ADSL services, a nearby customer employing a simple analog loop on a home run link can be shifted onto the DLC, freeing up this loop for easy ADSL deployment. Naturally, the feasibility of this would depend on the DLC/straight home run local loop service mix, proximity of the "exchanged" customers, resource commitment on the part of the service provider, and several other factors. The simplicity of this plan is appealing, but very dependent on the existence of home runs near the DLC, which is true enough in or near older neighborhoods, but less likely near newer "greenfield" construction without older dwelling units nearby.

Yet another possibility is to provide another local loop pair for ADSL alone and forget about analog voice support altogether. This should not be a serious liability because most people would be using ADSL for Internet access and not to make telephone calls, just as they use ISDN today. The issue is now whether or not ADSL is the best solution, because DSL services provided on "dry" (no voice) pairs would no longer be constrained by the need to support analog voice.

Of course, the most likely way that DLC-based customers will get ADSL will be to eventually deploy a DSLAM of some shape, manner, and form at the DLC remote terminal (RT) cabinet and backhaul the traffic on SONET/SDH fiber optic cables.

FROM ISDN TO ADSL

The limitations of analog modems when used almost exclusively for Internet access have been apparent to users for years. The situation was even worse several years ago when the fastest analog modems in common use were only 14.4 Kbps. As recently as 1992, a Hayes external 9600 bps modem cost $799 in the United States, and modems cost a lot more in other countries. Throughout the 1990s, therefore, users who have been especially sensitive to analog modem limitations for Internet access (mostly Internet Web site designers and other networking personnel working from home) have employed ISDN lines where available in an attempt to speed up their Internet access.

ADSL provides much higher speeds than the 128 Kbps (achieved by bonding the two 64 Kbps bearer channels) or 144 Kbps (achieved by adding the D signaling channel at the cost of dropping switched services) that can be used over the residential and small business 2B+D Basic Rate Interface (BRI) of ISDN. Also, using ISDN almost exclusively for Internet access unnecessarily ties up PSTN switches and trunks for long periods of time when the line is inactive, but still connected. Moreover, because ISDN is a switched service, a call setup delay occurs when accessing an ISP through ISDN (although this delay is usually quite small). Newer versions of ISDN, such as "always on" ISDN, seek to get around this problem, more or less successfully.

The real problem with ISDN when it comes to ADSL is that ISDN is a Digital Subscriber Line (DSL) in its own right. ISDN lines might already support analog telephones at the premises through ISDN terminal adapters (TA). In addition, some telephones could be compliant ISDN devices, and more exotic ISDN terminals might be present for true ISDN services such as videoconferencing. Any migration from ISDN to ADSL must address all of these possibilities. One possible "merging" of ISDN and ADSL is shown in Figure 13-2.

As shown in the figure, it is possible to carry ISDN BRI service under ADSL. However, this seems to be asking a lot from supposedly inexpensive ADSL equipment. In this way, the ISDN bandwidth is located somewhere under the starting ADSL frequency.

It is also possible to embed the existing ISDN service channels inside some range of the total 1.1 MHz ADSL bandwidth. That way, any analog phones on the premises would still be handled by normal splitter arrangements, while the 2B1Q signals traveled elsewhere. In this arrangement, the option is to embed the 2B1Q somewhere "inside" the 25 KHz to 1.1 MHz range and modulate the 2B1Q exactly as if it were just another digital input. Neither option is ideal, but some consideration must be given to ISDN users in an ADSL scenario.

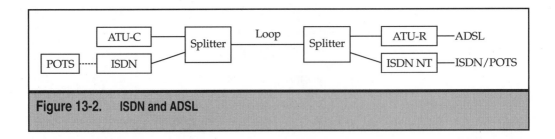

Figure 13-2.   ISDN and ADSL

Regardless of the approach considered, all ISDN-to-ADSL network scenarios must address ISDN standard issues. ISDN has been an international standard for years. Many ISDN documents specify that some interfaces operate, be provisioned, and be administered in a certain standard fashion. Therefore, violations to the ISDN standards in the name of ADSL should be approached with caution, although all standards can be revised or extended.

There is an even more important issue with regard to a migration from ISDN to ADSL. ISDN is a standard that defines more than DSL access to an ISDN switch. ISDN defines services and applications that can be accessed over the ISDN DSL. ISDN video servers and videoconferencing services are in place and in use already. Now totally replace an ISDN BRI line with ADSL. How do ISDN applications make the leap to ADSL? They don't. Customers that use ISDN for more than high-speed Internet access must be educated that ADSL is not just a faster form of ISDN.

A few variations on the whole ISDN and ADSL theme should be mentioned. These issues come up because both ISDN and ADSL are forms of DSLs, as was mentioned previously. One of the variations of the DSL family mentioned earlier in this book was something called IDSL. This form of "ISDN over ADSL" is not really the same as the migration from previously installed ISDN to ADSL considered here. In the case of IDSL, there is no ISDN switch to contend with. The ISDN link runs at 128 Kbps (usually), and the D-channel is essentially ignored (again, usually, but not always). IDSL allows the continued use of this speed and structure through the DSLAM by using the proper termination units at each end.

In many countries, especially outside of the United States, it is also common to deliver two analog telephone lines over one copper pair. This is done by digitizing the voice signals and transporting each individual conversation by means of a 64 Kbps digital channel.

This concept is known as *two channel pairgain* or *digitally added main line* (DAML). Manufacturers of this equipment were quick to notice that existing ISDN DSL chips offered an economic means of carrying the two 64 Kbps channels over the copper pair. A DAML system will therefore typically utilize ISDN transmission chips, and so the signals on the wire appear identical to an ISDN signal. However, the service is quite different from ISDN.

At least one ADSL equipment vendor supports ADSL with DAML "underneath" the ADSL instead of analog voice. It is, therefore, possible to deliver 2B1Q (two binary, one quaternary) coded ISDN-like service "under" ADSL.

INSIDE ADSL: USING IP FOR "FULL SERVICES"

ADSL is more than just a new way to adapt the analog local loop for digital information and services. As should be readily apparent, ADSL is part of a mostly complete network architecture extending from user device (PC client) to network server (perhaps an Internet Web site). But ADSL may be used and is indeed intended to be used to access a wide array of broadband, multimedia, and graphical services. Several of these services, which mostly remain potential services for ADSL, have been relatively routine for many ISDN users for years. However, if ADSL is to be "better than ISDN," then what happens to the services?

It is undoubtedly true that most users will employ ADSL for fast Internet access, and probably be happy with that. Sooner or later, though, users will want to do more, especially if those users have previously employed, or been exposed to, a complete range of ISDN services. Perhaps there will soon be a need to provide what is called a *full-service network* (FSN) with ADSL, not just faster Internet access.

However, it is also true that TCP/IP-based servers are the most common in the world. Many of these already serve up video and other broadband services (although in a somewhat limited capacity, to be sure). Nevertheless, most ADSL users will be running TCP/IP as client for their ADSL Internet access service.

Perhaps all broadband services delivered over ADSL networks will always be accessible only through TCP/IP. These services would ultimately include video-on-demand, broadcast TV services, and so on; they would be served up from TCP/IP–based servers, maybe still on the Internet, but not exclusively. For reliability and logistical reasons, the servers could just as well be in the service provider's central office location.

Of course, this move from Internet access to broadband services would most likely require massive changes to existing TCP/IP protocols and current routing practices. Some of these changes have already been made, but others will take years to complete. It is one thing to say that TCP/IP should have quality-of-service parameters and guarantees, have low and stable delays, and be as robust as it is now, but it is quite another thing to figure out exactly how to do it.

The wholesale offering of a TCP/IP and Internet-based full-service broadband network with high-quality video and audio would require massive changes to existing Internet structure, as well. The Internet backbone, as it is currently constructed, leaves a lot to be desired for handling massive flows of delay-sensitive and error-sensitive information. Detailing all the limitations cannot be done here, but a good example is the common "site not reachable" message that result from Internet congestion rather than a server being down. This obviously will not do for 911 voice services or even for routine television watching. Some changes are being made in this area, but much work remains. Furthermore, it is not yet clear who gets to build, pay for, and run this coordinated and upgraded Internet of the future.

USE ATM FOR "FULL SERVICE" ADSL?

Because ADSL will most likely be used simply for fast Internet access for the foreseeable future, it might be beneficial to try to deliver broadband services exclusively with TCP/IP. This approach, which was strongly suggested in the previous section, would require massive changes to the existing TCP/IP protocols and underlying Internet infrastructure. In fairness, many of these changes are underway, but they will take years to complete. However, early versions of protocols intended to add guaranteed quality of service parameters (such as for low and stable delays) and multicasting (for broadcast-quality video services) on the Internet have left a lot to be desired.

One possible alternative to IP-based and Internet-based ADSL full-service networks is asynchronous transfer mode (ATM) technology. ATM seems to be a better fit for "full-service" deployments, especially because ATM is part of the complete Broadband ISDN (B-ISDN) protocol stack, which outlines the form that most of these broadband services should take. So ATM is specifically designed for broadband services, not just a data delivery protocol with some voice and video features added on, as many accuse IP of being.

Internet-based broadband service proponents have always said things along the lines of "with a lot of hard work and effort, TCP/IP could be as good as ATM is when it comes to mixed forms of traffic like voice and video and data." This statement is startling for a number of reasons, but the aspect that startles when it comes to ATM and ADSL is the implication in the statement that ATM today is as good as TCP/IP might be tomorrow—but only after a lot of hard work and effort expended on TCP/IP. Why bother? Why not just use ATM instead of TCP/IP in ADSL (or any DSL) for all services and be done with it? After all, ATM cells are commonly used on the ADSL link from DSL modem to DSLAM, and to link DSLAMs to ISPs.

But things are never that simple. The fact is that there is no ATM hardware or software on the vast majority of customer premises, even business customer premises. The network protocol on the premises is most commonly IP, now bundled inside the majority of new PCs. The common argument is that the protocol of choice for ADSL must be IP for this reason, and it now seems that this will be true forever.

So the situation is bleak for ATM on DSL supporters. IP packets can be carried and switched inside a stream of ATM cells. In fact, this is one of the allowed ADSL transport modes of operation, as shown in an earlier chapter. If there is more than one TCP/IP device or PC on the premises, then the ATM Forum's *LAN emulation* (LANE) might be needed to shuttle the IP packets from user client to service provider server effectively. The ATM Forum also has a specification known as *multiprotocol over ATM* (MPOA), which essentially makes a multiprotocol router out of an ATM switch. Such a switch can route not just IP, but other protocols.

If IP is the only protocol of concern, then a group of protocols known collectively as *IP switching* or *multiprotocol label switching* (MPLS) can be used very effectively. Although implementations of IP switching have been proprietary to date, MPLS is a standard and has generated a great deal of interest.

At one point in time not too long ago, the use of ATM as the basis for full-service network deployment around the world had been seen as the only viable, long-term broadband solution. This is due mostly to ATM's roots in the B-ISDN service arena.

Moreover, using ATM with ADSL for this purpose requires only relatively minor changes to B-ISDN/ATM standards. Furthermore, many of these would merely involve redefining the speeds and media at which B-ISDN services and ATM networks are allowed to function.

It is worth noting that some of the architectures for implementing such full-service networks start with IP routers and others start with ATM switches. Consider the ATM switch first. ATM cells flow in and out on all ports. Now add LANE and MPOA (LANE is a kind of prerequisite for MPOA). The result is a device that sends and receives ATM cells and routes IP and other protocols. If only IP matters, add only IP switching.

Now start over again with an IP router. Add ATM adapters for input and output ports, mainly for speed and delay reasons. Now add LANE and MPOA, as before. The addition can still be made because both LANE and MPOA are simply standard documents that can be implemented in almost any programming language on almost any hardware platform. The result is a device that sends and receives ATM cells and routes IP and other protocols (maybe), but this is almost exactly what the ATM switch became! So which is it, an IP router with ATM switching, or an ATM switch with IP routing? It does not matter because the result is the same.

However, adding MPLS to an IP router can yield the same results without needing ATM at all. This is the direction most vendors and service providers are going today. And in spite of the glowing words about LANE and MPOA earlier in this section, these methods of making ATM more IP-aware and friendly have fallen out of favor as just more complicated and expensive ways of doing what MPLS does much more easily. All in all, the collision course between IP and ATM, Internet and telephone company, and router vendor and switch vendor has been a disaster for ATM proponents, in spite of the continued presence of ATM components inside a network using ADSL.

FROM ADSL TO NGDLC

Whether the PSTN is used with analog modems or ISDN, the transport of packets on circuits intended for short voice conversations makes congestion a problem. The problem actually runs deeper than this: At the root of the entire switch and trunk congestion issue is the very central office or local exchange (LE) concept upon which the PSTN is founded. Although the official term is "local exchange," which is what ISDN and B-ISDN documentation calls them, these switching offices are almost always called central offices (COs) in the United States. Local exchanges or central offices, by definition, aggregate traffic for switching and trunking purposes. Even if the local exchange is optimized for packets, if there are enough packets, congestion will result. After all, the Internet is optimized for packets, and it gets congested all the time, as telephone companies with lots of circuits are fond of pointing out to the discomfort of the ISPs.

Local exchanges (or central offices, the point is the same) exist mainly for historical reasons. Aggregating traffic makes sense when equipment is fragile, expensive, or both. Local exchange switches could not simply be plunked down anywhere and left alone. Even today, when PSTN switches are more robust than ever, they remain a huge expense to the service provider. These large, central switches do, however, provide the focal point for access to services.

Today, PSTN switches closely resemble large computers. The advances in modern computing power and methods make distribution, rather than centralization, attractive. Robust modern switches can be made smaller, consume less power, and require less local maintenance than ever before. A local exchange switch used to be one huge device handling 10,000 (or as many as 40,000) local loops and hundreds of trunks. Today, the local exchange switch might be a grouping of exactly the same small units handling 1,000 loops each, linked together by a 100 Mbps Ethernet LAN or fiber ring. Trunks may all emanate from a specialized trunk unit. In addition, the services no longer need to be located right next to the switch if the trunking network is reliable and fast enough.

Obviously, a local exchange switch on a LAN has already taken a big step from centralized to distributed. Why not distribute the services and their servers around the LAN, also? Why stop there? And the possible move from central to distributed access to services works hand-in-hand with the move to more widespread DLC systems. DLCs will soon comprise the most common loop architecture, due mainly to changing customer living patterns (they live farther and farther away from obvious local exchange sites in central areas) and the falling cost of fiber-based DLC equipment. As was pointed out earlier, most of the newer DLCs will access some form of SONET/SDH ring through a small SONET/SDH add/drop multiplexer (ADM) to support this robust NGDLC equipment.

The next step seems clear. Why not move the SONET/SDH ADM, NGDLC, and DSLAM technologies to a remote NGDLC site? Some switching functions could be included to avoid aggregating traffic into a funnel, which would lead to a congested local exchange or single-point-of-failure site. After all, how do you call to report that your full-service network is now acting empty of services?

Perhaps even IP routing and some Web services will follow. After all, why bring the packet traffic all the way back to the central location just to route them? Maybe there should be not only community Web site servers in the local exchange, but neighborhood Web site servers for a few hundred homes, tops. Whether this is feasible and whether this distributed approach occurs quickly or slowly all depends on how quickly local exchanges can be converted to participate in this new, distributed architecture.

FROM ADSL TO VDSL

In the near future, it may be common for the public network to include fiber feeder systems, as a matter of course, to offer widely used Internet services and to connect the increasing numbers of SOHO users who need not then congest the roads as well as the networks. Such a fiber-based access network will be common in the near future as end device computing power and networking capability continues to grow. Naturally, applications will demand more and more bandwidth, as they always have.

The point is that internal network computers and applications on servers must keep up with the demand placed on the network by "external" (client) user computers and applications—this is the reason for VDSL.

Given these simple network growth factors, VDSL seems as rational as the progression of Ethernet from 10 Mbps to 100 Mbps to 1 Gbps. However, where the migration from 10 Mbps to 100 Mbps Ethernet took some ten years and is still not always a given on inexpensive network cards and hubs, migration from ADSL to VDSL could take place more rapidly.

Furthermore, VDSL speeds align very nicely with the speeds needed for true broadband services. That is, the top VDSL speed of 51.84 Mbps (although VDSL can run marginally faster) is exactly the same as the OC-1 SONET/SDH interface that can be used for all broadband networks. By now it should be apparent that the relationship between ADSL and VDSL is not quite the same as the relationship between ADSL and other DSL technologies because of all the alphabet soup of DSLs, only VDSL is one of the ADSL target networks. This only makes sense. It would not be of any benefit to anyone to take existing ADSL links and turn them into HDSL, SDSL, or anything else. It does, however, make sense to convert existing ADSL links and turn them into VDSL. Maybe the time has come to take an even closer look at VDSL itself.

CHAPTER 14

Very High Data Rate Digital Subscriber Line (VDSL)

Very high data rate Digital Subscriber Line (VDSL, sometimes seen as Very High-bit-rate DSL) offers the highest upstream and downstream rates of any DSL designed to date. The downstream speeds range as high as 52 Mbps, depending on distance. The upstream speeds start at about 600 Kbps for the longest local loops, but most implementations should start at 1.5 Mbps (where ADSL essentially tops out for longer distances). Upstream speeds end at about 34 Mbps for short loops. It should be noted that in some forms, VDSL is symmetrical rather than asymmetrical, and there are different downstream and upstream speed ranges defined by the DSL Forum for symmetrical and asymmetrical forms of VDSL. Current VDSL trials use *frequency division multiplexing* (FDM) for line code signals, but echo cancellation may be needed ultimately for symmetrical or very high bit rates.

Figure 14-1 repeats the basics of the VDSL architecture. With one obvious exception, the figure could easily be detailing ADSL instead, but the presence of the *optical network unit* (ONU) shows that this is not the ADSL world at all. VDSL requires the use of a fiber feeder system of one form or another. Note that if the ONU were a *digital loop carrier* (DLC), plain and simple, the figure would simply be a variation on the ADSL theme, one that was mentioned previously in the ADSL migration scenarios. However, the ONU is not just a DLC. It is a special DLC, one that must employ fiber-optic cable to the local exchange. And this use of fiber takes the fiber closer to the home than other architectures—an important point. Of course, if the fiber were in a ring configuration, the ONU could also potentially become a *next generation DLC* (NGDLC), but for the purposes of this discussion, the ONU is similar to but not quite a NGDLC. In fact, a SONET/SDH ring can be used to branch off fibers to feed a VDSL ONU closer to the customer than the ring. The continued presence of the splitter is to support previously existing analog voice in the newly installed ONU/DLC world. Certainly nothing in VDSL mandates the presence of the analog voice service if digitization has already occurred.

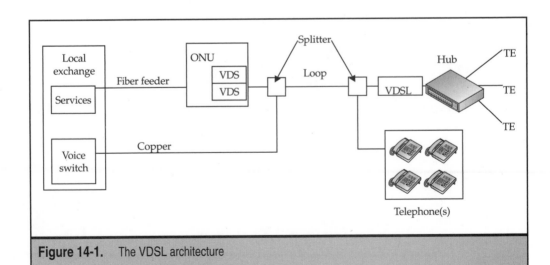

Figure 14-1.   The VDSL architecture

Two issues about VDSL should be addressed: First, how can VDSL offer such high speeds when other DSLs seem to struggle to achieve 1.5 Mbps and (with Asymmetric DSL [ADSL]) a mere fraction of that speed upstream? Second, is all of this blazing speed and its associated expense necessary for faster Internet access?

With regard to the first issue, VDSL is intended for the widespread fiber feeders or DLC systems of the near future. In other words, VDSL starts out by assuming that there will be fewer and fewer straight runs of twisted-pair copper from local exchange to customer premises, and that fiber-based DLC systems will become more prevalent—these are not bad assumptions. A sharp increase in the number of fiber-based DLCs is expected due to the combined effects of decreasing fiber costs and increasing condominium, apartment complex, and housing development construction in many parts of the United States and around the world. VDSL documents often refer to both *multidwelling units* (MDUs; usually residential) and *multitenant units* (MTUs; usually businesses) as the targets for VDSL services. With the current carrier serving area (CSA) rules limiting local loop runs and gauge changes, the only alternative to a world with many DLCs serving MDUs and MTUs is a world with many small local exchanges every few miles or so. And so VDSL anticipates no more than about 6,500 feet (2,000 meters) of copper on a loop, with the rest of the loop back to the central office being fiber.

Four main standards organizations are hard at work (along with the DSL Forum, of course) to create stable and interoperable versions of VDSL. These organizations are ETSI in Europe, ANSI committee T1 in the United States, and of course the ITU-T, which tends to wait until many aspects of a new technology are thrashed out by the other groups mentioned here. The fourth group is an organization known as FSAN, the *Full Service Access Network* group. This initiative was founded by leading telephone companies in Europe and the United States to create and define standards for broadband digital services delivered over both all-fiber and fiber–copper hybrid networks, as described earlier in this section. FSAN envisions the use of VDSL with passive optical network units which do not require remote electrical power feeds, a very desirable feature, of course. These *passive optical networks* (PONs) are not required to use VDSL, but it does make sense to use VDSL in an FSAN using a PON (in VDSL, as in other DSLs, the use of multiple acronyms is unfortunate but unavoidable).

Related standards work on the video aspects of VDSL and other broadband technologies has been done by the *Digital Audio Video Council* (DAVIC), based in Switzerland and now active only through its Web site at www.davic.org; the Full Service VDSL (FS-VDSL) Committee at www.fs-vdsl.net, which concentrates on using VDSL to deliver the video services defined by DAVIC; the VDSL Alliance, which promotes the use of DMT and related line codes for VDSL at www.vdslalliance.org; and the VDSL Coalition, which champions the use of versions of CAP/QAM as a VDSL line code (www.vdsl.org).

At present, there is a real difference in the line speeds envisioned for VDSL by ETSI and ANSI. ETSI has defined four asymmetric speed categories, called Class I, and five symmetrical speed categories, called Class II. For each category, there is also a best and worst case *reach* (the length of copper at the end of the fiber feeder) established depending on wire gauge, other impairments, and the power allowed on the loop (which might be limited if the cable bundle has to support other non-VDSL services or not). The seven speeds defined by ETSI, along with maximum (best case) and minimum (worst case) reach in meters, are shown in Table 14-1. Note how the reach increases as the speeds decrease. To go farther, go slower.

ANSI's VDSL services are also divided between asymmetrical and symmetrical, but there are eight asymmetrical categories and seven symmetrical categories, all distinct from the ETSI categories. In contrast to ETSI, the main distinguishing feature of these categories is the loop length, which is either long, medium, or short. If this sounds confusing, ANSI's plan should be clearer as presented in Table 14-2. Note the same "go farther, go slower" general aspects of this table, although at the same reach, a lower speed would result from actual physical line conditions.

Regarding the second issue—the purpose and use of these appreciable VDSL speeds—the intent of VDSL is not for mere Internet access, although pure high-speed Internet access is certainly possible with VDSL. VDSL is really intended for widespread asynchronous transfer mode (ATM)/Broadband ISDN (B-ISDN) service deployment. Whereas ADSL allows ATM to be used to provide services, for example, VDSL more or less expects ATM to be used for this purpose. These broadband services would include a wide range of multimedia and video services, including broadcast TV services. The video services envisioned for VDSL include multiple MPEG-encoded video channels, which currently require about 6 to 8 Mbps for broadcast quality video channels. In order to compete with cable TV, at least two independent channels will be needed for VDSL, making the minimum bandwidth required for VDSL video services about 13 Mbps downstream and about 1 Mbps upstream.

Also, VDSL is intended for widespread *small office/home office* (SOHO) premises networks. As housing spreads into the "exurbs" and formerly rural areas, the commuting pressure puts added stress on the highway infrastructure and adds to concerns about air

VDSL Service Type	Downstream Speed (Mbps)	Upstream Speed (Mbps)	Best/Worst Case Reach
Asymmetric (A4)	23.268	4.096	995/453 m (3,264/1,486 ft)
Asymmetric (A3)	14.464	3.072	1,344/729 m (4,408/2,391 ft)
Asymmetric (A2)	8.576	2.048 (E1)	1,691/789 m (5,546/2,588 ft)
Asymmetric (A1)	6.4	2.048 (E1)	1,791/843 m (5,875/2,765 ft)
Symmetric (S5)	28.288	28.288	298/212 m (977/695 ft)
Symmetric (S4)	23.168	23.168	397/261 m (1,302/856 ft)
Symmetric (S3)	14.464	14.464	845/575 m (2,771/1,886 ft)
Symmetric (S2)	8.576	8.576	1,294/820 m (4,244/2,690 ft)
Symmetric (S1)	6.4	6.4	1,444/876 m (4,736/2,873 ft)

Table 14-1.   VDSL Speeds and Reach Defined by ETSI

VDSL Service Type	Downstream Speed (Mbps)	Upstream Speed (Mbps)	Reach
Asymmetric Short	52	6.4	300 m (984 ft)
Asymmetric Short	38.2	4.3	300 m (984 ft)
Asymmetric Short	34	34	300 m (984 ft)
Asymmetric Medium	26	3.2	1,000 m (3,280 ft)
Asymmetric Medium	19	2.3	1,000 m (3,280 ft)
Asymmetric Long	13	1.6	1,500 m (4,920 ft)
Asymmetric Long	6.5	1.6	2,000 m (6,560 ft)
Asymmetric Long	6.5	0.8	2,000 m (6,560 ft)
Symmetric Short	34	34	300 m (984 ft)
Symmetric Short	26	26	300 m (984 ft)
Symmetric Short	19	19	300 m (984 ft)
Symmetric Medium	13	13	1,000 m (3,280 ft)
Symmetric Long	6.5	6.5	1,500 m (4,920 ft)
Symmetric Long	4.3	4.3	1,500 m (4,920 ft)
Symmetric Long	2.3	2.3	1,500 m (4,920 ft)

Table 14-2.   VDSL Speeds and Reach Defined by ANSI

pollution, energy consumption, and so on. Perhaps the best way to deal with this issue is to tell everyone to stay home, but still go to work. VDSL makes a lot of sense for business or office parks and college campuses as well. These places tend to have a lot of copper around and some fiber. If VDSL could be used to make maximum use of mixed fiber/copper environments, so much the better. And within an office park or on a college campus, runs tend to be much shorter than local loops.

All in all, VDSL is designed and intended for the network of the next 10–20 years in many countries around the world. Oddly, countries with less-developed copper infrastructures than in the United States may turn out to enjoy an edge when it comes to VDSL due to the higher percentage of fiber in their feeder systems.

VDSL AND ATM

VDSL is intended to take advantage of the presence of fiber in modern DLC systems, the need for more and more bandwidth for "converged" residential services with both Internet access and video, and the promise of ATM cell transport as a unifying transmission method for mixed-media broadband services. The whole package is neatly tied together.

What keeps it from falling apart? Each component is a vital part of the whole. If one is weak, it weakens the whole basis of VDSL.

On first glance, the "weak link" in this plan may seem to be asynchronous transfer mode. ATM has struggled to gain acceptance in all but a few well-defined markets, and depending on ATM to popularize VDSL might seem like a mistake. This is not all that VDSL offers, however.

VDSL documentation defines five main transport modes across the basic copper/fiber VDSL architecture. Most of these closely mirror their ADSL cousins. These VDSL transport modes are shown in Figure 14-2.

In its simplest form, VDSL offers *synchronous transfer mode* (STM), or what most people would just call *time division multiplexing* (TDM). This is the same type of "bit pipe" that bit sync mode in ADSL supports. Here, bit streams to and from different services and devices are assigned fixed bandwidth channels for the duration of the interaction.

VDSL also supports packet mode. In this transport mode, all bit streams to and from different services and devices are organized into individually addressed packets of varying lengths. All packets are sent on the same "channel" of maximum bandwidth. Naturally, the packets would be expected to be IP packets, but IP packets are not tied to VDSL in any way.

Additionally, VDSL supports ATM mode. In fact, ATM can be used in three distinct ways in a VDSL network. ATM mode is similar to packet mode in the sense that each bit unit is individually addressed and sent on the unchannelized line, but in this case the packets form small, fixed-length units called *cells* rather than variable length packets. ATM can be used to access service network in combination with STM and packet mode VDSL. At its logical limit, VDSL forms an end-to-end transport between ATM devices.

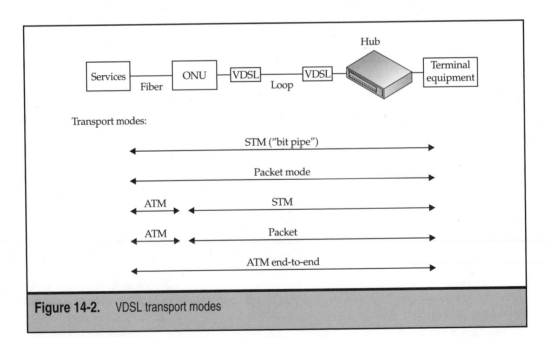

Figure 14-2. VDSL transport modes

So VDSL is also allowed to function with ATM at the service side to the ONU and STM (that is, TDM) on the loop, which makes sense because service providers are more likely to employ ATM switches and servers than users are. In fact, ATM is an ideal solution here, where mixing voice, video, and data services is common; and the ONU can take care of any conversion chores. A combination of ATM services to the ONU and packets on the loop is also supported; this combination takes advantage of the widespread use of IP packets on most types of distribution networks.

Please note that most implementers of early VDSL pilot systems intend to focus on the ATM end-to-end transport mode, which seems rather astonishing at first. However, the more mixing of services and types of traffic there is, the more ATM makes sense, both technically and financially, and VDSL is certainly intended for this mixture of traffic types.

VDSL/ADSL DOWNSTREAM SPEEDS AND DISTANCES

ADSL is designed for a world where most analog local loops consist of simple "home-run" pairs of copper wires extending from local exchange to customer premises. A large percentage of this local loop plant, however, is already on some form of feeder system, such as a DLC, and DLCs will be even more common in the future. And more and more often, the DLC feeder portion of the loop is carried on fiber-optic cable. This is most likely SONET/SDH, but not necessarily in a ring configuration. Figure 14-3 shows how the speeds and distances for typical ADSL and VDSL loop lengths complement each other.

Sometimes there is debate about carrying the fiber all the way to the home (called *fiber to the home*, or FTTH). It is a little known fact that this is supposed to be called *fiber to the basement*

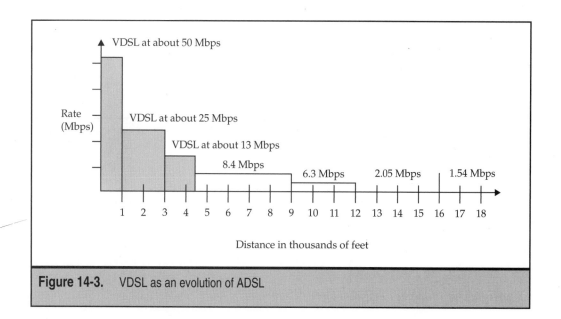

Figure 14-3.   VDSL as an evolution of ADSL

(FTTB) when a small office building is involved. Anyway, this premises deployment of fiber to all homes and businesses is probably too expensive and not absolutely necessary, and so there are variations for carrying the *fiber to the curb* (FTTC) and fiber to the neighborhood (FTTN, about 10 to 100 homes). The whole point is that VDSL is designed for FTTN, but nothing rules out the use of VDSL for other forms of fiber feeder networks.

In fact, it could be said that VDSL takes over where ADSL tops out (this is a generalization, of course: some VDSL speeds are lower than ADSL). The figure compares some classes of VDSL speeds and distances with those of ADSL. Note that VDSL may be used in conjunction with ADSL, depending on twisted-pair distance. Various standards groups have suggested that a combination of VDSL and ADSL offers the best opportunity to provide access to almost any PC server-based application, with interactive TV services on the same system.

VDSL IN DETAIL

VDSL is a challenging technology to implement. ADSL increased the useful bandwidth on the copper local loop from the usual 4 KHz used for analog voice to about 1 MHz. VDSL will in most cases extend this useful bandwidth on shorter copper runs to 10 MHz and even 12 MHz in some specifications. Working higher into normal radio frequency ranges means that many VDSL bands must be shut down to avoid interference with police and fire communications, because VDSL aerial loops are very effective transmitters as well as antennas at those frequencies. And once inside the premises, residential *Home Phone Network Architecture* (HomePNA) devices use the 5 to 10 MHz bandwidth on the same inside premises wiring that VDSL would ordinarily use to reach the VDSL modem near the PC or TV. All VDSL proponents freely admit that these HomePNA "squatters," although not sanctioned to use these ranges by any standards body, are not going to go away and must be acknowledged and dealt with in VDSL. In other words, it is a given that HomePNA and VDSL use will conflict. Customers might have to choose between or the other, or accept very restricted VDSL speeds and services.

Moreover, because VDSL is intended to be an international standard, VDSL frequency use in various countries must be adjusted to respect the police, fire, and other bands in use in those countries for public services. So a VDSL modem made to operate with, but avoid, the police and fire communications bands in the United States would not be appropriate for use in another country with different bands in use for these public services. These might make VDSL modems much more expensive to build and deploy than their ADSL modem cousins. An example of a VDSL frequency plan is shown in Figure 14-4.

Although VDSL is intended for use in a mixed fiber and copper environment, VDSL specifications cover only the copper portion of the entire path from service provider to customer. VDSL makes a lot of sense in a world in transition from an all-copper distribution network to an all-fiber distribution network. The asymmetric and symmetric aspects of VDSL service speeds offer additional flexibility for a wide range of both business and residential applications.

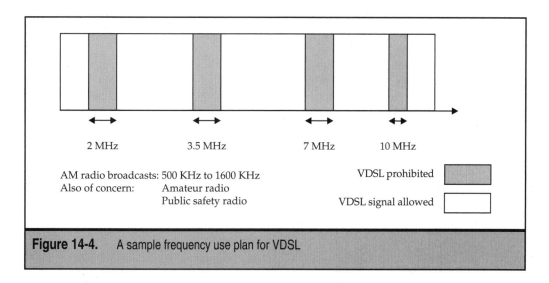

2 MHz   3.5 MHz   7 MHz   10 MHz

AM radio broadcasts: 500 KHz to 1600 KHz   VDSL prohibited
Also of concern: Amateur radio
                          Public safety radio   VDSL signal allowed

Figure 14-4.   A sample frequency use plan for VDSL

Another feature of VDSL that makes it unique and differentiates VDSL from simple ADSL service offerings is the fact that the higher bandwidths available with VDSL make VDSL attractive for a bundle of services rather than just higher speed Internet access. Video is the most common added service offered through VDSL service trials. Multiple digitally encoded video streams can be delivered at the same time that VDSL supports high speed Internet access and multiple digital voice channels (just another form of VoDSL) as well. For business environments, VDSL can support and extend the reach of multiple 10 Mbps Ethernet LANs, at least in some symmetrical offerings where the cable lengths are appropriate.

VDSL competes directly with FTTH schemes and various wireless local loop technologies. The success of VDSL will depend on these related factors:

▼   The willingness on the part of incumbent LEC (ILEC) companies, which control the local loop, to use existing copper for new services rather than to run fiber directly to the customer premises.

■   The existence of viable and financially secure competitive LECs (CLECs) to purchase and use unbundled local loops to reach their customers rather than trying to bypass the ILEC in some way.

▲   The overall cost/benefit tradeoff of using VDSL in a "fiber to the cabinet" scheme rather than some non-VDSL method to reach the customer from the fiber feeder cabinet.

So far, only the second item on this list has been called into question. And this uncertainty is not based on ILECs using something other than VDSL on unbundled loops, but the continued viability of the ILEC industry offering other than simple voice services at all. There will always be a small pool of dissatisfied ILEC customers willing to use any CLEC they can find for voice services, but whether this same pool of customers can justify the continued existence of "data CLECs" for Internet access services is another matter altogether. But even without this CLEC support for VDSL, the ILEC must still compete with cable TV and wireless local loop services, and VDSL seems to offer a very cost-effective way to do so.

The cable TV competition is one reason that ILECs have concentrated on trials and early service offerings using VDSL for residential communities. Business service overlays are possible with VDSL, of course, especially with relaxed rules about where SOHO business units may operate. The key to VDSL is that installation costs for services that use at least some existing copper are always lower than installation costs for services based on new fiber everywhere a service is needed. Think of VDSL as the ILEC equivalent of the cable TV companies' push for hybrid fiber-coax (HFC) architectures, but this time as hybrid fiber-copper. And as digital compression techniques continue to improve, the bandwidth limitations of VDSL versus all-fiber solutions will become less and less of a factor in choosing a broadband access technology.

One more factor should never be forgotten when assessing the viability of VDSL with regard to cable TV HFC and wireless local loop solutions such as LMDS. VDSL, like all other DSLs, is a simple point-to-point technology, not a bus architecture technology like HFC. So there is some measure of security already present in VDSL at the physical layer that is absent in the bus technologies. Of course, this does not imply that security at the higher layers of the architecture is not a concern with VDSL. It is. But VDSL starts with an inherent security absent in bus technologies.

CURRENT STATUS OF VDSL SPECIFICATIONS

Mention has already been made of the standards organizations involved in making VDSL a stable and interoperable standard technology for use around the world. This section details some of the current work in progress on VDSL by each organization.

ANSI

Work on VDSL is being done by the T1 Committee. The ANSI VDSL specification for use in North America is currently divided into three parts:

▼ Part 1 defines the overall VDSL functional requirements and specifications. These functions are independent of the line code used for VDSL. VDSL line code proposals mirror the types of line codes used in ADSL. So there are *multicarrier modulation* (MCM) line code proposals for DSL similar to DMT and using *frequency division duplexing* (FDD), not echo cancellation, to establish separate frequency ranges for downstream and upstream data flows. But there are also *single-carrier modulation* (SCM) line codes proposed for VDSL, similar to CAP/QAM. So the battle between DMT and CAP/QAM, won by DMT in ADSL, is now being fought all over again with VDSL.

■ Part 2 is the technical specification for a VDSL modem (transceiver) using an SCM line code such as CAP/QAM.

▲ Part 3 is the technical specification for a VDSL modem (transceiver) using an MCM line code such as DMT. There is an appendix covering a related MCM line code known as *filtered multitone* (FMT).

There is no need to be concerned over the presence of two types of line code for VDSL. All this means is that at present, neither SCM nor MCM line codes have been shown to be superior in each and every aspect of their operation. So in the interest of speeding up VDSL standardization, both types of line codes may be used. ADSL did quite well with two types for a long time. There were even dual-mode ADSL modems when interoperability was an issue. And there are many more compatibility issues among DSL equipment vendors than just the line code used. Dual-mode capability is not required for VDSL modems, of course. And even though VDSL modems using different VDSL line codes will not interoperate, the use of a common frequency plan will make sure that VDSL equipment using different line codes can coexist on the same physical cable plant.

All of the line codes in the ANSI specification use simple FDD and a common plan for frequency use (to avoid interference with other services). The use of the 25 KHz to 138 KHz range is optional in the ANSI specification for VDSL. There is considerable debate about the need to reduce power in the upstream direction, similar in function to the power back-off performed by HDSL2. In VDSL, this is known as *upstream power back off* (UPBO) and intended to minimize VDSL interference on the premises.

ETSI

Work on VDSL for use in Europe is being done by ETSI, and is currently divided into two parts:

▼ Part 1 is the functional requirements for all VDSL equipment, regardless of line code used.

▲ Part 2 is the transceiver (VDSL modem) specification, including the line codes.

Like the ANSI version of VDSL, ETSI envisions both SCM and MCM line codes for VDSL. Both will use FDD and a common frequency use plan. Initial spectrum use is from 138 KHz to 12 MHz, but there are plans to explore the use of frequencies above and below these limits. There are issues regarding not only UPBO, as in ANSI, but also the use of additional frequency use plans, given the extremely diverse European environment that VDSL must work in.

ITU-T

The ITU-T is also working on VDSL. The intent, as usual with the ITU-T, is to develop an interoperable international standard for VDSL. However, there has been little progress to date on "G.vdsl" given the controversies about issues as fundamental as line codes. However, it seems likely that both MCM and SCM line codes must be supported, at least initially. FDD will be used, perhaps with the 25 KHz to 138 KHz range as well. UPBO will be supported, and the general guidelines have been established. Many of the details about UPBO operation remain, though.

In addition to these standards organizations, several regulatory bodies have been keeping a close eye on VDSL. This scrutiny is necessary because VDSL goes into frequencies on the local loop between 1 MHz and 10 (or even 12) MHz where no service has gone before.

For example, the European Community (EC) will establish uniform requirements for radio disturbances due to VDSL use. Not only will the EC fix limits of this radio disturbance, but the EC will describe the methods to be used to measure the VDSL disturbance and to standardize the operating conditions for VDSL and for the interpretation of test results. In addition, two countries, the UK and Germany, concerned that the EC might not go far enough in dealing with potential radio interference with VDSL, have proposed additional regulatory restraints on VDSL.

In the United States, the FCC is mainly concerned with the effects of the high VDSL frequencies on nearby telephony services.

VDSL SERVICE CHARACTERISTICS

Regardless of the details of VDSL operation, all versions of VDSL require that the *bit error rate* (BER) be better than 1 bit error for every 10 million bits (10^{-7} BER). VDSL modems are essentially "error free" with this bit error rate, although performance details vary based on service type. Data services, such as IP-based Internet access, can resend errored packets and segments, and video signals should not be adversely affected by this bit error rate.

VDSL defines two types of delays, or latency parameters. With single latency mode, all VDSL system payloads flow on one channel. Single latency mode provides for programmable burst error protection, such as provided by using a *forward error correction* (FEC) scheme. Delay is then completely determined by the data rate provided and the added delay for FEC processing. In dual latency mode, there are two VDSL channels, not one. There is a "low delay" fast channel (similar in concept to the fast data buffer in ADSL) and a "higher delay" slow channel (similar in concept to the interleaved data buffer in ADSL). The slow channel adds a FEC, as described in the single latency mode. The details, such as the bandwidth on the VDSL link allocated to the low and higher delay channels, are left up to the equipment vendor's network management method.

A concern in VDSL is the presence of other types of service nearby, perhaps even within the same 50 pair cable binder or bundle. VDSL specifications are expected to minimize interference with and from adjacent services such as ISDN or analog voice. Work on VDSL in this regard is complicated by the fact that VDSL might have different spectral characteristics on the local loop, within the DLC cabinet, and on the feeder back to the central office. VDSL proper is concerned only with the transmission on the copper portion of the local loop, of course.

The intention of VDSL specifications with regard to adjacent wire pairs is to allow both symmetrical and asymmetrical versions of VDSL to operate within the same cable binder, as long as they use the same frequency ranges. As long as these frequency ranges are respected, no special care need be taken with regard to wire pair selection. In addition, VDSL is expected to allow operation of analog voice and ISDN on the same wire pair as VDSL. A splitter is used to separate the services, of course.

MIGRATION TO VDSL

VDSL documentation from the DSL Forum envisions two major types of VDSL service deployments: *Green Field* and *existing plant*. Green Field deployments connect new customers without any previous telecommunications service to broadband services provided by VDSL. A typical Green Field VDSL installation would be for a new housing development, strip mall, or office park. Existing plant deployments basically evolve analog voice, ISDN, or even ADSL services to VDSL. The concerns for each type of VDSL service deployment are slightly different and deserve some mention.

Green Field deployments take into account the fact that in the United States and many other places around the world, new homes and business offices are built far from city centers and parallel major road systems. These sites are often beyond the reach of ADSL using a central office DSLAM, and placing ADSL DSLAMs into remote DLC cabinets might prove to be self-defeating, costly, or just impossible. So it might be better to use NGDLC or a PON with a fiber feeder and VDSL to reach these customers. Not only are new housing units and business parks appropriate for Green Field VDSL, but high-rise residential and business units and extensions to campus networks are good candidates as well.

Now, this could also be an opportunity for ADSL, but in many cases the higher speeds of VDSL might be a better and more cost effective alternative to ADSL. For about the same cost, VDSL offers much more bandwidth than ADSL, the promise of additional services, and no need to "evolve" the access network to anything else beyond VDSL. In order for this attractive VDSL strategy to work, however, the "VTU-R" or VDSL modem must be able to support the same types of service (fast Internet access) as ADSL, and the same types of premises equipment and technologies (Ethernet) as ADSL. In fact, VDSL can be deployed in this Green Field to look and feel very much like ADSL or SDSL, if the VDSL modem supports the proper modules and the fiber feeder has access to the proper types of DSLAMs and networks. Distinctly VDSL services (video) can be added at a later time. In places where competition from all sorts of services from cable TV operators is strong, Green Field VDSL can give the ILEC a way to compete. In the future, the fiber could be extended right to the premises, once a secure customer base (and revenue stream) is established.

To sum up, here is why service providers should consider VDSL instead of ADSL for Green Field deployments:

▼ DSL Forum guidelines can be used to support ADSL modems and ADSL speeds on VDSL, even if full VDSL speeds and services are not initially offered.

■ Copper loop lengths from feeder cabinet to customer are typically 1,500 feet (about 500 meters) or less in most new construction. Yet ADSL cannot run faster than 6–8 Mbps downstream on even the shortest loops.

■ VDSL can offer all the bandwidth needed for even the newest services, and even VDSL speeds can evolve as equipment such as VDSL modems evolve as well, so pressure from cable TV competitors is reduced.

▲ There are fewer compatibility and crosstalk concerns with older, existing broadband services.

However, at least three concerns apply even for Green Field VDSL deployments:

▼ The issue of interference with radio transmissions

■ The availability of standard and stable VDSL equipment

▲ The interoperability of VDSL modems from different vendors with the VDSL modem in the remote cabinet

For VDSL deployments on the existing cable plant, the major difference is that customers already have access to a number of different types of broadband access services. There must be strong incentives, both for the service provider and for the customer, to deploy and use VDSL instead of something else. One major difference from Green Field VDSL deployments is the need to seek permission from authorities to run fiber to remote cabinets with only copper feeders in order to upgrade these existing cabinets for VDSL. Many authorities have imposed a moratorium on trenching of any kind for this type of "upgrade" work. For Green Field VDSL, this permission is assumed, or no services are available at the new locations at all.

There is also more of a need for VDSL to be compatible with existing services such as ISDN, analog voice, or ADSL. As before, it is possible for a single VDSL modem to be compatible with existing ADSL services so that VDSL services can be added in the future, but this is not a requirement.

In summary, here is why service providers should consider VDSL instead of ADSL for existing plant deployments:

▼ VDSL can allow support for ADSL services, if DSL Forum guidelines are followed.

■ VDSL can be used with analog voice or ISDN on the same loop.

▲ A VDSL modem can support the same types of premises equipment (PCs) and networks (Ethernet) as ADSL.

There are more concerns with existing plant VDSL deployments than in Green Field VDSL, however. VDSL standards and interoperability are more of an issue than with ADSL, given the newness of VDSL. Radio interference is a more acute concern, because other services pre-exist to a site, by definition. The time and permissions required to deploy remote cabinets with VDSL is also a major concern.

Typically, a VDSL deployment scenario would proceed as follows. First, ADSL will be used to service customers close to city center. As ADSL deployment moves to remote DSLAMs with fiber feeders, the remote cabinets will serve fewer customers and be closer to those customers. At some point, there will be ample opportunities for Green Field VDSL. If service offerings exceed the ADSL capabilities, existing plant VDSL will be the answer, as long as the loop lengths allow.

VDSL SERVICES

The DSL Forum has identified two major VDSL service types. These are asymmetrical services primarily aimed at residential customers, and symmetrical services primarily aimed at small- to medium-sized businesses (what the DSL Forum calls SMEs: *small/medium enterprises*). The main difference between the type of service is the emphasis on video services in the residential package. The rates set for the services in the residential package assume three independent and simultaneous digital video (and audio, of course) channels using at least 5 Mbps per channel. The services for small- and medium-sized businesses assume symmetric transport over VDSL. The asymmetrical residential service speeds are shown in Table 14-3.

The downstream speeds listed are greater than the minimum required, and the upstream speeds listed are less than the maximum required to allow the service to work properly. So online radio, or listening to broadcast radio stations over VDSL, requires more than 1 Mbps downstream, but less than 100 Kbps upstream. Note how the presence

Service	Downstream Speed	Upstream Speed	Guarantee
Video services (3 channels assumed)			
Switched digital video	>15 Mbps	<200 Kbps	Yes
Video on demand	>15 Mbps	<200 Kbps	Yes
Near video on demand	>15 Mbps	<200 Kbps	Yes
Audio services			
Audio on demand	>1 Mbps	<100 Kbps	Yes
Online radio	>1 Mbps	<100 Kbps	Yes
Internet/corporate intranet access			
Multimedia downloads	>10 Mbps	<100 Kbps	Best effort
Downloading applications	>10 Mbps	<100 Kbps	Best effort
Virtual reality (games)	>10 Mbps	<1 Mbps	Best effort
Shopping online	>10 Mbps	<100 Kbps	Best effort
Web server on-site, or Web hosting	>400 Kbps	>2 Mbps	Best effort
Voice services (derived voice)			
VoDSL with ATM (less than four channels)	<32 Kbps	<32 Kbps	Yes
VoIP (less than four channels)	<32 Kbps	<32 Kbps	Best effort

Table 14-3.   VDSL Asymmetric Service Speeds for Residential Services

of a Web server on-site, or a Web hosting service, effectively reverses the upstream and downstream speed requirements. The voice services require less than 32 Kbps, and can use ATM or IP to deliver the digitized voice packets.

There is one column in the table that deserves further explanation. The column represents whether the service has any guarantee with regard to quality of service parameters such as bandwidth or delay, or whether the service is a simple "best effort" on the part of the IP protocols and routers to get information to the customer and back. When VoDSL with ATM is used, the underlying ATM network provides the voice with a service guarantee. This is not possible when pure VoIP is used.

The three video service offerings are related, but slightly different. Switched digital video services use content provided by the DSL service provider, usually all of the channels that a typical cable TV service provider would broadcast. Video on demand is usually limited to movies that start when and only when the customer is ready to view them, but there are no firm definitions of either term. Near video on demand (also called "stagger cast" in the VDSL documents from the DSL Forum) offers movies starting not exactly when the customer likes, but perhaps every 15 minutes or so—near enough to when the customer wants.

Table 14-4 shows the symmetric services established by the DSL Forum for small and medium businesses using VDSL. This table may be contrasted with the residential services table.

In this case, the derived voice is assumed to be 64 Kbps digital voice, and there are more than 16 channels (simultaneous voice calls). The video service is for interoffice

Service	Downstream Speed	Upstream Speed	Guarantee
Interoffice communications			
Derived voice (>16 channels @ 64 Kbps)	<2 Mbps	<2 Mbps	Yes
Video conferencing (high quality)	<8 Mbps	<8 Mbps	Yes
Internet/corporate intranet access			
Large file transfers	>10 Mbps	>10 Mbps	Best effort
Downloading applications	>10 Mbps	<2 Mbps	Best effort
Access to virtual reality Web sites	>10 Mbps	<2 Mbps	Best effort
Webcast media hosting	<2 Mbps	>10 Mbps	Best effort
Web hosting	<2 Mbps	>10 Mbps	Best effort
Remote learning	>10 Mbps	<2 Mbps	Best effort

Table 14-4.   VDSL Symmetric Service Speeds for Small/Medium Business Services

videoconferencing and high quality digital video at Ethernet speed or higher. Both of these services have quality of service guarantees, and the rest of the Internet and intranet services are best effort IP-based.

Webcasting is the practice of using a Web site and the Internet to distribute information, with or without multicast, to Web browsers that wish to view the content (such as an online fashion show or music concert). Again, note how the presence of a Web server reverses the upstream and downstream speed requirements. And because these are service speed requirements, not transport speed requirements, the asymmetrical nature of the application or service is listed in the table. VDSL for business is always symmetrical, so the higher bandwidth requirement will determine the speed of the VDSL link in both directions.

For business services, VDSL adds a service protection feature. This means that there should be redundant links and devices in place to guard against failure of any individual component of the service. This type of protection is absent in residential VDSL services. Neither table implies that all VDSL service providers will support all of these services.

VDSL DEPLOYMENT

Ideally, high bandwidth, or broadband, services would be delivered to customers directly on fiber all the way to the customer premises. Today, it makes economic sense to deploy fiber to a feeder close to the customer, but not all the way to the customer. This is mainly due to the fact that because the customer (in almost every case) has no equipment on the premises that can support a direct fiber interface, some optical-to-electrical conversion is necessary between the feeder and the premises. The current conversion devices are very expensive, and generally much too fragile to expose to the elements when mounted on the side of a building. So it makes sense to group these converters together and shelter them inside the remote feeder cabinet. When the "last mile" (or "last 1,000 feet") to the customer premises is provided on copper, it makes sense to deploy VDSL.

For asymmetrical residential services, where the highest bandwidth needs are about 15 Mbps downstream and 1 Mbps upstream, the reach of VDSL is about 1 km (3,280 feet). For symmetrical business services, where the highest bandwidth needs are about 10 Mbps downstream and upstream (the common Ethernet LAN speed), the reach of VDSL is about 0.7 km (2,296 feet). Rate and reach specifics depend on the actual service mix and line conditions.

Just as ADSL, VDSL may support analog voice in the 4 KHz frequency range with a splitter. However, VDSL splitters will be very different than their ADSL cousins due to the higher frequencies used in VDSL (about 10 MHz in VDSL as opposed to about 1 MHz in ADSL). When ADSL is already used to service a customer with a home run local loop to a central office DSLAM, migration to VDSL is complicated by the need to install and provision VDSL in a remote cabinet, and perhaps change the central office DSLAM and customer's DSL modem.

VDSL can also be used in a business campus environment with symmetrical services. A fiber feeder cabinet can be installed in a central location, and new or existing copper can be used to provide essentially 10 Mbps Ethernet LAN speeds between endpoints located 1.4 km (4,592 feet) apart, because the fiber feeder supports a 0.7 km radial distance. If the copper is already in place, no major rewiring of the campus is needed. New end equipment is needed, but this new equipment would be needed no matter what high speed solution was chosen for the campus. If all that is needed is on-campus connectivity, all that is needed is the VDSL "hub." A fiber feeder link can be added to a VDSL service provider when appropriate (or available). VDSL has the further advantage of being a standard solution.

The business environments that are appropriate for VDSL include hospitals and medical centers, company or college campuses, office parks, office towers with different companies on every floor, and factories. In multitenant environments, VDSL might service both asymmetrical residential units and business units in an adjacent strip mall (for example). In this scenario, a SONET/SDH ring feeds various SONET/SDH units around the ring. Point-to-point fibers branch out to either individual buildings or NGDLC for VDSL services. In the case where fiber runs to the building, the VTU-O (VDSL Termination Units – Optical) is located in the basement. For NGDLC scenarios, the VTU-O is located with the analog voice interfaces in the DLC cabinet. These possibilities are shown in Figure 14-5.

The major difference between the MTU (mostly business) and MDU (mostly residential) environments is that MTUs require symmetrical speeds from 1.5 Mbps to more than

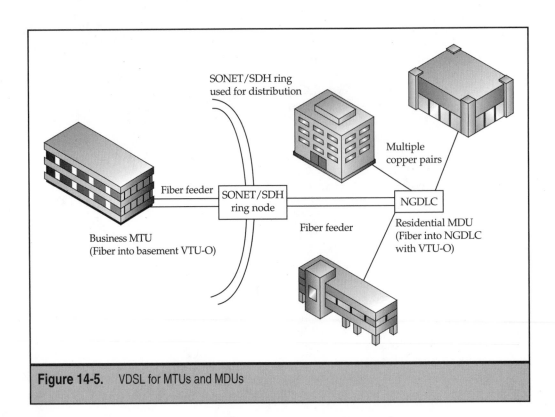

Figure 14-5. VDSL for MTUs and MDUs

10 Mbps. There are not only service quality guarantees but service protection offered as well. For MDUs, asymmetric rates can be higher than 15 Mbps downstream to 1 Mbps upstream. Voice, video, and data support is necessary to compete with cable TV offerings, but only the voice and video require service quality guarantees. There is no service protection offered for these services.

The places where mixed MTU and MDU VDSL would be found are places where a small distance separates high-rise commercial office buildings from high-rise apartment towers, condominium and townhouse complexes intermixed with strip malls, medical centers and clinics near small office parks, and college dormitories near hotels and motels. None of these situations are particularly uncommon.

VDSL is normally a point-to-point technology. However, even in symmetrical service offerings, there is always a distinct central component (VTU-C) and remote component (VTU-R) to VDSL. This is because VDSL uses FDD to isolate upstream and downstream frequency ranges, the risk of crosstalk precludes the co-location of VTU-Cs and VTU-Rs in close proximity at the end of a shared cable. Therefore, the most common VDSL architecture will be a star network with VTU-Cs required to be at the hub (usually in the remote cabinet, not the central office) and VTU-Rs at the spokes. Where VDSL is used in a less structured point-to-point environment, care must to taken to make sure that VTU-Cs and VTU-Rs are sufficiently isolated from each other to allow the links to operate properly. In a partial or dense mesh of point-to-point links, this is especially important and requires careful network engineering and planning.

VDSL ISSUES

VDSL has not achieved anywhere near the maturity of ADSL, which is arguably immature itself. In fact, VDSL is still only in the definition stage. Few VDSL products even exist, and trials have only been underway for a few years. A lot of work needs to be done on local loop characteristics, noise effects, upstream multiplexing techniques, and other aspects of VDSL operation to establish a set of firm standard properties. The biggest and most important VDSL question mark involves the maximum distance that VDSL can operate reliably for a given bit rate. This may sound odd for a new technology such as VDSL, but it really reflects the unknowns of real local loops at VDSL frequencies. Furthermore, short bridged taps or unterminated extensions in homes that have no effect at all on analog voice, ISDN, or even ADSL, may have very disruptive effects on VDSL. Some early trials have made a practice of terminating even the shortest of bridged taps, and there may be no other way to allow VDSL to work properly.

One important issue with VDSL is that both ADSL and VDSL use the frequency range below 1 MHz. If ADSL and VDSL are both viable technologies is a given area, care must be taken to make sure that ADSL and VDSL do not interfere with each other on an unbundled local loop. Cynics might note that one solution to this problem appears to be in the struggles of data CLECs and ISPs to compete with the ILEC with DSL services. But regulation and a firm frequency plan will be needed in any case, even if the ILEC is the only DSL provider around.

VDSL also is susceptible to radio interference, just like ADSL. A long local loop acts as both an antenna and a transmitter at these frequencies. If VDSL signal levels are very low, the antenna effects inject amateur radio noise into the line, but VDSL reach is too short to be an effective technology. But once the signal levels are increased, the transmitter effects begin to dominate and interfere with the radio signals. Standards bodies have tried to strike a balance between inadequate and high VDSL power levels, and power back-off is an important feature in VDSL. However, as mentioned previously, the UK and Germany have added regulations for those countries intended to minimize any chance of VDSL interference. There will be both a standard way to measure the electromagnetic emissions of VDSL and a standard "intervention" threshold for taking action. The communities most concerned about VDSL are the AM radio broadcasters and civil aviation groups. Faced with the possibility of VDSL interference from homes and businesses close to airport runways, authorities might have no choice but to shut down VDSL systems even if the power used is within the technical limits of the specifications. This would have a chilling effect on VDSL deployment. No service providers would like to spend large amounts of capital to deploy a service in an area where it might be unlawful to use. But few would have it any other way when the choice might be between a plane crash or the availability VDSL video services. In these cases, ADSL might have to do for these areas, but in at least one country (UK), the level is set so low that even ADSL runs the risk of being shut down if interference complaints are filed.

Another related issue is the presence of untwisted wires on the drop between cable binder and home (common in the United States) and the possible presence of untwisted wire carrying VDSL signals on the premises. Even poorly run premises wiring that has lost some twist properties have been a concern. VDSL filtering can help in many cases on the premises, especially when the HomePNA frequency range between 5 and 10 MHz is also in use. Studies have shown that VDSL emissions increase on the untwisted drop wire leading to the premises, which in turn places pressure on the need to lower VDSL power levels everywhere. VDSL has even been shown to interfere with G.lite ADSL running next door with Home PNA, and vice versa. This effect is much worse with CAP/QAM VDSL than with DMT, and just might tip the balance in favor of DMT forms of VDSL. It all depends on the popularity of G.lite and HomePNA.

It would be a mistake to assume that shielding or bonding the cable, or simply burying it, will overcome any ingress or egress radio interference concerns about VDSL. Bonding to maintain system balance might be done poorly, and a few inches of earth have been proven not to do much to reduce VDSL interference. Radio frequencies often penetrate the ground easily, especially when it is wet.

Also, the AM broadcast and civil aviation groups have only been the most vocal communities of interest regarding VDSL. The spectrum between 1 and 20 MHz is used by many groups. Some are very important and include not only civil aviation, but military communications, diplomatic services, ship communications systems, and health services. All in all, VDSL interference issues are very important not only for VDSL, but all the services that are affected. Debate over the best way to handle these relationships between VDSL and other services will continue for quite some time.

Another VDSL issue involves the services that VDSL is designed to deliver. VDSL more or less needs ATM cell streams at some as-yet unspecified bit rates to carry information with firm delay and bandwidth guarantees. But what about non-ATM system using IP packet for everything? VDSL, like anything else at this low level, cannot be completely independent of upper layer protocols. This is particularly true in the upstream direction where multiplexing information from potentially many pieces of customer equipment may require VDSL to be aware of, and handle properly, a variety of protocols, not just ATM.

Another issue also is a concern in ADSL—the area of the premises distribution network and the interface between the telephony portion of the network and customer premises equipment itself. There are many considerations, including cost, that favor a passive network interface device with VDSL on the premises installed in a customer premises equipment package. In this form, the upstream multiplexing is handled very much like multiple access in LANs. However, other factors exist, including system management, reliability, regulatory constraints, and migration considerations that would favor an active VDSL network interface. The active VDSL interface could act just like a LAN hub and use either point-to-point or shared media distribution to multiple premises equipment devices over premises wiring. This wiring would be physically isolated and totally independent from the network wiring. But when all is said and done, the VDSL modem should essentially mirror the operation of the ADSL modem.

However, as with most things, cost may be the overwhelming concern. The fact is that only a few customers are serviced by small fiber feeder cabinet arrangements, at least compared to HFC cable TV systems. The common equipment costs in the feeder cabinet, such as fiber links, interfaces, and equipment cabinets, must therefore be spread over a relatively small number of customers. VDSL has a much lower target cost than ADSL because ADSL can be supplied directly from a wire center in a local exchange. In spite of the obvious benefits of VDSL, deployment will be totally driven by the cost and rate of return on the VDSL investment. Pricing VDSL services low to compete with cable TV is fine, but few service providers—or regulators—will approve a service that will not begin to make money after about three years. This break-even point might not be possible for VDSL (many ADSL deployments had a 5- to 10-year loss cycle before they were projected to make to make money even with impressive buy rates).

A VDSL SERVICE SAMPLE

What would VDSL serving a residential community actually look like? One ongoing trial in the Southwestern United States has begun to generate answers to the issues raised by VDSL. This section briefly describes this VDSL trial.

Users of the VDSL service have a choice of three independent television channels, drawn from the same approximately 160-channel pool as cable TV users in the same community. The video service is multicast *switched digital video* (SDV) using ATM directly from the DSL service provider (the ILEC) with FEC and so slightly higher delays. Fast Internet access is also included, as is support for analog voice with a remote splitter. The line code used is QAM, but mainly due to cost, availability, and maturity considerations rather than a belief that QAM is somehow better than DMT for VDSL.

This was a trial using existing pairs, but has had few compatibility issues. Feeder installation was simple and direct and had no overwhelming permission problems. Downstream speeds range between 20 and 26 Mbps, and upstream speeds between 2 and 4 Mbps. Distance has been a real concern, with crosstalk between VDSL pairs themselves limiting reach much more than anticipated. Any bridged taps had to be located and terminated to improve performance.

All in all, VDSL has worked as advertised in the trial, but has had the typical struggles of a young technology just starting out.

CHAPTER 15

Outstanding DSL Issues

o far, all of the major components of the DSL network family in general, and ADSL in particular, have been explored in detail. However, the approach taken has been mainly descriptive, without mentioning many of the numerous controversies and issues that divide DSL proponents. Notable exceptions were the ADSL line coding controversy, where aspects of CAP and DMT solutions were covered in some detail, and the discussion on DSLAM form and location. However, the same type of debate goes on over many other aspects of DSL technologies. The time has come to consider some of these issues in more detail. The answers to these questions are crucial in determining just what DSL technologies will look like, how they will perform, and how much they will cost.

This chapter considers the current issues and controversies surrounding DSL technology in four main areas:

▼ **Overall DSL network issues** Issues in this category include items that affect the entire operation and functioning of an DSL network.

■ **Equipment issues** Issues in this category include items that affect the various product packages in which DSL equipment will be available.

■ **Service issues** Issues in this category include items that affect the various service packages that DSL systems will deliver.

▲ **Active and passive network interface device or network termination arrangements**

DSL NETWORK ISSUES

Any DSL technology, from HDSL to VDSL, is a relatively new technology. As with any new technology, from ATM to Gigabit Ethernet, a good number of network issues should be resolved before anyone grabs a couple of boxes and begins plugging them together. These issues address overall networking concerns. The following lists some of the more critical network issues regarding DSL. Each issue is then explored in more detail.

▼ How should loop impairments such as loading coils and *Digital Loop Carriers* (DLCs) be addressed?

■ What will be the network of choice on the service side of the *DSL access multiplexer* (DSLAM)?

■ How are DSL links to be tested, repaired, and managed?

■ What should be done with non-DSL telephone company solutions?

■ How should premises issues about splitters and inside wiring be addressed?

▲ How should varying DSL bit rates be billed over time and among customers?

How Should Loop Impairments Such As Loading Coils and DLCs Be Addressed?

Obviously, because a large number of local loops have loading coils, and the percentage of local loops on DLC systems is constantly rising due to population movements away from city centers, something must be done about these loops before any DSL can be deployed, but what? In order to use any DSL on local loops with loading coils, the loading coils must be removed. This was a big stumbling block to ISDN deployment. The process of removing the loading coils is labor intensive and time consuming. It will be no easier with DSL. Moreover, this removal must be done by the entity that controls the local loop, almost always the ILEC. When loops are sold unbundled to competitors, any loading coils present must be removed by the ILEC, not the CLEC or ISP buyer of the loop. This process can often take weeks before crews are scheduled and dispatched. Meanwhile, the potential data CLEC or ISP customer is left without the promised service, even though the CLEC or ISP has paid for the unbundled loop. It is small wonder that in many cases, the DSL arena has been all but abandoned by the CLECs and ISPs and left to the ILECs.

The trouble with DLCs is that they digitize and pass along only the analog voice passband of 300 to 3.3 KHz, so that ADSL cannot take advantage of the 1MHz usually used on the DSL twisted pair without doing something about the DLC system. And VDSL is even worse, because the frequencies range up to 10 and even 12 MHz in proposals. The simplest answer would be to use spare home runs to DLC customers, but this only shifts the issue. Perhaps these users' best short term bet would be satellite services or cable modems, although this sounds strange to telephone company service providers.

Actually, in spite of the issues addressed here, there are clear answers to the loading coil and DLC issues today. First, loading coils will be dealt with as they were in ISDN: They will be removed. This is not a DSL show-stopper—it's just a pain. Second, with regard to DLCs, the DSLAM functions will eventually make their way out to the *remote terminal* (RT) cabinet, and separate trunks will handle the extra loads, which is another unavoidable pain in the neck.

What Will Be the Network of Choice on the Service Side of the DSL Access Multiplexer (DSLAM)?

Basically, what this boils down to is the choice of using either the *Internet Protocol* (IP) or ATM. Both positions have merit and neither can be totally ruled out, even on grounds of economics. ATM may be more expensive, but perhaps the performance aspects of ATM networks and services in terms of quality of service guarantees for delay and bandwidth would justify this approach. On the other hand, it is an IP world, especially where the Internet and Web are concerned. The entire Internet protocol suite (TCP/IP) is evolving to a more adequate transport for voice and video, but this transformation is by no means complete. Perhaps the latest advances in IP routing, which make the IP router as fast as an ATM switch and offer at least some service quality guarantees, will doom ATM use in the DSL arena once and for all.

How Are DSL Links to Be Tested, Repaired, and Managed?

It took a long time for outside plant personnel to be trained and comfortable with ISDN DSL links. Personnel took a while to become familiar with *Basic Rate Interface - Terminal Equipment* (BRI-TE) cards and so forth. It was quickly apparent that this was no longer an analog world and that radically different testing, repair, and management techniques were needed. By the same token, DSL testing and maintenance procedures will take a while to work out. One complication is that in spite of their similarities, the DSL technologies are all different. Testing ADSL is different than testing HDSL or HDSL2, and these in turn are different than testing SDSL or VDSL. Test equipment might be able to handle the differences between ADSL and G.lite, or HDSL and HDSL2, but it will be difficult if not impossible to combine testing for all DSLs in one affordable and portable test equipment package.

Other questions along these lines exist. What is the standard procedure for splicing a break in an ADSL link? Where are the uniform network management packages for multivendor DSL environments? Any DSL might prove to be too immature for the constant day-to-day pounding that working networks demand. ADSL has already been shown to be susceptible to Internet worms and viruses. How are these effects to be found and risks to be eliminated?

It is easy to take a pessimistic attitude regard this testing and repair issue, but it need not be. HDSL links are being readily maintained today. And adding ADSL electronics at the local loop might actually help with management, due to the increased intelligence and capabilities of the devices.

What Should Be Done with Non-DSL Telephone Company Solutions?

DSL is not the only game in town, which is painfully obvious at many trade shows and conferences. In fact, DSL solutions are not always the solution of choice within the telephone companies in spite of DSL's promise to maximize use of existing twisted-pair copper wire. Some ILECs seem almost neutral about DSL, especially if the perceived deployment costs are high. One reason might be that the telephone companies fought for years for the right to build and operate cable TV networks. At least one telephone company decided to practically abandon their twisted-pair infrastructure in favor of building the whole thing over again as a full-service *hybrid fiber-coax* (HFC) network that will be almost indistinguishable from any other cable TV network. The company has since backed off this extreme position, but just the fact that it could be considered by a traditional telephone company is startling to many. And abandoning ISDN customers in favor of ADSL or VDSL might be a traumatic experience for many telephony service providers, not to mention their ISDN customers. Also, the *multichannel multipoint distribution system* (MMDS) technology is essentially a telephone company initiative. What is the future of these alternative telephone company technologies in what might yet turn out to be a DSL world?

How Should Premises Issues About Splitters and Inside Wiring Be Addressed?

Exactly where should the splitter be located and how should it be packaged when full ADSL is used? If packaged with the ATU-R, as some splitters still are, the concern is maximized that ATU-R failures will affect analog telephone service, although there are ways to address this possibility. Unfortunately, these vary from vendor to vendor. If the splitter is separated from the DSL modem, as most are today, there may be long parallel runs of wire that cross-couple noise from the analog voice, which, of course, is always a major concern. How should those entrusted to run the premises wiring be monitored for standards compliance? Inspections of inside wiring are rare outside of new commercial construction, and even here the inspection is often limited to a visual inspection of splices, connectors, and other non-electrical characteristics of the wiring system. Although these physical features affect electrical performance of the inside wiring, more precise electrical testing should be conducted. But electrical testing is time consuming and expensive and so rarely done at all.

Of course, G.lite makes a lot of these issues easier to address. Nothing could be simpler than using splitterless ADSL, and in fact this will be very desirable from the customer point of view in the majority of cases. There will still be inside wiring concerns, but these will be lessened as G.lite microfilters become more and more effective.

How Should Varying DSL Bit Rates Be Billed Over Time and Among Customers?

Money is always a crucial issue both to customers and service providers. Because DSL line rates vary from installation to installation, what are customers really paying for? Obviously, it is not pure bandwidth. Also, packet-based services are bursty, so a customer might not want to pay for pure connection time either (or else customers might want their old circuit back!). When PPPoE is used, the required login makes it easy to track customer's connection times, but this is harder to determine when other encapsulations are used or when the DSL modem or home security device handles the user login even when the PC is turned off.

So just how should DSL be best billed? Right now, a flat monthly fee is paid to the DSL service provider, and usually a connection charge to the ISP. But is this really fair? Dial-up modem users do not complain. After all, if a dial-up modem connects at 14.4 Kbps on a noisy local loop, a customer pays the same as a neighbor that consistently connects at 33.6 Kbps. What if neighbors with ADSL or G.lite have service that varies even more drastically? What if a customer can get only 640 Kbps while a neighbor consistently gets 1.5 Mbps, about three times more bandwidth for the same price? That is just the way it will be, it seems. Service providers can always try to justify equal charges, and it is debatable whether the customers will ever have any easy way to compare DSL line rates with others.

There is a related bandwidth issue as well. In many cases, DSL service providers would like to be able to minimize, rather than maximize, the bandwidth available to DSL users. Now, it would seem to be in the DSL service provider's best interest to offer the maximum upstream and downstream bandwidth possible. However, this maximum bandwidth just stresses the backbone network linking the DSLAMs to the ISP, especially when the DSL service provider is a small CLEC. In many cases, the DSL throughput is nowhere near what the DSL line rate can actually support. So not only do DSL bit rates often vary, in many cases the DSL line rate in no way accurately reflects the bandwidth actually available to the customer.

DSL EQUIPMENT ISSUES

The technology exists today to combine a security firewall, LAN hub, and router all in the same package. Many reasons to do so exist because most of these devices are used together in most cases anyway. The resulting device is about the size of a LAN hub, and costs much less than buying three separate devices. Why aren't these devices universally used? There are many reasons, among which are that people like having separate devices for these functions, different vendors have different strengths, and so on. There are similar issues regarding DSL equipment, as shown in the following list. Each issue is explored in more detail in the sections that follow.

▼ How should different DSL line coding methods and technologies relate to each other?

■ How should DSL premises equipment be packaged to ensure customer acceptance?

■ What types of customer equipment interfaces should DSL devices support?

■ What is the best ATM or IP interface for DSL access multiplexers (DSLAM)?

■ Where are the end-to-end, turnkey DSL systems?

■ How much will all this equipment cost?

▲ What about compatibility issues?

How Should Different DSL Line Coding Methods and Technologies Relate to Each Other?

If both *carrier amplitude/phase* (CAP) modulation and *discrete multitone technology* (DMT) had continued to coexist as DSL line coding methods, what would, or should, happen if an ATU-C was DMT-based and ATU-R (which is, after all, customer premises equipment) was CAP-based? Is it logical to expect that DSL equipment would have eventually supported both? Before DMT became much more common in ADSL, some vendors, faced with trying to distinguish their products in an increasingly standards-based, commodity market, had announced dual CAP/DMT support in an effort both to separate themselves from the crowd and to allow the maximum use of their products in all situations.

This controversy is currently being played out not in ADSL, where DMT currently rules, but in other DSLs such as SDSL and VDSL. The previous chapter mentioned that VDSL standards groups, in order to move VDSL along, have endorsed both CAP/QAM- and DMT-based VDSL line codes. Should this dual support be a routine part of all DSL equipment standards? What about vendors who wish to decide on one line code or another? Will the risk of guessing wrong mean one vendor less for that DSL? What will this do to costs, interoperability, and vendor presence in the crucial deployment phases?

How Should DSL Premises Equipment Be Packaged to Ensure Customer Acceptance?

The plain old telephone service splitter can be located in one of several places to support the analog voice in full ADSL. Which splitter form and position will users prefer? Which will they tolerate? Which will they reject? The ATU-R itself can be a PC card, a set-top box attached to a TV, a stand-alone device, or something else entirely. Right now, DSL modems are basically standalone devices with an Ethernet port. Some have built-in LAN hub, routing, and even firewall filter protection, while others are just plain DSL modems and no more. But how does a vendor or service provider decide which packaging the customer would prefer? This issue involves considerations of economics, marketing, psychology, and just plain consumer intangibles. The art of packaging is hard enough to do just right for packaging food, let alone technology!

G.lite is a bet that customers will embrace splitterless arrangements and handle microfilters properly. So far, arrangements without a splitter seem more or less preferred for all forms of DSL, not just ADSL.

What Types of Customer Equipment Interfaces Should DSL Devices Support?

Customers do not really care about the technical aspects of their equipment; they just want to plug things in and get the system up and working. But what kind of plug is being plugged in? The more types and choices on the customer premises, the higher the potential cost of the device. Just give the users an Ethernet port? In what form? A 10Base-T RJ-45 only, as many ATU-R vendors do now? What about the other types of connector? PCs now support the *Universal Serial Bus* (USB), a sort of "serial SCSI" interface with multiple fast serial devices hooked up to a PC. Many DSL equipment vendors just put in the 10Base-T RJ-45 and that is that, but this means users have to buy an Ethernet card just to be able to plug in to the DSL modem. USB is much simpler for a single PC, but a DSL modem on a PC card is even simpler. But for now, the DSL modem with Ethernet RJ-45 is all but universal, even when USB support is included.

What Is the Best ATM or IP Interface
for DSL Access Multiplexers (DSLAM)?

This is not a question of using ATM or IP behind the DSLAM, which has already been addressed. This is rather a question of what kind of ATM or IP interface or device should be employed. Look at the IP choices first. Should the IP router inside or adjacent to the DSLAM be accessed at 100 Mbps? What about Gigabit Ethernet? Is even more speed needed? Should the service provider consider something more exotic such as Fibre Channel, which offers speeds into the multiple gigabit ranges? Should the DSLAM support the *Small Computer Systems Interface* (SCSI)? Something else here as well? What about the High Performance Parallel Interface (HPPI), which runs close to 6 Gbps. ATM interfaces are pretty much in the same boat, but choices could be more restricted due to ATM Forum and B-ISDN standards. SONET/SDH is pretty much it for ATM on the DSLAM.

Where Are the End-to-end, Turnkey DSL Systems?

Customers may like simply to plug it in and run it—whatever the "it" may be—but so do service providers in many cases, believe it or not. Why buy equipment and software from four different vendors and spend months performing quality assurance and interoperability testing? Just buy everything—from the ATU-R to the DSLAM to the backend distribution network to the ATM server (or IP server)—from one vendor or *value-added reseller* (VAR), rely on them to install the system from customer to ISP, and this lets the DSL service provider just go out and sign up users. These are called *turnkey* solutions. Common enough in Web site, LAN, and even some ATM applications, common turnkey solutions seem to be somewhat lacking in DSL, although there are now several system integrators in the field. There seems to be a precedent in MMDS systems. Telephone companies do not have the in-house expertise to engineer and build MMDS networks, so they rely on outside contractors to turnkey the whole thing for them. The same thing appears to be happening with VDSL trials, which have often been built by the equipment vendors under contract to the DSL service provider. The whole issue of handling a DSL in-house or relying on turnkey providers is a key one.

How Much Will All This Equipment Cost?

This is the stickiest issue. It is even possible to use subsidies to get DSL off the ground in rural areas, a process that worked quite well with cellular phones, at least in the early stages of deployment. Early cellular phones that cost about $700 were commonly sold for $350 or less. How could this be? Simple. The cell phone had to be used with one service provider or another. Whoever signed up the user would rebate a portion of the subscriber's monthly bill to the manufacturer. When the difference in cost and consumer price was even, it was all the same in the long run. Could DSL modems (ATU-Rs) themselves and other aspects of DSL services be marketed the same way? Should a service provider be eligible for a subsidy to deploy DSL services in areas that they otherwise would not? Perhaps, but this is pure speculation. Attempts in the United States to pass laws in the U.S. Congress supposedly to help ILECs deploy DSL outside of compact city centers and suburbs have been stalled and criticized as a form of "corporate welfare."

What About Compatibility Issues?

Cost issues for ATU-Rs are one thing. But compatibility issues can be just as important. Without full and complete standards for the management of DSL systems, the lack of interoperability for operations, administration, and maintenance (OAM) purposes tends to lock both users and service providers into single vendor solutions. This can be a situation that stifles growth, especially for a new technology like DSL that needs all the help it can get.

For example, when the new 28.8 Kbps analog modem standard was first available, one major national ISPs' dial-in ports that could use the new 28.8 Kbps technology were incompatible with one of the least expensive consumer 28.8 Kbps modems on the market. Because the low-priced modems were so popular, frantic efforts were made to patch the ISP's modem operation so that the modems on each end would train (connect initially) properly. The alternatives, forcing either the entire ISP to change vendors, or all customers who had the low-cost modems to buy new ones, appealed to no one and would hardly help the new 28.8 Kbps technology. And there was nothing really "wrong" with these consumer's modems—they just didn't work with this ISP's equipment—so returning them for credit towards a new modem was sometimes an issue.

There are many compatibility issues to consider when it comes to ADSL. There are still CAP ADSL modems around, and the types of encapsulations supported vary from vendor to vendor. Also, there are several distinct implementations of PPPoE to consider.

Today, it is common for the DSL service provider to require the customer to buy the DSL modem (ATU-R) from them in order to assure that the DSL modem is compatible with the DSLAM that the DSL service provider is using. That is a little heavy-handed, but as long as the customer agrees to this, there is no interference with free trade or customers' rights. Some DSL service providers just give the DSL modem away and build its cost into the monthly price for the services with an appropriate payoff period.

But all agree that in order for DSL to be as popular and common as a dial-up service, customers should be able to walk into a store and buy a standard DSL modem that they can use with confidence with any DSL service. Unfortunately, until DSL management standards for OAM are completed and then implemented (which can take 6 to 12 months), this day will remain in the future.

DSL SERVICE ISSUES

People make money from technology in two ways. The first is to package and sell the technology itself, such as with automobiles or VCRs. The second way is to package and sell the service that the technology represents, such as leasing a limousine for the evening or renting videos. Obviously, the two approaches can be combined to some degree (such as buying a modem, the technology itself, or accessing the online service, which the modem technology represents). When it comes to DSL technologies and computer/network technologies in general, people are not so much impressed by the technology (which remains a semimystical, black box device to users and even to many in the industry), but are more often impressed by the service ("I need that and I want that now!"). Many of the service issues surrounding DSL concern pricing and packaging, but not exclusively. The issues are listed here, and more detail is provided in the sections that follow:

▼   How should DSL services be priced to maximize use and minimize payback periods?

■   What marketing mechanisms exist to identify DSL target customers?

■   Will DSL services simply shift bottlenecks elsewhere in the network?

■   What impact will DSL services have on other provider revenue streams?

■   Will slow deployments mimic the glacial ISDN deployment pace?

▲   Can I change my voice provider and keep my DSL service provider?

How Should DSL Services Be Priced to Maximize Use and Minimize Payback Periods?

A payback period is the length of time that it takes a service provider to recover the initial costs necessary to start offering a new service from scratch. The new service may involve totally new technology, but not always. The initial AT&T coast-to-coast, direct-dialed calls (DDD, or direct distance dialing) cost $12 per minute (in 1960 dollars). This amount was too expensive for most people to afford, but based on the cost of the equipment, pro-rated over a certain payback period, AT&T figured that this was a fair price to charge. No one used it initially. AT&T dropped the price to $6 per minute and made money hand over fist. Long distance is now deregulated, but the LECs that are considering DSL services are still closely watched by the state regulators. The point is that the same type of cost, revenue, and payback period curves for DSL have yet to be widely publicized, and there is debate over the ones initially released.

Some studies have concluded that in order to compete with other technologies such as cable modems, ADSL services will have a payback period of 5 to 10 years, given current rates that new users sign up for the service when available. But many regulators will not approve a service that does not pay for itself in three years. The idea is that a new service must be profitable on its own in order to avoid the issues of "cross subsidies" that for years kept local telephone service unnaturally low and long distance service artificially high. The goal here is to encourage competition by keeping costs equal for all. But given the cost of unbundled loops, the payback period for a data CLEC or ISP appears to be even higher than 5 to 10 years! How can a data CLEC or ISP compete with the ILEC's $40 a month ADSL service when the unbundled loop from the ILEC is costing $50 or more a month in many cases? It is perhaps not surprising then that the ADSL field seems to have been abandoned by data CLECs and ISPs and left to the ILECs.

What Marketing Mechanisms Exist to Identify DSL Target Customers?

Marketing is the art of creating a perceived need for a product. The key term here is *perceived*. How many people have found themselves saying, "I really shouldn't have bought this, but it was such a bargain…"? To be most effective, marketing efforts should focus on those who are most likely to recognize the product's benefits. Why waste time marketing to users who have no interest in the service or could not use it even if they wanted to, such as people without home PCs? The benefits could be real or imagined, but good marketing makes them known in either case. However, outside of obtaining a list of current Internet users (and where exactly would that be?), how are DSL service providers to decide whether one neighborhood has more potential buyers than another? Right now, the most common marketing tool for DSL service providers is household income. Studies have shown that the more disposable income (money) a household has, the more likely they are to purchase DSL services. But this practice is open to very real charges of elitism, or worse. How much DSL should there be in low-income housing projects and neighborhoods? And when?

Will DSL Services Simply Shift Bottlenecks Elsewhere in the Network?

DSL will alleviate the bottlenecks caused by restricted local loop bandwidths, true enough. What about other potential bottlenecks outside of the access bottleneck in the network? Will the DSLAMs now be flooded with traffic, or the IP routers or ATM switches? Will the backbone network now grind to a halt, which has been predicted for the Internet for some time now? Will the servers on the Internet at the remote end of the DSL link and system be able to keep pace with the traffic demand? It is one thing for a heavily visited Web site to have to stream out information over a 100Base-T Ethernet LAN to a router feeding 64 Kbps downstream to 10 users (640 Kbps aggregate), but will the same server be able to handle the load over the same LAN to the same router when users now get 1.5 Mbps downstream (15 Mbps aggregate)? The concept of a "shifting bottleneck" is not unique to DSL, of course, but it is always important.

What Impact Will DSL Services Have on Other Provider Revenue Streams?

Service providers, especially telephone companies, do not often put all of their eggs in one basket. A variety of voice services are available to fit almost any user need. Services include call forwarding, call waiting, caller ID, and so on, all of which complement each other nicely. Users do not need a wireline phone or a cellular phone—they need both! But DSL is different. If DSL becomes common for SOHO usage, what will that do to T-carrier revenues? What about users of ISDN frame relay or ATM services? The success of DSL could cause a drastic drop in other revenue streams.

Will Slow Deployments Mimic the Glacial ISDN Deployment Pace?

The ISDN has been around since the mid-1980s, and yet, as recently as October 1996, there was no ISDN available in New Mexico, Alaska, Washington, D.C., nor in several other major portions of the United States. Ironically, as soon as ISDN became almost universally available, DSL became the hot, new technology. ISDN deployment in the United States was just too slow to be really popular with the public. The kiss of death for almost any technology has frequently been the familiar "we will deploy services based on customer demand" refrain from the service providers. Customers do not often demand services, any more than people demand bus stops. The bus does not just cruise the boulevards and look for knots of people waiting on corners ("better put a bus stop here..."). However, if the bus stops at a convenient location, and there is a bus stop sign there, people will climb aboard. And when service providers begin to ask for special rules to "encourage" their own service deployments, this is never a good sign.

Can I Change My Voice Provider and Keep My DSL Service Provider?

Deregulation in many countries enables customers to shift service providers almost effortlessly. However, if DSL is delivered to a user with a splitter or even a microfilter, how can a customer change DSL providers and keep their analog voice service provider? Not many telephone service providers with DSL plans have even considered this possibility.

The LECs, naturally, see the Internet customer as a way to help hold and protect their voice customers. And given the recent abandoning of the DSL field by data CLECs and ISPs, there might an exodus from competitive voice services as well, leaving no place for ILEC customers to go.

ACTIVE AND PASSIVE NIDS

The *customer premises equipment* (CPE) remote unit in many DSL arrangements (especially ADSL) supports both newer services and analog voice services. The CPE in ADSL consists of an ATU-R and splitter, sometimes in the same device, but not often. The boundary between the CPE and the network is determined by a *network interface device* (NID). Sometimes the term *network interface unit* (NIU) or even *network termination* (NT) is used in the same context. It all depends on who or what is talking about this network access point. In any case, there must be some form of NID involved in the DSL link, which is the term used in this chapter.

As it turns out, there are two major types of NID: active and passive. The differences are mainly due to differing regulatory environments and have little to do with straightforward technological issues. These are sometimes referred to as the *wet wire* NID and *dry wire* NID, respectively. The terms *wet* and *dry* refer to the ability and willingness of the service provider to deliver electricity to the DSL device over the copper loop itself. In a wet arrangement, electricity is part of the service package, but in a dry arrangement, electricity must be provided separately by the customer.

In the United States, where deregulation is considered a worthwhile goal and widely pursued, the NID is almost always a passive, "wires-only" interface, which means that the NID delivers no power to the premises and works by just plugging things in to the proper jacks on the NID. There are exceptions, but not using a passive NID for a service in the United States often involves the permission and cooperation of the regulating authority, typically the individual state in DSL arrangements.

It is more common in a deregulated environment, particularly in the United States, for the ATU-R to be the CPE and contain all of the active DSL electronics. That is, the customer could in theory purchase the ATU-R from any approved vendor, install it themselves, and the service provider has no control over the situation at all. In practice, the DSL modem must be compatible with the DSLAM, so choices are limited, in many cases to exactly one.

In many other countries, however, where deregulation is not as prevalent, both the NID and ATU-R are part of the network. This is an active NID because the electronics can be split between the NID and ATU-R. In these countries, the ATU-R can be bundled with the NID device. The customer must purchase or lease the ATU-R from the service provider. The service provider arranges for installation, and so the service provider exercises a fair amount of control over the situation.

The question is, how should DSL vendors package their products for these environments? Should a vendor manufacture passive NIDs and active CPEs for sale in the United States and other deregulated countries? Should the vendor then repackage the product as an active NID/ATU-R combination for sale elsewhere? This is not an ideal situation, but there seems to be no way around it.

Figure 15-1 shows how both the passive and active NID arrangements work. In the passive NID scenario, as will be typical in the United States, the ATU-R is packaged separately from the NID. There is a "wires-only" interface for DSL signals to and from the ATU-R unit, which has its own separate alternating current (AC) power supply (or shares power with another device, such as a PC or set-top box). The NID sits at the end of a dry (no electrical power) DSL link.

In the active NID scenario, also shown in the figure, the ATU-R can be bundled within the NID itself. There is no need for a wired DSL interface anywhere else. After all, the user interfaces on the ATU-R are not governed or determined by DSL standards. These non-DSL interfaces might be 10Base-T, USB, or something entirely different. This arrangement is quite attractive for that very reason. The service provider is not responsible for the non-DSL interfaces. The active NID/ATU-R unit sits at the end of a possibly wet

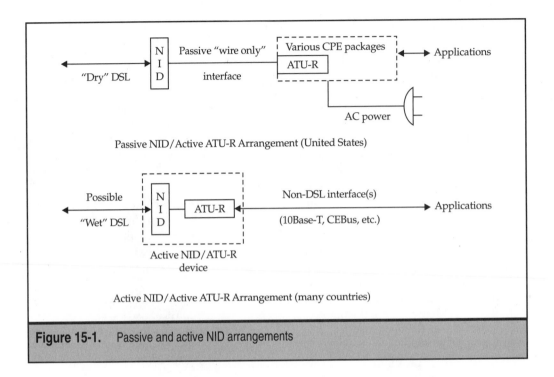

Figure 15-1.   Passive and active NID arrangements

(powered) DSL link, although power distribution methods are subject to more than just active/passive considerations (for example, battery power is an option).

PASSIVE NID/ACTIVE CPE ISSUES

Although mandated by deregulation initiatives, the passive NID is simple, and it minimizes service provider involvement, which can be seen as a direct benefit to both the customer and service provider. The passive NID fits in very well with the traditional "modem" purchase and installation philosophy. That is, a customer can buy whatever modem they like; if it complies with certain common standards, the customer can use it with confidence, without concern for service provider support and compatibility.

The passive NID delivers no power to the CPE. However, if packaged as a PC board, the DSL CPE needs no additional power supply. In addition, even if packaged as an external modem with separate power supply needs, the DSL CPE can be deployed as the customer pleases, without regard for PC location.

One distinct benefit to vendors exists in the passive NID environment: There is ample opportunity for vendor distinction based on feature set. In other words, because customers have a wide range of choices for DSL CPE equipment, competition might be fierce, and prices will drop rapidly at the same time the features become richer.

ACTIVE NID/PASSIVE CPE ISSUES

Regulated countries can use a bundled, active NID and ATU-R. When the active NID/ATU-R issues are considered, this arrangement guarantees maximum service provider control over the entire network loop. The service provider might have an easier time configuring, troubleshooting, and maintaining the loop, although this is not necessarily a given.

This arrangement means that the service provider has no premises wiring concerns at all when it comes to DSL. The interfaces on the ATU-R are governed by standards other than DSL and are beyond the scope of DSL.

Also, there is no ATU-C-to-ATU-R (or any DSL devices) interoperability concern at all in this configuration. Interoperability can be a problem in deregulated environments, where there is no guarantee that a customer's CPE will function properly with the service provider's equipment if they come from different vendors. Of course, if they both comply with all relevant standards, they are supposed to work together flawlessly. However, this is not a perfect world.

The active NID presents a good opportunity for the service provider to perform effective remote management and configuration. Knowing exactly what the device is at the premises will help this situation. Moreover, the active NID/ATU-R presents a simple and familiar interface(s) to users (such as 10Base-T). There is no need to worry about whether premises wiring will carry DSL signals because there will be none within the premises itself.

Naturally, future deregulation trends in many countries could alter the applicability of active NID solutions. Many nations see deregulation as an ultimate goal, although projected timelines vary greatly.

Please note that in some circumstances, even in the United States, an active NID configuration arrangement is still possible. It all depends on the willingness of the relevant regulating bodies to go along with service providers' plans and tariff agreements.

CHAPTER 16

International Issues
and DSL

ADSL and all other DSL technologies in general are not intended solely for use within the United States, of course. The goal of maximizing use of the existing copper local loop plant is a worthy one, and not only a concern of service providers in the United States. All developed countries, telecommunications-wise, have a relatively extensive infrastructure of copper local loops. Not all are as old as the loop infrastructure in the United States, but there are still challenges in terms of usage and increasing network access speeds on these loops. Many of the concerns about local loops are truly global in nature. It should be noted that although this chapter specifically references the United States, there is little difference between the local loop infrastructure in the United States and the local loop infrastructure in Canada. The inventor of the telephone had close ties to Canada, and the Canadian telephone system is in many ways identical to the telephone system in the United States. Some sources use the term *North America* to group the United States and Canada together, but this book just uses the terminology "United States" and "rest of the world" for consistency with related topics discussed throughout this book.

This is not to say that there are not significant differences between the copper local loop infrastructure within the United States and throughout the world. Some appear to be simply cosmetic, such as the difference between measuring distances in kilometers instead of miles and kilofeet, but others are more profound and should be addressed. This chapter considers the issues surrounding the global use of DSL, including not only the obvious differences in wire gauges and distances, but also considering the degree of pre-existing wiring infrastructure.

It is a given that the more extensive the copper local loop infrastructure is in a particular country, the more interest there will be in DSL products, equipment, and services. The differences among the various countries and regions around the world and their approaches to DSL and high-speed services in general are certainly worth pointing out in a chapter of their own.

Before looking at local loops in particular, a few general comments about telephone networks around the world may be in order. First and foremost, tariffs for telephony services vary widely around the world. Many countries do not have monthly flat rate services, but they may have flat rate service on a per-call basis, which is at least similar in concept. Some countries have a monthly maximum on measured service, which amounts to a kind of price cap. The point here with regard to DSL is that the switch clogging effects of Internet access due to flat rate services might not be as severe around the globe as they are in the United States. It is worth noting, however, that the traffic engineering practices (such as the use of Erlang B tables) used to design telephone networks is universal.

Also, 911 is not a universal emergency number. For example, some countries use 999 for emergencies. In business environments, activation digits for special services or outside lines may be different. For instance, an outside line in many countries is obtained by dialing a 0 and not a 9, as is common in the United States.

Cable TV service, which is almost as universally available in the United States as telephone service, is not found all over the world. Much of the world's television is delivered by way of government-controlled networks. Most countries only have three or four television stations anyway, making 60-channel cable TV networks somewhat useless.

However, satellite systems are becoming popular in many countries with developed tele-communications infrastructures, which is an important point because cable modems are frequently mentioned as an alternative to DSL technologies, but when the marketing considerations broaden to include non-U.S. markets, the applicability of cable modems as a broadband access solution falls off dramatically.

Oddly, ISDN varies widely around the world in spite of its position as an international standard. The United States follows something known as *National ISDN* (NISDN) whereas many other countries have followed a more international standards-based flavor of ISDN. As a result, much of the ISDN deployed in the United States is incompatible with other countries' forms of ISDN. Even when it comes to connectors, the world is not totally on the standard modular connectors known as RJ-11 or RJ-45, as the United States is. There may be an RJ-11 in addition to the "national connector" on the telephone handset, and so even the simple act of plugging something in can be a challenge.

There are other differences, of course, but the differences of most concern here involve the local loop itself.

LOCAL LOOPS OUTSIDE THE UNITED STATES

Naturally, there are many types of local loops outside of the United States that do not necessarily correspond to the common local loop architectures developed and deployed in the United States. All other countries must adapt local loop technology for their local environments. Some of the changes are more or less straightforward, such as the changes from feet to kilometers or from *American wire gauges* (AWG) to international wire gauges, measured in millimeters, but other differences are more profound. Because almost all countries are interested in higher-speed access lines and DSL in general, this topic is an important one.

It is always good to keep in mind when dealing with international situations that the telecommunications infrastructure in the United States is somewhat unique. After all, the telephone was invented in the United States, and many other countries are not as developed as the United States given this built-in head start. (Then again, many other countries are not as limited as the United States by network decisions made in the late 1800s or early 1900s.)

One of the most significant differences in local loop infrastructures is that not all countries make extraordinary efforts to reach distant customers. Although it is true in many countries—especially developed telecommunications countries like the United States—that governments and telephone companies have a firm commitment to the concept of universal service, where everyone who can pay for a telephone can get one, this is not always the case.

If a potential subscriber cannot generate enough revenue to cover the cost of extending the local loop to the premises, so much the worse for the customer. Many other countries' local exchanges therefore serve more compact areas than in the United States. This makes the use of loading coils almost unheard of except on long trunk circuits between local exchanges. Oddly enough, trunk loading coils are never used in the United States today, due mainly to the nearly universal use of digital trunks in the United States.

Of course, outside of the United States, mixed gauges and bridged taps still exist. In fact, mixed gauges are even more common in many countries than in the United States because the rest of the world compensates for signal loss with creative wire gauge changes rather than loading coils in most cases. But most European countries do not have bridged taps at all because the access network never assumed that each and every building in an area would need one or more telephone lines.

Some countries, especially developed telecommunications countries, have a high percentage of digital access lines. Some of this is due to ISDN deployment, but not always. Germany has a significant percentage of its access lines digitized already, and had about half of the world's ISDN lines a few years ago, because backward compatibility was not so much an issue in Germany as new deployment after the reunification of East and West Germany. South America has many digital access lines, but few ISDN switches. And so on.

Given the enormous variability of local conditions for telecommunications infrastructure in the world outside the United States, it is hard to make valid generalizations. But this does not mean there are not at least some distinguishing guidelines that can be used so that service providers and equipment vendors and interested users can get an idea of how and when DSL would come to a particular area or region. One of the main things to look for is to divide the countries into emerging, developing, and developed telecommunications infrastructures, as does the United Nations and ITU-T. This grading is based mainly on the infrastructure maturity of the country in question. The internationally accepted infrastructure maturity indicator has been telephone penetration: the ratio of the number of telephones to the population of a given country. This idea is explored more fully in the "Infrastructures Around the World" section later in this chapter.

INTERNATIONAL WIRE GAUGES

In the United States, the copper wire used in the local loop is manufactured according to the AWG sizes. Naturally, the rest of the world feels no urge to conform with this specification, and indeed other countries do not. It is universal outside the United States to measure wire gauges in simple terms of millimeters (one-thousandth of a meter, which is 3.28 feet). Officially, the United States, like every other country around the world, is on the metric system. But while the American *government* might be on the metric system, the American *people* are still on the older systems they understand and are familiar with. The same is true about the companies they work for and the people these companies sell to within the United States.

Table 16-1 shows the differences between wire gauges used in the United States and in the rest of the world. The AWG is shown in the left column. The center column translates the AWG to an individual wire's diameter in millimeters. The rightmost column shows the closest international gauge to the AWG in question.

The most common gauges in use internationally are 0.4 mm and 0.5 mm, which corresponds to 26- and 24-gauge wire in the United States. Note that AWG numbers grow larger as the size of the wire grows smaller. Legend has it that these numbers reflect the number of wires that can be passed through a standard-sized hole: the smaller the wire, the more could be placed through the hole. Be that as it may, it is common in many local

American Wire Gauge (AWG)	Diameter (mm)	Closest International Gauge (mm)
19	0.912	0.9
22	0.643	0.63
24	0.511	0.5
26	0.404	0.4
28	0.320	0.32

Table 16-1. International Wire Gauges

loop configurations to see 0.62 mm (22-gauge) or 0.9mm (19-gauge) wire outside of the United States, as well. Even 0.32 mm (28-gauge) wire is used on some local loops. Keep in mind that the rest of the world can be much more creative than the United States when it comes to mixing wire gauges and constructing bridged taps (if they are needed).

KILOFEET TO KILOMETERS

As mentioned in the preceding section, the people and businesses in the United States are one of the few groups left in the world that insist on using the traditional English system of measurement (inches/feet/miles). The rest of the world uses the more rational metric system, based as it is on tens and hundreds instead of twelves and other multiples and fractions of this basic factor. There is nothing critical about technology when seeing DSLs in terms of feet instead of meters. It is just quite awkward to try to convert units mentally.

Table 16-2 shows the most common local loop lengths used in loading coil situations in the United States. The rightmost column translates these figures to kilometers. Keep in

Feet	Kilometers
12,000	3.66
14,000	4.27
16,000	4.88
18,000	5.49
30,000	9.15

Table 16-2. Kilofeet and Kilometers

mind that the United States is actually officially on the metric system, so many standards organizations, such as ANSI, use metric units in their specifications.

For local loop purposes, the table shows local loops of 12,000 feet (3.66 km), the point at which creative steps are taken to extend its reach. The table also lists 18,000 (5.49 km), long considered the ISDN limit. Even 30,000 feet (9.15 km) is listed to show the extent of some local loops in some parts of the United States (legends of 20-mile [105,600 feet, or 32 km] local loops running on barbed wire fences are common in the Far West, and similar distances must be spanned in Alaska and other remote locations).

For all other conversion purposes, keep in mind that 1 km equals 3.28 kft or 3,280 feet. For going the other way, 1,000 feet equals 0.305 km.

SOME SAMPLE GLOBAL LOCAL LOOPS

Those concerned with local loops globally, and not unique to any particular country or region, are often faced with a bewildering array of local practices and variations in local loop construction. Fortunately, standards organizations, long concerned with testing methods for compliance among equipment vendors, have established a set of standard international local loops in common use around the world. For instance, a set of eight test loops is included in the ANSI T1.413 specification for ADSL. The four samples shown in Figure 16-1 are based on this document. These local loops are part of the complete *carrier serving area* (CSA) local access architecture that has been used as the standard since the mid-1980s in the United States and around the world. The loops are numbered CSA 1 through CSA 8, and a special CSA 0 loop is used for testing purposes.

In its simplest form, a CSA local loop might consist of 2 to 4 km of either 0.4 or 0.5 mm twisted-pair copper wire. This is the first configuration shown in the figure.

The second configuration shows a mixed gauge situation where up to about 2.5 km of 0.4 mm wire is followed by 1.5 km of 0.5 mm wire. This is quite common. The 0.4 mm section can vary from 1.5 to 4.55 km in length under most conditions, extending the reach to about 6 km (18,000 feet), which is about the ISDN limit.

The third configuration shows a more complex mixture of gauges. It is common to take 0.32 mm (28 AWG) wire about 0.2 km (only some 656 feet) from the local exchange and then travel from 1 to 5 km on 0.4 mm wire. At that point, 0.9 mm (19 AWG) wire travels the remaining distance of up to 4.0 km to the customer premises. Note that this mix extends the reach of the loop without the need for loading coils, at the cost of reduced voice quality.

Finally, a bridged tap configuration is shown. In this case, all of the wire is 0.4 mm. However, the bridged taps can extend up to 0.5 km (about 1,600 feet) from the loop itself. Note also that it is common to have one of the bridged taps right at the customer premises.

The other four configurations not shown are basically variations of gauge and distances on the third loop type in the figure. Those in need of more detail are referred to the ANSI document itself, or any other standards document that references the CSA architecture.

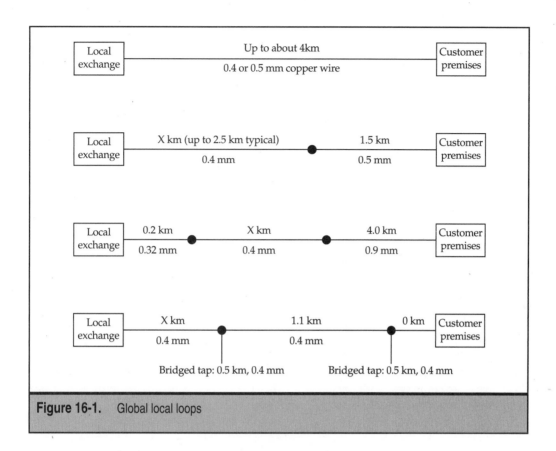

Figure 16-1. Global local loops

GLOBAL LOCAL LOOPS IN GENERAL

It is hard to summarize and generalize the local loop architectures outside of the United States. Considerable variation exists due to local conditions, both regulatory and with regard to existing infrastructure. Nevertheless, it is possible at least to attempt to make some valid statements about local loops around the world.

First, local loops in excess of 5 km are relatively rare, with 2 to 4 km as the typical local loop length. This was one reason why the maximum access line length for ISDN was established at 5.49 km (18,000 feet). In most parts of the world, this is more than adequate.

In spite of the high degree of modest length local loops, signal quality on local loops in many places around the world is quite poor. One reason is that loading coils are rare, and efforts are made to extend loops with creative gauge mixing, often with poor results. Oddly, loading coils are often used on analog trunks outside of the United States, due to their longer distances between local exchanges. This practice is no longer done in United States.

Also, mixed gauges are even more common around the world than in the United States, which is at least partly because of the effort to extend loop distance without the need for loading coils.

Outside of the United States, there are usually fewer switches and access lines per unit area. For example, the United States has more than 22,000 switches of all types (some 13,000 of them local exchanges) to serve about 270 million people. This works out to about 12,000 people per switch. In contrast, Australia, considered to be as developed a country as the United States with regard to telecommunications, has some 5,000 switches to serve about 18 million people. This works out to only 3,600 people per switch. Part of the difference is because of regulation, but another part is due to the smaller reach of the serving area.

As a final note, some countries are not seeking to preserve their copper local loop infrastructure, but to more or less abandon it. These countries are moving right into wireless local loop or fiber-based loop services. There is more talk of wireless broadband than DSL in these countries, which even include many countries considered to have developed infrastructures, as defined in the next section.

INFRASTRUCTURES AROUND THE WORLD

DSL technologies are of interest to service providers not only in the United States, but around the world, and yet each individual country has its own distinct approach to handling their telecommunications infrastructure. These differences reflect local conditions that vary from country to country and are to be expected. For example, a country without a concept of eminent domain (and there are many) to pre-empt private rights-of-way for public use might favor undersea cables for coastal trunks, whereas a country comprised of islands might favor as many wireless links as possible.

Whatever the circumstances, it turns out that there are three recognized categories of nations with regard to telecommunications: emerging, developing, and developed. This structure is established by the United Nations and ITU-T, and is in no way intended to portray any individual country as in some way better than any other. Many countries lacking in modern telecommunications infrastructures openly embrace these categories to measure their own progress and justify efforts at modernization.

▼ Emerging nations have only rudimentary telecommunications infrastructures, and services are typically offered only in major metropolitan areas.

■ Developing countries have more mature telecommunications infrastructures based on advanced technology, and services are typically offered beyond the major metropolitan areas.

▲ Developed countries have fairly state-of-the-art telecommunications infrastructures, and services are typically offered almost everywhere.

The point is not to be critical of various countries, but rather that countries have much more in common within each of these three categories than they have with other countries in the same geographical area that are not in their category. And one of the most commonly used indicators of telecommunications category is a measure known as *service*

Country	Population (M)	TVs (M)	Penetration Rate	Telephones (M)	Penetration Rate
United States	266.5	215.0	80.7%	182.5	68.5%
Canada	28.8	11.5	39.9%	15.3	53.1%
Germany	83.0	85.0	100%	44.0	53.0%
Australia	18.2	9.2	50.5%	8.7	47.8%
Malaysia	20.0	2.0	10%	2.5	12.5%

Table 16-3.   Penetration Rate of Television and Telephones

penetration, which is the ratio of a service device to a number of people. The faults of this type of measure have been frequently pointed out. It is true that raw math cannot take into account many of the widespread variations in coverage and local conditions. However, this has not stopped this form of statistic from being used almost universally, especially by the countries that it concerns the most.

Table 16-3 shows the penetration rate of basic telephony service not too long ago for a variety of countries around the world. Note that telephony penetration closely follows other technologies such as television, but there are exceptions (Germany, for example). In fact, countries such as Germany point out the hazards of using pure numbers like this to establish categories along these lines. Of course, this stops no one from using statistics of this nature and such categorization is done all the time.

DSL AND INTERNATIONAL CONDITIONS

Due to the large degree of variation among countries when it comes to the telecommunications environment, sweeping generalizations are difficult to make. Nevertheless, some statements still apply to countries that share a category of telecommunications infrastructure maturity.

The United Nations considers a nation *developed* if the penetration rate for telephony is 45 percent or more. The nation is *developing* if the penetration rate is between about 20 to 30 percent, and it is *emerging* if the penetration is less than 20 percent or so. Many nations fall into this emerging category. Others are considered to be developing. Few have telecommunications infrastructures mature enough to be considered developed.

What has this to do with DSL? It is mostly developing and developed telecommunications nations that are most interested in DSL due to the presence of a mature infrastructure. Other nations have more to explore from a telecommunications perspective than the need to preserve existing copper-wire local loops.

Also, wireless services are popular in some environments, especially in countries consisting of islands or those with densely forested interiors or uplands. For these countries, DSL becomes less of an issue except in major metropolitan areas.

Regardless of category or geographical circumstance, all countries are interested in deregulation and broadband access services. Both are desirable from a DSL perspective. Deregulation spurs competition and service deployments, whereas broadband access is always a given for things such as video and Internet access.

DSL IN EUROPE

When it comes to talk about DSL outside of the United States (and Canada), most conversation revolves around DSL deployments and plans in Europe. Many DSL equipment vendors are European companies, even those with strong American market presences. So a section on the status of DSL in Europe is certainly appropriate.

Generally, the deployment of DSL services in Europe has been slow, often puzzlingly so. Everyone agrees that higher access speeds are needed, and in many places there are few obvious obstacles to DSL deployment, yet DSL does not appear. A number of reasons have been proposed to explain this lack of DSL, the least charitable one being that the ILEC is Europe does not want to cut into their E1 revenue streams.

But it is also expensive to deploy DSL. Not only will revenues due to the sales of E1 private lines inevitably fall with HDSL/HDSL2 and SDSL as an option (as they have in the United States in some cases), but rolling out ADSL in an area is not inexpensive. For example, British Telecom (BT) put the cost of DSL rollouts nationwide as "hundreds of millions" of UK pounds sterling, money that has to come from somewhere.

One complication is the EU plan to unbundle access to all local loops, despite fierce resistance in the UK and France. DSL in the United States was available for a long time in many areas only through data CLECs and ISPs with access to the local loop, and European ILECs seem unwilling to lose control of DSL in this way. But unbundling is inevitable, especially given the real political pressure throughout Europe for high-speed Internet access, although Internet access is not as indispensable in Europe as it seems to be in North America. Some European ILECs have been charging data CLECs and ISPs metered (time-sensitive) rates, whereas the data CLECs and ISPs must charge their own users flat rates to compete, much to the woe of the CLECs and ISPs.

Co-location rules concerning power and space have been an issue in Europe as well, but these issues closely mirror such controversies in the United States and need no further discussion here. It is easy to sympathize with the competitors in such cases, but consider the plight of one European ILEC that announced an ADSL trial in an area, only to be faced with 14 requests from competitors for co-location space. Granting all would be impossible, yet limiting participation to a few could be deemed "anticompetitive."

It is also true that many Europeans already have a form of high-speed Internet access available in the form of ISDN. Germany in particular has embraced ISDN and there is not the pressure to adapt ADSL as there is when Internet access is limited to dial-up modem speeds.

However, DSL deployment in Europe must increase, or else the EU risks being left behind other regions of the world. DSL is the only way to upgrade an infrastructure for higher speeds without running fiber everywhere.

Initial European interest in DSL is expected to come from the business community, not residential users as in the United States. As mentioned, the need for Internet access is often less among the general population in Europe, but businesses using the Internet as a substitute for expensive international circuits and private lines of all sorts are tired of waiting for file transfers over low-speed Internet access lines. SDSL will be a popular choice here, mainly for its symmetric speed, flexibility, and cost-effectiveness. But a revived interest in work-at-home might drive ADSL access to corporate LANs.

The business aspect of DSL will require service providers to do more than just link premises to the Internet. Remote LAN access will require security in the form of VPNs or the like.

DSL IN ASIA/PACIFIC

Perhaps surprisingly, the region of the world with the most DSL lines is not the United States (North America), or Europe, but the Asia/Pacific region. However, this section is much shorter than the European section on DSL. This is not to slight the importance of DSL in the Asia/Pacific region. It is just an acknowledgement that many of the same issues that apply to European DSL also apply to DSL in the Asia/Pacific region and so do not need to be examined all over again.

Although ADSL might appeal to almost every country, most of the interest in ADSL services to date has come from compact, developing/developed countries. This is not surprising because many emerging countries do not have the infrastructure in place to be overly concerned about preserving existing local loops, although there are exceptions. The Asia/Pacific region has a concentration of smaller, developed countries, and so DSL makes a lot of sense and is quite popular in some countries.

One important difference is that in many cases deregulation is not as important in Asia/Pacific as it is in other places. Governments in this region of the world are generally accustomed to exercising tighter control on the public aspects of people's lives, and to be frank, the people do not seem to mind all that much unless the government is perceived as oppressive. So for the most part, the ILECs, which are often a bureau or branch of the government, are much more in control of DSL deployments than in many other locations.

This is not to say that DSL deployment in Asia/Pacific has been slow. There are many DSL links, but then again the population is higher in Asia/Pacific than in many other areas. So percentage-wise, DSL lags deployments in many other areas around the world. This section briefly mentions two examples of DSL initiatives taken by countries in the Asia/Pacific region. In Singapore, Singapore Telecom Limited has ADSL equipment in 27 local exchanges. More equipment is to be added, and essentially all of the population of Singapore has ADSL service available to them.

And just to show it is not all an ADSL world, Hong Kong Telecom Limited is seriously looking at running ATM cells on VDSL links in Hong Kong. The bandwidth and services offered will be unparalleled but limited to access lines of about 1 km or so. Luckily, the 1 km limit is no problem at all in Hong Kong.

APPENDIX A

Case Studies in DSL Installation

I t is all well and good to talk about DSL technology in the abstract. But specifications and standards and white papers can only take a person so far. Sooner or later, if DSL technology is as good as this book claims it is, people must grapple with DSL directly, as they come to grips with cable TV, VCRs, and camcorders. So this appendix deals directly with all aspects of my experience with DSL installation: from ordering it, to having the phone line installed, to connecting the DSL modem to the computer.

Perhaps understandably, when the first edition of this book was prepared in late 1997, I had no direct experience with DSL at all. Few people had, and DSL deployment was miniscule at the time this book first appeared in the spring of 1998. Trials were about the best one could hope for at that time, and early services were plagued by unexplained outages, flaky equipment barely more stable than prototypes, and numerous bugs of one sort or another. It is a measure of the success that service providers have had in smoothing out the rough edges of DSL that the first edition of this book was as accurate as it turned out be with regard to the popularity of DSL. More than I like to admit today was based on faith in what others told me.

But no longer do I have to rely on the words of others to write about DSL installation. In fact, between April of 2000 and June of 2001, I have had four separate DSL links installed into various homes and offices I have occupied. I regret that I kept incomplete records about the first three installations. Happily for the second edition of this book, however, I had a chance to fully document every phone call and conversation I had regarding the installation of a DSL link ordered in May of 2001.

Before reading on, I have to say that I have no real horror stories to relate. No tales of hung computers or links that just never seemed to work at all. The best I can do is tell of a few missed due dates, but nothing that I considered to be more than the normal inconveniences that people have to put up with every day. The same type of thing happens with pizza delivery, dry cleaning, and all types of other services.

One conclusion I have come to based on my own personal experiences is that over the roughly year and half prior to June 2001, DSL installation has gotten a lot easier not only for the customer, but for the service provider as well. It should be noted that on each of these DSL lines, I use additional security hardware and software not detailed here for obvious reasons.

DSL CASE STUDY #1: BELL ATLANTIC, APRIL 2000

My worst experience was the first, when I tried to get DSL from Bell Atlantic (now Verizon), starting in mid-1999. Now, I just said that the first DSL I had was installed in April of 2000. That's right: it took eight months from the time that the service was supposedly available in my area until I finally got the link installed. If there is a horror story in this appendix, this is it. But let me start at the beginning.

In early 1999, I began checking the Bell Atlantic Web site because they had announced a big push to bring DSL to my area of New York state, just north of New York City. I would type in my telephone number, and then see a Web page telling me if DSL was available or not at that central office (CO). Through the beginning of 1999, the response was always something like "DSL is coming soon to your area…." Then in June of 1999, there was a breakthrough, or so I thought. The Bell Atlantic Web site response to my telephone number was something like "DSL is coming to your area in July!" I was thrilled, and waited until right after the Fourth of July holiday to check again. This time it said "DSL is coming to your area in August!" I simply figured that it was harder to deploy those DSLAMs than they thought. I waited some more. In August, it was "September/October." They were obviously getting smarter with their responses. I waited until mid-September, but by then the message had gone back to "DSL is coming soon to your area…" Clearly, someone who knew a thing or two about real customers had taken over the Web page.

I later read that Bell Atlantic had severely underestimated the amount of time and effort it would take to install a DSLAM and related equipment in a CO. So deployment schedules had to be scaled back dramatically. I would just have to wait.

I still checked off and on, however, and in January of 2000 my wait was apparently over when I got a different type of message when I entered my telephone number. DSL was available, *if* my line qualified in terms of distance, bridged taps, loading coils, and in short, all of the things mentioned in this book that can bring DSL dreams back to earth. All I could do at that point was request DSL service and wait for the qualification process to be completed in a week or two.

I heard nothing, so I called the toll-free number that appeared on the Web page. I could not have DSL after all, they said. My line was not qualified for DSL service. I asked why. The answer made no sense to me, so I asked again. To make a long story short, over the next two months I heard three different stories as to why my line was not good enough for DSL. At each point I was able to draw upon my experience and knowledge of the telephone network to counter these claims. But I wondered about people without the expertise to realize that they were being handed an excuse and not a reason.

First, Bell Atlantic said I was on a digital loop carrier (DLC). I knew this was untrue, because when the line was installed to my home office, the technician told me that the copper pair bundle was installed in 1932, but still seemed fine all the way back to the central office. There was no DLC in 1932, and because most of the houses in my neighborhood were built around that time, I saw no reason that one should have been installed since then. I knew how conservative the local access upgrade pace could be.

It took some calls to clear this one up. My line went back into the qualification hopper, I guess. Then I heard that my line was an ISDN line. Now, at the time, ADSL, which is what I was getting, required an analog local loop and could not run on a digital ISDN line at all. This one was easily refuted through the telephone company's own records. I was not paying for ISDN, therefore I had no ISDN. Qualification time again.

Third and last, I was informed that my local loop was longer than 18,000 feet and therefore could not have DSL under any circumstances at all. Now, I had driven the distance from my home to the CO where my line ran and it was about a mile and a half on the car odometer, or about 9,000 feet, I figured. The lines were above ground on telephone poles, so they were easy to follow. Even allowing for inaccuracies, I did not think there was any way I could be more than 10,000 from the CO.

I got nowhere with that argument. Only the telephone company knew for sure where the cables ran, I realized. I tried a weaker argument: if I had ISDN, as they had claimed a week or so earlier, I had to be less than 18,000 feet from the CO. This got me nowhere.

By now it was around February of 2000 and I despaired of ever getting DSL. Then I got a call from someone at Bell Atlantic's Community Relations division. He was conducting a survey on how satisfied customers were with Bell Atlantic service. The New York State public utility commission had recently fined Bell Atlantic due to poor service and made them return millions in rate hikes. As part of the penalty, or maybe to prove that they were not as bad as that, Bell Atlantic was calling people all over the state to let them have their say about the service.

Again to make a long story short, with the help of Bell Atlantic's Community Relations department, I was able to get my line "hand qualified" by a technician instead of relying on the automated tools that showed my line more than 18,000 feet from the CO. I later heard that the tool that Bell Atlantic was using at the time was famous for giving "false highs" when it came to loop length. I was about 12,000 feet from the CO, they said in March of 2000.

I also heard much later that Bell Atlantic was so swamped with requests for DSL that they routinely told people they were too far away from the CO to get the service. Then they handled only the requests that were from people who persisted or were smart enough to know that this reason could not be valid. I hope, of course, that this story is as bogus as I suspect it is.

Once the qualification logjam broke, things moved along swiftly. I was sent some paperwork with user ID and password, which was required for the PPPoE arrangement that Bell Atlantic was using. The paperwork had a date in March on it that I assumed was the installation date. When that date came and went with no installer, I called and was informed that the date was for internal work, not customer premises work. That was eventually scheduled for April 8, 2000.

The technician showed up as planned and we got the DSL modem installed and running in about an hour. My PC already had an Ethernet card and TCP/IP installed and configured for dial-up, so that helped. All I really had to do was install the PPPoE software. But as soon as everything was plugged in, I was up and running at between 300 and 400 Kbps downstream, depending on time of day.

I was billed a one-time charge for the Westell DSL modem, about $200. The basic cost of the phone line is about $30 per month. The cost of the DSL link and access to the ISP is $40 per month additional, making the total $70 per month, plus any toll calls made on the phone line. The only outages to date have been for a day or so, but in some cases I suspect that the presence of my personal firewall connected to the DSL modem might be contributing to the problem.

DSL CASE STUDY #2: US WEST, AUGUST 2000

In July of 2000, I ordered DSL for an office in a suburb of Minneapolis, Minnesota. US West (now Qwest) had a Web site similar to Bell Atlantic's, and this time I qualified right away (in fact, my line was prequalified, and all the Web site did was display this result from a database). I was 9,000 feet from the CO, and so I thought I should be able to get a fairly high DSL data rate.

There is not much more to tell. US West billed their service as "customer installable" and it went very smoothly. Once the DSLAM was configured, US West shipped a small box with the DSL modem and all the analog telephone filters, cables, and software required. The box arrived a day or two ahead of schedule, which pleased me.

However, the instructions were somewhat confusing initially. I was installing the DSL link to be used with a 10Base-T Ethernet LAN for multiple PCs, and so I had a LAN hub connected to the DSL modem, not a simple direct connection to a single PC. But there was nothing in the box that clearly spelled out the differences, if any, between the two configurations, and I was unable to locate the proper steps to follow. I installed some drivers and executed some commands that I thought would do the trick, but when I plugged everything in, I could not access the Internet.

There was a toll-free number to call with questions, so I called it. I was only on hold for a few minutes. I worked with the technician for about an hour, and the basic problem was that I had configured the DSL modem incorrectly. This DSL software version required a serial link from PC to DSL modem, and the use of some fairly complex commands entered from the terminal interface on the PC. I was certain that few people could have installed the DSL properly without a call.

The DSL modem is configured for 640 Kbps downstream and 272 Kbps upstream. There is also an "auto" speed option, but I have not experimented with this. Download throughput is usually around 550 Kbps. There is often a delay in DNS lookups, especially in the late afternoon and early evening. That is, when initially accessing a Web site, there is sometimes a 10- to 20-second delay while the browser displays a "looking up site…" message. However, once the initial page of the site is loaded, and its IP address is stored in the browser cache, access to other pages at the same site is very fast. But this type of delay is hardly unique to DSL. Many ISPs simply have overburdened DNS servers.

I was billed a one-time charge of about $240 for the Cisco 675 DSL modem (which also included network address translation [NAT], the dynamic host configuration protocol [DHCP], and some filtering support as well as being a router). The basic cost of the phone line is $32.95 per month. The cost of the DSL service is $29.95 per month. Internet access to the ISP is an additional $17.95 per month. So the total cost is $80.85 per month, without any toll calls. If there have been any outages at all on this link in the past year, I am unaware of them. However, the Cisco 675 DSL modem was hammered in July of 2001 by the Code Red virus (my security hardware and software were "behind" the DSL modem, and therefore useless to protect the DSL modem itself). Even after Qwest's "fix" to the DSL modem was applied, the DSL service was out sporadically into August 2001, until a second "fix" finally solved the problem. Bringing up the Web browser during this period was an adventure, requiring a power off/on several times a day, but the link was technically intact and in service. Qwest has refused to give refunds for outages that were, after all, not any fault of theirs.

DSL CASE STUDY #3: COVAD, OCTOBER 2000

The third DSL installation was even less eventful. In fact, because the whole operation was handled by my employer, all I had to do was be around when the line went in and the router was installed. I was not part of the ordering or qualification process, but the distance of 9,000 feet was not a problem.

The access line used by Covad was provided by Qwest (formerly US West). The line was ordered on October 12, 2000, and ready to go by October 20, 2000. On that date, the Covad technician came and ran the line the last few hundred feet to my office. He also installed the DSL modem, a Cabletron Systems Flowpoint 2200 SDSL modem. The speed was set to 768 Kbps, symmetrical, of course. The DSL modem doubled as a LAN hub, so there was little to do except plug in the PC through the Ethernet port, and that was that. The PC came preconfigured from my company with the proper software, so there was nothing to do in that area either.

I do not know the cost of the SDSL modem. The SDSL service cost my employer $334.35 per month. Throughput was about 605 Kbps, not much better than the US West DSL in Example #2 in the preceding section, but a whole lot more money. However, the SDSL link could upload files from my computer much faster than the asymmetrical DSL.

DSL CASE STUDY #4: QWEST, JUNE 2001

The best documented DSL case study I can provide was my latest experience with DSL. I had to install a second DSL line for business use in the spring of 2001. I was already planning the second edition of this book, so I kept careful notes of dates and telephone calls. I never told any of the people I spoke to or wrote to that I was an author or was writing a book specifically about DSL. I did not know if I would or could get preferential treatment, but for the purposes of this book I did not want it anyway. I was starting from scratch, meaning that I did not even have a second telephone line to install DSL on when I began the process with Qwest in early May of 2001.

In fact, that was my first surprise. The Qwest Web site would not even allow me to order DSL until I had a working telephone line to put it on. So my first call, on Thursday, May 10, was to order a new analog telephone line for DSL. I emphasized that I wanted to put DSL on the line, so I would prefer not to have a DLC or other arrangement that would make the process of adding DSL more difficult or impossible. I had the advantage of knowing that my original DSL line (see Case Study #2, earlier in this chapter) worked fine at essentially the same distance (and the jacks on the wall are no more than a foot apart). I expected some resistance along the lines "we-can't-guarantee-anything," but the person I spoke to was understanding and helpful. I was told that it would take about 10 business days after the DSL was ordered on the new line to install the DSL. I got an analog line installation appointment for Monday, May 14.

I did get a call the next day from the installation organization at Qwest. There was apparently a damaged cable to the crossbar (an element of the central office switch), and although they could install the analog line for Monday, this new line would not support DSL. If I could wait until Wednesday, May 16, a new cable could be used and all would be well to add the DSL. I was impressed that the information about this new line being mainly for DSL had spread to the "new line" group. The analog line installation on May 16 was on time and uneventful. It cost about $100 to install and the basic service was $22 a month including all taxes and fees.

I immediately revisited the Web site and ordered DSL for my new telephone line. I had to choose the Qwest ADSL "Professional" service for $60 per month, because that was all my line qualified for. Apparently, Qwest knew that the line was for DSL, but could not order it for me. This was no big deal to me, and I received a call saying that my DSL would be ready to go on Friday, June 1, about two weeks later. I would be shipped the DSL modem to install myself—a Cisco 678—along with some software for the PC. The DSL modem, which included network address translation (NAT), the dynamic host configuration protocol (DHCP), and some filtering support as well as being a router, cost $195 (a special discount from the usual $295). I was told, but I already knew, that I would be sent the various IP parameters such as a login ID that I needed separately.

By June 1, I had neither the paperwork for the IP settings nor my DSL modem. I called the number that the DSL installation group gave me when they verified my order, and was told that the DSL installation had been postponed until June 8. I was supposed to have been called, but was not. The person I spoke to was courteous and apologetic, and disarmed any thoughts I had of giving them a hard time over the missed date. I was given a number to call if the DSL modem did not show up by Friday, June 8.

By June 8, I had my IP setting paperwork, but still no DSL modem. I called again and found out that the date had again been pushed back, this time another week until June 15. This time I did point out that the date had already been postponed once and that I did not want this to happen again. However, the person I spoke with was so relentlessly understanding and apologetic that I found it impossible to become very outraged. The person seemed as exasperated at the delays as I was. If this attitude was purely due to training and marketing, it was brilliant training and marketing. In fact, I was eventually given the DSL modem tracking number so that I could trace the progress of the shipment. This impressed me, it really was useful for tracking my shipment, and the DSL modem arrived on June 12, along with the usual filters, cables, instructions, and so on.

I was not able to install the hardware and the software until the morning of Friday, June 15, but this went off without a hitch. I was up and running in about an hour. There was no hub this time, so things were quite simple: telephone wire to the jack, and Ethernet crossover cable to my PC.

Total monthly cost is about $156.00—$15.00 for the analog line that the service runs on (any toll calls add to this basic charge, of course), $66.00 for the Qwest DSL Pro Deluxe service, and about $55.00 for the DSL connection to the ISP (Qwest). Various fees and taxes make up the rest of the $20.00 or so. One-time DSL "installation" charges were about $180.00. The total DSL timeline was from May 10 to June 15, or 36 days from start to finish, and required me to make four telephone calls, two for orders and two for order status. All in all, not a bad experience except for the installation delay, although I did receive a very welcome $25.00 "delay credit" on the first monthly bill. At every step of the way, I was impressed by the information available to the people I talked to, and their relentlessly cheerful attitudes in what must be a very difficult customer service position.

This line runs at 553 Kbps downstream, comparable to the other Qwest DSL line. This might seem the same as the speed of the initial DSL link detailed in Case Study #2, but for a lot more ($156 compared to about $80 per month). However, I have found significant differences between the "Pro Deluxe" link and the basic DSL service on the first line. Unlike the link in Case Study #2, I have never seen any delay I could relate to DNS lookups or initial Web page loads regardless of time of day or day of week. The additional cost might not show up in line rate, but whatever resources in and behind the DSLAM that are included in the price easily makes my throughput much higher than on the link in Case Study #2. For business use, the added throughput is very welcome and necessary.

I have not been able to determine the upstream speed easily, but it seems very fast. Unlike the Cisco 675, the 678 has no speed settings to examine. Both Qwest offerings run PPP over ATM, but it is the DSL modem that maintains the login information (that is, I never have to log in to the service provider as a separate step, as with some forms of PPPoE). This DSL modem was unaffected by the Code Red virus during the summer of 2001.

APPENDIX B

Sources of Standards and Specification Information

DSL FORUM (FORMERLY ADSL FORUM) DOCUMENTS

For the latest information, see www.dslforum.org or www.adsl.com.

Document Number	Document Name	Date
PR-001	ADSL Recommendation Number 1 - Splitter mode operation	December, 1998
PR-002	ADSL Recommendation Number 2 - Splitterless operation	December, 1998
PR-003	ADSL Recommendation Number 3 - ATM compliance	December, 1998
TR-001	ADSL Forum System Reference Model	May, 1996
TR-002	ATM over ADSL Recommendations	March, 1997
TR-003	Framing and Encapsulation Standards for ADSL: Packet Mode	June, 1997
TR-004	Network Migration	December, 1997
TR-005	ADSL Network Element Management	March, 1998
TR-006	SNMP-based ADSL Line MIB	March, 1998
TR-007	Interfaces and System Configurations for ADSL: Customer Premises	March, 1998
TR-008	Default VPI/VCI Addresses for FUNI Mode Transport: Packet Mode	March, 1998
TR-009	Channelization for DMT and CAP ADSL Line Codes: Packet Mode	March, 1998
TR-010	Requirements & Reference Models for ADSL Access Networks: The "SNAG" Document	June, 1998
TR-011	An End-to-End Packet Mode Architecture with Tunneling and Service Selection	June, 1998
TR-012	Broadband Service Architecture for Access to Legacy Data Networks over ADSL ("PPP over ATM")	September, 1998

Document Number	Document Name	Date
TR-013	Interface and Configurations for ADSL: Central Office	March, 1999
TR-014	DMT Line Code Specific MIB (see TR-024)	March, 1999
TR-015	CAP Line Code Specific MIB	March, 1999
TR-016	CMIP-based Network Management Framework (see TR-028)	March, 1999
TR-017 ATM over ADSL	Recommendation (see TR-042)	March, 1999
TR-018	References and Requirements for CPE Architectures for Data Access	May, 1999
TR-019	ADSL Forum Recommendation for Physical Layer of ADSLs with a Splitter	May, 1999
TR-020	ADSL Forum Recommendation for Physical Layer of ADSLs without a Splitter	May, 1999
TR-021	ADSL Forum Recommendation for ATM Layers of ADSLs	May, 1999
TR-022	The Operation of ADSL-based Networks	August, 1999
TR-023	Overview of ADSL Testing	August, 1999
TR-024	DMT Line Code Specific MIB (update of TR-014)	August, 1999
TR-025	Core Network Architecture for Access to Legacy Data Network over ADSL	November, 1999
TR-026	T1.413 Issue 2, ATM-based ADSL ICS	November, 1999
TR-027	SNMP-based ADSL Line MIB	November, 1999
TR-028	CMIP Specification for ADSL Network Element Management (update of TR-016)	December, 1999

Document Number	Document Name	Date
TR-029	ADSL Dynamic Interoperability Testing	February, 2000
TR-030	ADSL EMS to NMS Functional Requirements	February, 2000
TR-031	ADSL ANSI T1.413 – 1998 Conformance Testing	May, 2000
TR-032	CPE Architecture Recommendations for Access to Legacy Data Networks	May, 2000
TR-033	ITU-T G.992.2 (G.lite) ICS	May, 2000
TR-034	Proposal for an Alternative OAM Communications Channel Across The U Interface	May, 2000
TR-035	Protocol Independent Object Model for ADSL EMS-NMS Interface	May, 2000
TR-036	Requirements for Voice over DSL (see TR-039)	August, 2000
TR-037	Auto-Configuration for the Connection Between the Broadband Network Termination (B-NT) and the Network using ATM	August, 2000
TR-038	DSL Service Flow-Thru Management Overview	March, 2001
TR-039	Addendum to TR-036 Annex A; Requirements for Voice over DSL (update of TR-036)	March, 2001
TR-040	Aspects of VDSL Evolution	June, 2001
TR-041	CORBA Specification for ADSL EMS-NMS Interface	June, 2001
TR-042	ATM Transport over ADSL Recommendation (update to TR-017)	August, 2001
TR-043	Protocols at the U Interface for Accessing Data Networks Using ATM/DSL	August, 2001

SOME MAJOR IETF DOCUMENTS RELATING TO DSL

For the latest information, see www.ietf.org.

RFC	Title	Description
791	Internet Protocol	IP Standard
793	Transmission Control Protocol	TCP Standard
768	User Datagram Protocol	UDP Standard
826	Address Resolution Protocol	MAC to IP address resolution
1034	Domain Names - Concepts and Facilities Host Extensions for IP Multicasting	Description of the DNS; outlines IP multicasting
1142	OSI IS-IS Intra-domain Routing Protocol	Using IS–IS with IP
1155	SMI for TCP/IP	Structure of Management Information for SNMPv 1
1213	Management Information Base for Network for Management of IP-based Internet	SNMPvl Management of TCP/IP-based Internets: MIB-11
1332	The PPP Internet Protocol Control Protocol (IPCP)	Control for PPP
1483	Multiprotocol Encapsulation over ATM Adaptation Layer 5, 1993	Describes encapsulations for carrying network interconnect traffic over ATM AAL5
1631	The IP Network Address Translator (NAT)	Basic NAT definition
1662	PPP in HDLG-like Framing	PPP over frame connections
1723	RIP Version 2 - Carrying Additional Information	Intra-domain routing protocol
1771	A Border Gateway Protocol 4 (BGP-4)	Interdomain routing protocol

RFC	Title	Description
1883	Internet Protocol, Version 6 (IPv6) Specification	New version of IP
1901	Community-based SNMPv2	Simplified SNMPv2
1918	Address Allocation for Private Internets	Private address spaces defined
1932	IP over ATM: A Framework Document Dynamic Host Configuration Protocol (DHCP) Automated IP Address Assignment	Describes the various proposals of mapping IP into ATM
2138	Remote Authentication Dial In User Service (RADIUS)	User authentication and authorization standard
2205	Resource Reservation Protocol (RSVP)	Application QoS Support
2211	Specification of Controlled-Load Quality of Service	(see next entry)
2212	Specification of Guaranteed Quality of Service Specification of Controlled Load Quality of Service	Describes the network element behavior required to deliver guaranteed service in the Internet as previous for Controlled Load
2225	Classical IP and ARP over ATM Update	Support for SVCs over ATM Updates RFCs 1577 and 1626
2251	Lightweight Directory Access Protocol (v3) Architecture for Describing SNMP Management Frameworks	User information SNMPv3 management
2272	Message Processing and Dispatching for SNMPv3	Management
2273	SNMPv3 Applications User-based Security Model for SNMPv3	Management Management

RFC	Title	Description
2328	OSPF Version 2 Intra-domain Routing	Intra-domain routing protocol
2331	ATM Signaling Support for IP over ATM UNI Signaling 4.0 Update	Describes support for IP over ATM signaling. Replaces RFC 1755
2362	Protocol Independent Multicasting (PIM) Sparse	Multicast routing
2363	PPP over FUNI	Frames in frames
2364	PPP over AAL5	PPP over ATM
2401	Security Architecture for the Internet Protocol	IPSec
2402	IP Authentication Header	For IPSec
2406	IP Encapsulating Security Payload (ESP)	For IPSec
2408	Internet Security Association and Key Management Protocol (ISAKMP)	For IPSec
2409	Internet Key Exchange (IKE)	For IPSec
2427	Multiprotocol Interconnect over Frame Relay	Updates RFC 1490
2516	PPP over Ethernet	PPPoE
2597	Assured Forwarding PHB Group	For Diffserv
2598	An Expedited Forwarding PHB	For Diffserv
2661	Layer Two Tunneling Protocol - L2TP	Secure tunneling
3031	Multiprotocol Label Switching Architecture (MPLS details are the topic of many other RFCs)	Basics of MPLS

SOME RELEVANT ATM FORUM DOCUMENTS RELATING TO DSL

For the latest information, see www.atmforum.com.

Document Title	Date	Document Number
LAN Emulation over ATM 1.0	1995	af-lane-0021.000
LANE v2.0 LUNI Interface	1997	af-lane-0084.000
Multi-Protocol over ATM v1.0	1997	af-mpoa-0087.000
Frame-based ATM Transport over Ethernet	2000	af-fbatm-0139.000
Frame-based ATM Interface (Level 3)	2000	af-phy-0143.000 P-NNI V 1.0
	1996	af-pnni-0055.000
PNNI Addendum - ABR Support	1997	af-pnni-0075.000
PNNI Addendum – Path and Connection Trace	2000	af-pnni-0141.000
FUNI2.0	1997	af-saa-0088.000
Circuit Emulation 2.0	1996	af-vtoa-0078.000
VTOA to the Desktop	1997	af-vtoa-0083.000
Dynamic Bandwidth Utilization	1997	af-vtoa-0085.000
ATM Trunking for Narrowband Services	1997	af-vtoa-0089.000
ATM User–Network Interface Specification V3.1	1994	af-uni-0010.002
UNI Signaling 4.0	1996	af-sig-0061.000
ILMI v4.0	1996	af-ilmi-0065.000
Traffic Management 4.0	1996	af-tm-0056.000
Signaling ABR Addendum	1997	af-sig-0076.000
Traffic Management ABR Addendum	1997	af-tm-0077.000
Traffic Management 4.1	1999	af-tm-0121.000

Document Title	Date	Document Number
Addendum to TM 4.1: Differentiated UBR	2000	af-tm-0149.000
RBB Architecture Framework	1998	af-rbb-0099.000
ATM Security Specification 1.0	1998	af-sec-0096.000
ATM Security Specification 1.1	2001	af-sec-0100.000
Loop Emulation Service Using AAL2	2000	af-vmoa-0145.000

MAJOR RELEVANT ITU-T STANDARDS RELATING TO DSL

For the latest information, see the International Telecommunications Union at www.itu.int.

Number	Document Title and Date
I.361	BISDNATM Layer Specification, November, 1995 (a more detailed description of the information in 1.150)
I.362	BISDN Asynchronous Transfer Model Adaptation Layer, October, 1993 (the principles of the AALs corresponding to the B-ISDN service classes)
I.363	Asynchronous Transfer Mode Adaptation Layer Specification, February, 1994/1996 (a more detailed description of the AALs from 1.362, including the division of the AAL into the Segmentation and Reassembly Sublayer and the Convergence Sublayer)
I.363.1	BISDN ATM Adaptation Layer (AAL) Specification, Type 1, August, 1996
I.363.5	BISDN ATM Adaptation Layer (AAL) Specification, Type 5, August, 1996
G.991.1	High Bit-rate Digital Subscriber Line (HDSL) Transmission System on Metallic Local Lines, October, 1998
G.991.2	Single-Pair High-Speed Digital Subscriber Line (SHDSL) Transceivers, February, 2001

Number	Document Title and Date
G.992.1	Asymmetrical Digital Subscriber Line (ADSL) Transceivers (G.dmt), June, 1999
G.992.1	Annex H: Specific requirements for a Synchronized Symmetrical DSL (SSDSL) System Operating in the Same Cable Binder as ISDN as Defined in G.961 Appendix III, October, 2000
G.992.2	Splitterless Asymmetric Digital Subscriber Line (ADSL) Transceivers (G.lite), July, 1999
G.994.1	Handshake Procedures for Digital Subscriber Line (DSL) transceivers, February, 2001
G.995.1	Overview of Digital Subscriber Line (DSL) Recommendations, February, 2001
G.996.1	Test Procedures for Digital Subscriber Line (DSL) Transceivers, February, 2001
G.997.1	Physical Layer Management for Digital Subscriber Line (DSL) Transceivers, June, 1999
H.320	Narrow-band Visual Telephone Systems and Terminal Equipment
H.321	Visual Telephone Terminals over ATM
H.323	Visual Telephone Terminals over Non-Guaranteed Quality of Service LANs
Q.2010	B-ISDN Overview. Signaling Capability Set 1, Release 1, August, 1991
Q.2100	B-ISDN Signaling ATM Adaptation Layer (SAAL) Overview Description, February, 1995
Q.2110	B-ISDN ATM Adaptation Layer Service Specific Connection Oriented Protocol (SSCOP), April, 1995
Q.2119	B-ISDN ATM Adaptation Layer Protocols - Convergence Function for the SSCOP above the Frame Relay Core Service (1996)
Q.2120	B-ISDN Meta-signaling Protocol, September, 1995
Q.2130	B-ISDN SAAL Service Specific Coordination Function (SSCF), April, 1995
Q.2140	B-ISDN ATM Adaptation Layer - Service-specific Coordination Function for Signaling at the Network-node Interface (SSCF at NNI), November, 1995

Number	Document Title and Date
Q.2144	B-ISDN Signaling ATM Adaptation Layer (SAAL) - Layer Management for the SAAL at the Network Node Interface, December, 1995
Q.2210	B-ISDN Signaling Network Protocols. Message Transfer Part Level 3 Functions and Messages Using the Services of ITU-T Recommendation Q.2140 (1996)
Q.2931	B-ISDN Digital Subscriber Signaling System No. 2 (DSS2) User-Network Interface (UNI) Layer 3 Specification for Basic Call/Connection Control (modified by Q.2971), February 1995
Q.2971 (modifies Q.2931, Q.2951, and Q.2957)	B-ISDN DSS2 UNI Layer 3 Specification for Point-to-Multipoint Call/Connection Control, December, 1995

INDEX

L

M

N

O

P

Q

V

INTERNATIONAL CONTACT INFORMATION

AUSTRALIA
McGraw-Hill Book Company Australia Pty. Ltd.
TEL +61-2-9417-9899
FAX +61-2-9417-5687
http://www.mcgraw-hill.com.au
books-it_sydney@mcgraw-hill.com

CANADA
McGraw-Hill Ryerson Ltd.
TEL +905-430-5000
FAX +905-430-5020
http://www.mcgrawhill.ca

**GREECE, MIDDLE EAST,
NORTHERN AFRICA**
McGraw-Hill Hellas
TEL +30-1-656-0990-3-4
FAX +30-1-654-5525

MEXICO (Also serving Latin America)
McGraw-Hill Interamericana Editores S.A. de C.V.
TEL +525-117-1583
FAX +525-117-1589
http://www.mcgraw-hill.com.mx
fernando_castellanos@mcgraw-hill.com

SINGAPORE (Serving Asia)
McGraw-Hill Book Company
TEL +65-863-1580
FAX +65-862-3354
http://www.mcgraw-hill.com.sg
mghasia@mcgraw-hill.com

SOUTH AFRICA
McGraw-Hill South Africa
TEL +27-11-622-7512
FAX +27-11-622-9045
robyn_swanepoel@mcgraw-hill.com

**UNITED KINGDOM & EUROPE
(Excluding Southern Europe)**
McGraw-Hill Education Europe
TEL +44-1-628-502500
FAX +44-1-628-770224
http://www.mcgraw-hill.co.uk
computing_neurope@mcgraw-hill.com

ALL OTHER INQUIRIES Contact:
Osborne/McGraw-Hill
TEL +1-510-549-6600
FAX +1-510-883-7600
http://www.osborne.com
omg_international@mcgraw-hill.com